# Engineering of Functional Skeletal Tissues

Felix Bronner, Mary C. Farach-Carson and Antonios G. Mikos (Eds)

# Engineering of Functional Skeletal Tissues

Volume 3 in the Series
**Topics in Bone Biology**
*Series Editors:*
Felix Bronner and Mary C. Farach-Carson

Springer

Felix Bronner, PhD
Professor Emeritus
University of Connecticut Health Center
Farmington, CT, USA

Mary C. Farach-Carson, PhD
Professor of Biological Sciences
Department of Biological Sciences
University of Delaware
Newark, DE, USA

Antonios G. Mikos, PhD
Professor of Bioengineering
Department of Bioengineering
Rice University
Houston, TX, USA

British Library Cataloguing in Publication Data
A catalogue record for this book is available from the British Library

Library of Congress Control Number: 20069222753

ISBN-10: 1-85233-962-4 e-ISBN 1-84628-366-3 Printed on acid-free paper
ISBN-13: 978-1-85233-962-3



9 8 7 6 5 4 3 2 1

Springer Science+Business Media
springer.com

# Preface

The science of bone replacement has greatly advanced in recent decades, but replacing bone with bone tissue rather than with metallic components remains in early development. The current volume, third in the series *Topics in Bone Biology*, deals with problems inherent in inducing the body cells to accomplish bone tissue repair, to degrade devices introduced to provide initial mechanical support, and to attract and stimulate bone formation. It is therefore logical that Chapter 1, by Hicok and Hedrick, deals with stem cells, i.e., pluripotential cells that may differentiate into cartilage and bone cells. The chapter begins with a description of how stem cells may be harvested; the limitations of autologous, embryonic, and adult stem cells; and the need to expand the harvested cells in culture. The authors then discuss the influences of the body environment on implanted cells and on the scaffolds that need to be introduced. They emphasize the need for adequate oxygenation and for rapid integration with the vascular system of the host/patient. Stem-cell-engineered cartilage is discussed at some length, along with the need for stem-cell-engineered ligaments and tendons. The chapter concludes with an analysis of what needs to be learned to make stem-cell-engineered bone tissue a reality.

In Chapter 2, Gerstenfeld and colleagues review osteogenic growth factors and cytokines, soluble proteins that regulate postnatal bone repair. These molecules are of importance because many are targets of efforts to promote therapeutic bone healing and repair. Molecules discussed are the tumor necrosis factor $\alpha$ (TNF-$\alpha$) family of cytokines and their role in bone remodeling, the bone morphogenetic proteins (BMPs) and their role in signaling, and angiogenic factors such as the vascular endothelial growth factor (VEGF) and angiopoietin families, with detailed discussion of the role of angiopoietins in bone development and tissue healing. The authors then discuss parathyroid hormone (PTH) and parathyroid hormone-related peptide (PTHrP): the differences between their paracrine and endocrine effects, their signal transduction and nuclear effects, and their effects on endochondral development and bone repair. A concluding section deals with bone healing and the roles played by skeletal stem cells, cytokines, and morphogenetic signals. This chapter, like all the others in this volume, has an extensive reference list.

Transplantation of bone allografts is a common orthopedic practice, but unless great care is taken, the allograft may give rise to infection and its sequelae in the host/patient. Tuan and colleagues Moucha, Renard, Gandhi, and Lin, in Chapter 2, discuss the harvesting and processing of musculoskeletal grafts and the conditions that must be met for the graft to be safe, i.e., not to cause inflammation, disease, or other harm to the host. This means that the medical and social history of the donor must be known in order to avoid complications that might arise, for example, as a result of transmission of the AIDS virus through the donor. The graft itself must be sterilized, and the authors discuss the various possible procedures to achieve this aim. Freezing or gamma-irradiation may weaken the graft, preventing adequate weight bearing initially. Infections due to improperly sterilized grafts include human immunodeficieny virus (HIV), one of the most serious, other viruses such as hepatitis C virus (HCV), and bacteria such as the *Clostridium* species. Factors that may affect performance and mechanical properties of the graft are discussed at the end of the chapter.

Park, Temenoff, and Mikos, in Chapter 4, provide a general discussion of biodegradable implants and the functional characteristics and requirements of such implants. Implants must have high mechanical strength and stiffness if employed in sites subject to high loads, and the chapter discusses various materials suitable for that purpose. Discussed also are nano- and microparticles as means for delivering bioactive molecules to the site, the use of hydrogels to entrap and release drugs, and the kinds of cells that can be embedded in the scaffolds. The implants must be biodegradable and biocompatible, have biological functionality, have suitable mechanical properties, and be composed of appropriate materials, the requirements for which are discussed in detail in the second half of the chapter.

Biodegradable scaffolds are highly desirable, but, as discussed in Chapter 4 and also in Chapters 6 and 7, they are not sufficiently developed for universal use. In Chapter 5, van den Dolder and Jansen summarize results achieved with a nondegradable scaffold made of titanium fiber mesh. Titanium has excellent biocompatibility and, in spongelike form, has been used extensively for tissue-engineering purposes. The authors review in detail the properties that make for biocompatibility of titanium. They then discuss other nondegradable metals, including tantalum and stainless steel. Like biodegradable scaffolds, the nondegradable scaffolds are used to deliver cells or extracellular matrix proteins to the defect site. Van den Dolder and Jansen describe methods of cell seeding and review the effects of matrix proteins on osteoblast differentiation in the titanium fiber mesh scaffolds. The chapter concludes with a review of the cell-based and growth-factor-based approaches to in vivo bone engineering.

The next two chapters describe and review in detail the use of scaffolds in bone tissue engineering. Betz, Yoon, and Fisher, in Chapter 6, discuss the fabrication and properties of polymers used for scaffold construction, including descriptions of curing methods and of the surface and mechanical properties of these scaffolds, as well as their biodegradation and biocompatibility. Polymer entanglement and cross linking, two major curing methods, are described, as is polymer assembly. The chapter describes several conventional fabrication methods (fiber bonding, phase separation, and gas foaming, among others), as well as different types of prototyping, including sheet lamination and laser stereolithography. This is followed by a detailed analysis of the various polyesters and other synthetic polymers and an extensive description of the properties that are desired in scaffold design, as they relate to surface, macrostructure, and mechanical properties and their suitability in terms of biodegradation and biocompatibility.

Chapter 7 deals with injectable scaffolds, which ideally can be used to replace hard or soft tissues. Such materials minimize the need for invasive surgery and thus improve current methods. Migliaresi, Motta, and DiBenedetto discuss the properties that an injectable scaffold must have and then describe injectable scaffolds that are ceramic-based, i.e., hydroxyapatite, tricalcium phosphate, biphasic calcium phosphate, and bioactive glasses. These materials, developed some three decades ago, have porosity, so that cells can be attracted or proteins inserted into the scaffold; the materials therefore must be resorbable. To use these materials, the engineer must impart a setting rate that is not too slow, so that the scaffold assumes mechanical strength rapidly, but that allows the scaffold to be resorbed in a time adequate for replacement of the implant by cells from the host. Soft tissue can be effectively replaced by hydrogel-based scaffolds. The chapter describes the many synthetic and natural hydrogels that have been used for injectable scaffolds. As the authors state, for a scaffold to be injectable, composite technology must be used creatively and the viscoelastic properties of the material must be understood, as must be the effect of the biological environment into which the scaffold is to be placed.

In Chapter 8 on Motion and Bone Regeneration, Ko, Somerman, and An discuss the three stages of bone regeneration—healing, osteogenesis, and osseointegration—and how regenerating bone responds to the signals emitted by limb movement. Bone healing in turn involves three stages—inflammation, reparation, and remodeling—and much of the chapter is devoted to an analysis of how mechanical factors influence bone healing. The authors show the relationships between cellular and organ events, how movement is transduced to the bone cells, and how the resulting intracellular increase in mRNA of protooncogenes and bone matrix proteins in turn affects bone healing and bone repair. A section of the chapter is devoted to distraction osteogenesis, a technique for producing new bone, and its application in principle in dentistry, inasmuch as tooth movement is equivalent to distraction. The final section, on bone and tooth implants, building upon information presented in earlier chapters, analyzes the effects of mechanical loading and bone repair, emphasizing that the correlation depends on the synergy between general boundary conditions and specific bone properties.

In dentistry, functional tooth replacement has become a reality as a result of the development of dental implants. Oates and Cochran, in Chapter 9, describe the bone and periodontal ligament loss frequently encountered in individuals with periodontal disease, a chronic infection. Bone implants have been used, though not always successfully, to stop the fairly extensive resorption of alveolar bone that occurs after tooth extraction. The chapter discusses in detail bone formation around dental implants, methods for speeding the rate of bone healing, how to regenerate bone in areas unsuited for implants, and bone grafting materials. Traditionally implants have been inserted some time after tooth removal, but there is great interest, as pointed out by the authors, in implant therapy very soon after tooth extraction. This may be possible, because healing in the tooth socket does not appear to be significantly affected by implant placement. Because space in the posterior maxilla is limited, implant therapy at that site has been difficult. Sinus augmentation, as described at the end of the chapter, seems to be the solution. The authors conclude by pointing out that further progress in dental practice, as in the recent past, will come from continued progress in bone research.

Computers have found increasing use in two- and three-dimensional design. In the last chapter, Melissa Knothe Tate illustrates the strength of computational modeling to extend experimental findings to the design of implants. An important aspect of modeling is that a given design can be expanded in length or in mechanical properties with the help of the computer, and the resulting expanded design can then be tested. Knothe Tate describes how the theory of poroelasticity has been adapted to bone modeling and how pressure gradients that cause nutrients and waste to be moved to and from cells have become part of the modeling approach. Similarly, the need to take into account cyclic compressive loads in designing bone replacements can be most readily met by appropriate modeling. In the second half of the chapter, the author illustrates in figures and equations the resolution of a variety of design problems. For example, a stochastic model is shown that represents the exact conformation and organization of the pericellular network, as well as reflecting microporosity. Other examples deal with the delivery of drugs to bone, fluid velocity magnitudes in the pericellular space, and the calculated and model-predicted permeability of a specific scaffold. There can be little doubt that computational modeling will find increasing use in implant and scaffold design.

This book appears at a time when functional engineering of bone tissue is ready to play a growing role in orthopedic and orthodontic practice. The editors are grateful to the authors of this book for their critical and timely discussion of this topic and for sharing their perspectives, so important to the many patients in need of bone repair or replacement, whether the very young, athletes, or the elderly. We also thank Springer-UK for their interest, patience, and willingness to publish the needed illustrations.

Felix Bronner
Farmington, Connecticut

Mary C. Farach-Carson
Newark, Delaware

Antonios G. Mikos
Houston, Texas

October 2006

# Contributors

**Kai-Nan An, PhD**
Biomechanical Laboratory
Department of Orthopedics
Mayo Clinic Rochester
Rochester, MN, USA

**Martha W. Betz, BS**
Bioengineering Graduate Program
University of Maryland
College Park, MD, USA

**Felix Bronner, PhD**
University of Connecticut Health Center
Farmington, CT, USA

**David L. Cochran, DDS, PhD**
Department of Periodontics
University of Texas
Health Science Center at San Antonio
San Antonio, TX, USA

**Anthony T. DiBenedetto, PhD**
University Professor of Chemical Engineering, Emeritus
University of Connecticut
Storrs, CT, USA

**Juliette van den Dolder, PhD**
Department of Periodontology and Biomaterials
Radboud University Nijmegen Medical Center
Nijmegen, The Netherlands

**Cory M. Edgar, PhD**
Orthopaedic Research Laboratory
Department of Orthopedic Surgery
Boston University Medical Center
Boston University School of Medicine
Boston, MA, USA

**Thomas A. Einhorn, MD**
Orthopaedic Research Laboratory
Department of Orthopedic Surgery
Boston University Medical Center
Boston University School of Medicine
Boston, MA, USA

**Mary C. Farach-Carson, PhD**
Department of Biological Sciences
University of Delaware
Newark, DE, USA

**John P. Fisher, PhD**
Department of Chemical and Biomolecular Engineering
and Bioengineering Graduate Program
University of Maryland
College Park, MD, USA

**Ankur Gandhi, PhD**
Department of Orthopedics
New Jersey Medical School
University of Medicine and Dentistry of New Jersey
Newark, NJ, USA

**Louis C. Gerstenfeld, PhD**
Orthopedic Research Laboratory
Department of Orthopedic Surgery
Boston University Medical Center
Boston University School of Medicine
Boston, MA, USA

**Marc H. Hedrick, MD**
Cytori Therapeutics Inc.
San Diego, CA, USA

**Kevin C. Hicok, MS**
Biologics Research Laboratories
Cytori Therapeutics Inc.
San Diego, CA, USA

**Kimberly A. Jacobsen, MA**
Orthopedic Research Laboratory
Department of Orthopedic Surgery
Boston University Medical Center
Boston University School of Medicine
Boston, MA, USA

**John A. Jansen, DDS, PhD**
Department of Periodontology and Biomaterials
Radboud University Nijmegen Medical Center
Nijmegen, The Netherlands

**Sanjeev Kakar, MD, MRCS**
Orthopaedic Research Laboratory
Department of Orthopaedic Surgery
Boston University Medical Center
Boston University School of Medicine
Boston, MA, USA

**Melissa L. Knothe Tate, PhD**
Department of Biomedical Engineering and Mechanical & Aerospace Engineering and
Thinktank for Multiscale Computational Modeling of Biomedical and Bio-Inspired Systems
Case Western Reserve University
Cleveland, OH, USA

**Ching-Chang Ko, DDS, MS, PhD**
Department of Orthodontics
University of North Carolina at Chapel Hill
School of Dentistry
Chapel Hill, NC, USA

**Sheldon S. Lin, MD**
Foot and Ankle Division
Department of Orthopedics
New Jersey Medical School
University of Medicine and Dentistry of New Jersey
Newark, NJ, USA

**Claudio Migliaresi, PhD**
Department of Materials Engineering and Industrial Technologies
University of Trento
Trento, Italy

**Antonios G. Mikos, PhD**
Department of Bioengineering
Rice University
Houston, TX, USA

**Antonella Motta, PhD**
Department of Materials Engineering and Industrial Technologies
University of Trento
Trento, Italy

**Calin S. Moucha, MD**
Division of Adult Joint Replacement
Department of Orthopedics
New Jersey Medical School
University of Medicine and Dentistry of New Jersey
Newark, NJ, USA

**Thomas W. Oates, DMD, PhD**
Department of Periodontics
University of Texas
Health Science Center at San Antonio
San Antonio, TX, USA

**Hansoo Park, MS**
Department of Bioengineering
Rice University
Houston, TX, USA

**Regis L. Renard, MD**
Department of Orthopedics
New Jersey Medicine School
University of Medicine and Dentistry of New Jersey
Newark, NJ, USA

**Martha J. Somerman, DDS PhD**
Department of Periodontics
University of Washington School of Dentistry
Seattle, WA, USA

**Johnna S. Temenoff, PhD**
Department of Bioengineering
Rice University
Houston, TX, USA

**Rocky S. Tuan, PhD**
Cartilage Biology and Orthopedics Branch
National Institute of Arthritis and Musculoskeletal and Skin Diseases
National Institutes of Health
Bethesda, MD, USA

**Diana M. Yoon, BS**
Department of Chemical and Biomolecular Engineering
University of Maryland
College Park, MD, USA

# Contents

# 1. Stem Cells and the Art of Mesenchymal Maintenance

*Kevin C. Hicok and Marc H. Hedrick*

## 1.1 Introduction

The most promising emergent medical technology of the early twenty-first century is stem-cell therapeutics. Traditionally, stem cells possess two important characteristics: the ability to undergo nearly unlimited self-renewal and the capability to differentiate into many (multipotent/pluripotent) or all (totipotent) mature cell phenotypes. The existence of stem cells and their ability to generate every tissue of the body during embryonic development has been known for many years. Transplant experiments performed in the 1970s, in which single stem cells were injected into early-stage blastulas, produced a chimera of donor and recipient cells in each organ of the resultant animal [29, 47].

The isolation and propagation of human embryonic stem cells (hES), however, has been achieved only relatively recently [111]. Political, moral, and ethical concerns surrounding procurement of these cells from embryos have held back their development as a source of cells for therapeutics or tissue engineering. Research efforts in the field of therapeutic cell cloning have skirted these issues by providing alternative methods, such as somatic cell nuclear transfer (SCNT), that generate hES cells without the use of intact embryos [46]. Until recently, the success rate of SCNT was extremely low, and the characterization of these cells to determine similarity to hES cells generated via sexual reproduction has not yet advanced far. Furthermore, teratoma formation by hES cells remains a safety issue, so that large-scale clinical trials involving these cells cannot be undertaken until the safety issue is resolved.

Adults also have stem cells. Hematopoietic stem cells in the bone marrow that can reconstitute the immune system have been known and studied for many years [13]. Stem cells in the liver allow rapid regeneration after liver surgery; stem cells in the dermis undergo continuous cell division and differentiation to replace skin cells; and mesenchymal stem cells (MSCs) in bone provide osteoblasts for bone remodeling throughout life. Until the mid-1980s, these stem cells were thought to be committed to regenerating only the tissue in which they resided and were believed to be unable to differentiate toward cell fates not associated with their germinal layer of origin. Their potential as “true” stem cells was therefore not realized. In the 1990s, the molecular mechanisms involved in cellular differentiation began to be understood more fully. Moreover, development of in vitro differentiation assays helped cell biologists and tissue engineers realize the therapeutic potential of these cells. This chapter will review the successes, challenges, and future prospects of using stem cells in the tissue engineering of bone, cartilage, tendon, and ligament.

## 1.2 The Challenges of Mesenchymal Tissue Engineering

Unique challenges face those attempting to reconstruct or repair damaged bone, cartilage, ligament, or tendon. As the major support and connective tissues in the body, they must sustain high mechanical stress. All four tissues are largely devoid of cells and are made up mainly of extracellular matrix (ECM) proteins and minerals. Cartilage, tendon and ligament cells must all be able to survive in hypoxic conditions, because these tissues are largely avascular. As a result of this acellularity, these tissues, when damaged, often heal slowly, if at all. Moreover, the body's healing response diminishes with age [24, 28, 55, 73, 79, 84, 86, 100].

The tissues that orthopedic surgeons employ to repair damaged mesenchyme therefore have great demands on them. Success in using autologous or allogenic graft materials for mesenchymal tissue repair has been mixed, depending on the size and site of the wound or defect and the age and health of the patient. Autologous grafts for bone repair (the "gold standard" in orthopedics) have been more successful than autografts for cartilage, tendon, and ligament [15, 119]. For example, Brittberg and colleagues [15] reported positive results for 88% of patients in a clinical study of femoral condyle cartilage defect repair that used autologous chondrocyte-seeded grafts. On the other hand, the results for patellar transplants were less impressive, with only one third of the patients having a successful outcome [15].

Aside from the difficulties associated with harvesting autograft material due to donor-site morbidity and the difficulty of obtaining enough donor tissue, a major deficit of autografts has been their frequent failure to become integrated with the surrounding host tissue. Often the resultant chimeric tissue fails to attain the properties of the original tissue, so that secondary grafting procedures are needed. Loading the graft material with mature phenotype cells has increased the amount of graft integration; however, limitations in the number of available autologous donor cells restrict the size of the graft that may be used [106].

Stem cells constitute an exciting alternative to the limitations of the current repair therapies. Stem cells can undergo more than 50 rounds of replication and thus provide an abundant source of cells to repair or regenerate large regions of tissue. Because stem cells can differentiate into many different cell phenotypes, they can be used in situations where multiple tissues must be generated to restore organ function. In addition to providing a source of mature phenotypes, culture-expanded adult stem cells secrete paracrine factors that support vascularization of new tissue. Furthermore, in instances where the cells themselves do not differentiate or produce a requisite factor for endogenous tissue healing or ex vivo regeneration, stem cells can be used to deliver gene therapy that in turn may enhance regeneration of the endogenous host tissue [12, 31, 92, 123]. These characteristics provide the mesenchymal tissue engineer with an abundant, renewable, and flexible source of cells that are capable of generating adequate amounts of ECM and of providing the enzymes, cytokines, and growth factors for the remodeling processes that are needed for the integration of implanted tissue.

## 1.3 Stem-Cell Repair of Bone

Recently stem cells of both embryonic and adult origins have been utilized. However, most early constructs utilized either endogenous or culture-expanded bone marrow-derived mesenchymal stem cells (BM-MSCs). In fact, orthopedic surgeons have, for many years, unknowingly utilized endogenous stem cells for bone repair. Early autograft transplant studies revealed the healing potential of bone marrow, soon recognized to contain a therapeutically valuable mesenchymal cell population capable of generating osteoblasts [30, 88]. However, the identification and characterization of "stem" cells within this population has been accomplished only in the last 20 years [35, 90, 93].

When grown in vitro, these putative stem cells were found to reside principally within the adherent cell subpopulation. Researchers have taken advantage of this adhesive property to isolate and enrich the cells [16, 17, 30, 88, 90, 93]. This remains the principal way in which MSCs are enriched for use in tissue-engineering applications.

To establish that stem cells are as multipotent, researchers have isolated single adherent cell clones that were then expanded in culture [41, 90, 93, 99]. Initial in vitro studies demonstrated that the BM-MSCs divide more than 50 times and differentiate into smooth muscle, osteoblastic, adipocytic, and chondrogenic phenotypes [90, 93] and, later, into tendon, ligament, and even cardiomyocytes (Fig. 1.1) [94, 124]. These cells can also be induced to display characteristics of endoderm and ectoderm tissues such as hepatic, neuro/glial, endothelial, and epithelial markers [91, 97, 121]. Transplantation of labeled stem cells produced chimeric mice and demonstrated differentiation of the stem cells into cells of endodermal, ectodermal, and mesodermal origin [52].

The osteogenic potential of BM-MSCs has been extensively characterized in vitro [16, 17, 90, 93, 99]. Under appropriate conditions these cells express genes from osteoblasts and synthesize proteins including type I collagen, bone sialoprotein, osteocalcin, osteopontin, and osteonectin. In vivo, BM-MSCs that are loaded either onto allogenic demineralized or mineralized bone matrix or onto to synthetic hydroxyapatite matrices generate osteoid tissue [19, 18, 60, 61, 63, 85]. It was therefore obvious that stem cells derived from human bone marrow would be a source of osteoblasts for bone tissue engineering.

In order for BM-MSCs to be used for generating tissue-engineered bone, two important problems had to be solved. First, BM-MSCs are quite rare; by some estimates there are as few as one to two cells per million isolated mononuclear cells [19, 93, 99]. It was therefore necessary to develop methods to increase the number of stem cells, while maintaining their osteopotential. This was accomplished by Bruder and colleagues in 1996 [16, 17].

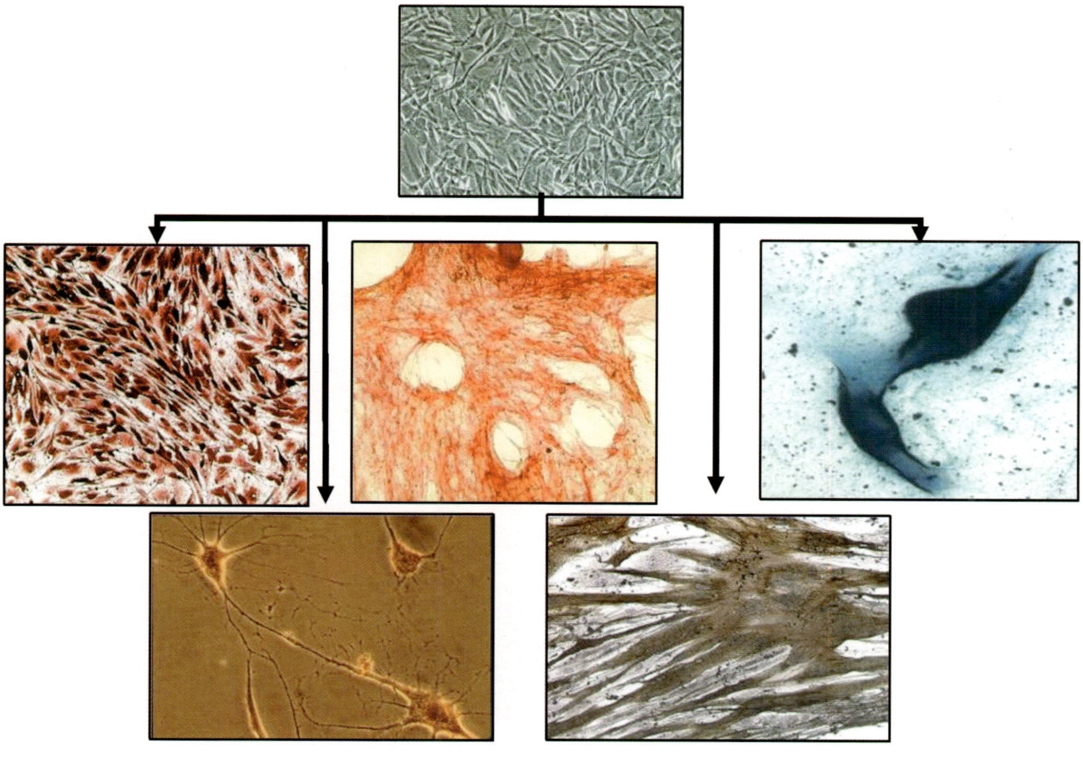

**Figure 1.1.** Adult adipose-derived stem cells (ADSCs) have the potential to differentiate into many different mature cell phenotypes. From top to bottom: Row 1: undifferentiated adult ADSCs in expansion culture, 10× phase objective. Row 2: ADSCs possess the ability to differentiate into adipocytes (left), osteoblasts (middle), and chondrocytes (right). Row 3: ADSCs also possess the ability to differentiate into neurons (left) and myocytes (right).

Others searched for osteogenic stem cells in tissues such as skin, muscle, and fat [5, 6, 33, 50, 51, 104, 112, 125]. Of these, adipose tissue appeared to be the most promising, both economically and practically. Adipose tissue is abundant and relatively easy to harvest, and the number of stem cells that can be harvested from it is two to three log units higher per number of isolated cells than is the case for BM-MSCs. When the number of these cells was increased, they differentiated into osteoblasts that synthesize bone matrix in vitro [37, 125] and in vivo [23, 38, 40, 92] (Fig. 1.2A). Even though adipose-derived stem cells (ADSCs) constituted an abundant alternative to BM-MSCs, it was necessary to do some ex vivo expansion prior to utilizing them in vivo [23, 38, 40, 92]. Recent data from our laboratories suggest that freshly isolated ADSCs may be loaded directly onto osteo-supportive matrices and can form bone in vivo (Fig. 1.2B). However, this treatment strategy still needs rigorous testing.

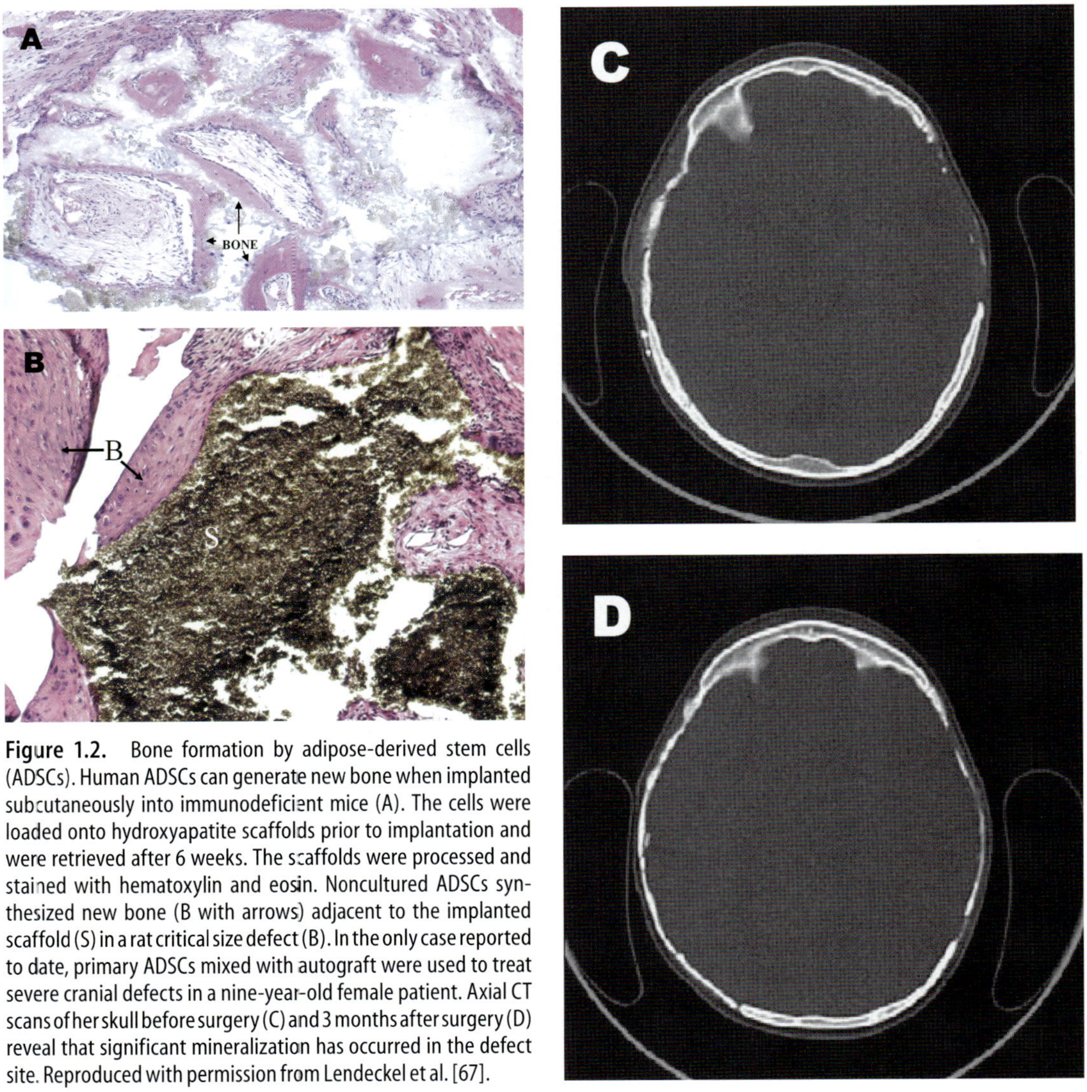


**Figure 1.2.** Bone formation by adipose-derived stem cells (ADSCs). Human ADSCs can generate new bone when implanted subcutaneously into immunodeficient mice (A). The cells were loaded onto hydroxyapatite scaffolds prior to implantation and were retrieved after 6 weeks. The scaffolds were processed and stained with hematoxylin and eosin. Noncultured ADSCs synthesized new bone (B with arrows) adjacent to the implanted scaffold (S) in a rat critical size defect (B). In the only case reported to date, primary ADSCs mixed with autograft were used to treat severe cranial defects in a nine-year-old female patient. Axial CT scans of her skull before surgery (C) and 3 months after surgery (D) reveal that significant mineralization has occurred in the defect site. Reproduced with permission from Lendeckel et al. [67].

## 1.4 Microenvironmental Influences on Bone Formation by Stem Cells

Several factors are critical to the successful formation of bone by stem cells. The osteopotential of a stem cell is significantly influenced by environmental signals that include soluble growth and differentiation factors, as well as cell–cell and cell–ECM interactions. When stem cells are delivered either ectopically or into a critical size defect model in a soluble vehicle solution such as physiologically balanced saline, relatively few stem cells are actually found to multiply and differentiate into osteoblasts [60, 92]. One reason for this may be a lack of adequate environmental signals to direct the stem cells toward osteogenesis. However, when the stem cells are allowed to adhere to a bone or bonelike matrix, either alone or in the presence of endogenous signals such as bone morphogenetic protein 2 (BMP-2), retinoic acid, dexamethasone, or 1,25-dihydroxyvitamin $D_3$, bone formation is increased [23, 38, 92].

Various natural and artificial scaffold materials have been utilized to serve as a delivery vehicle for stem cells and to provide the cells with appropriate cell–matrix interactions. Generally, these scaffolds are composed of either autologous or allogenically derived bone, demineralized bone matrix, coral, collagen, calcium salts, or composites of these. Typically, the more similar a scaffold is to natural bone, the better it supports new bone growth. Thus, scaffolds containing tricalcium or bicalcium phosphate salts and hydroxyapatite appear to be most effective in supporting stem-cell osteogenesis. Studies performed in a canine segmental defect model illustrate how adult stem cells loaded onto an appropriate scaffold, a hydroxyapatite/β-tricalcium phosphate ceramic, can repair large gaps within long bones [18]. Other groups have utilized collagen-based scaffolds, poly(lactic acid) (PLA) polymers, and hydrogels, with variable success; however, these appear to be more suitable for cartilage formation [26, 38, 40, 61, 113].

Determining the optimal combination of soluble factor and matrix signals that gives rise to stem-cell osteogenesis is complicated. Stem-cell response to these signals may be model- and species-dependent. The length of time for differentiation, the number of passages in culture, the health status of the stem-cell donor, and the tissue from which the cells have been obtained are all variables that influence the final differentiation potential of the stem cells. For example, culturing ADSCs on hydroxyapatite scaffolds in the presence of BMP-2 prior to implantation appears to aid in bone formation [23, 92], whereas in vitro stimulation of these same cells on the same scaffolds with other osteoinductive reagents, such as dexamethasone or 1,25-dihydroxyvitamin $D_3$, may or may not aid in bone formation [38, 40]. Hattori and colleagues [52] demonstrated that human ADSCs are superior to undifferentiated cells for the formation of ectopic bone when seeded onto atelocollagen matrices cultured in the presence of dexamethasone, ascorbate, and β-glycerol phosphate. Hicok and colleagues, however, found that when dexamethasone and 1,25-dihydroxyvitamin $D_3$ were used to predifferentiate ADSCs, there was no benefit to ectopic bone formation [40].

The age of the donor from which the stem cells are derived may be important in determining the extent of predifferentiation required for effective bone formation. Mendes demonstrated that MSCs derived from either young or old donors, when implanted subcutaneously into nude mice, formed ectopic bone without dexamethasone pretreatment. However, dexamethasone significantly increased bone formation in implants that contained cells from individuals older than 50 years of age [78]. Other age- dependent factors, such as advanced glycation end products from elderly or diabetic recipients, inhibit stem cells from proliferating and differentiating into osteoblasts [62].

The concentration of osteoinductive factors and the length of exposure to them affect stem-cell efficacy both in vitro and in vivo. ADSCs that were genetically modified to express either constitutive BMP-2 or BMP-7 demonstrated increased levels of osteoid formation in comparison with stem cells cultured with these molecules [92, 123]. Epigenetic modification of the stem cells may also be important, since compounds such as valproic acid, which has histone deacetylase inhibitory activity, have been shown to enhance osteogenesis of both adipose-derived and bone marrow-derived stem cells [22].

Species-specific differences in the responsiveness of stem cells to their environment further complicate our understanding of which combinations of extracellular signals are most effective in inducing bone formation. Species differences in the response of osteoblast progenitor cells to osteogenic stimuli are known [21, 65, 69], but they need detailed characterization. Rat BM-MSCs readily adhere to and proliferate on an alginate gel surface, whereas human cells fail to adhere, unless type I collagen or β-tricalcium phosphate is added to the gel [64]. Srouji and colleagues [63] reported that BM-MSCs, when predifferentiated prior to transplantation in a rabbit tibia defect, gave rise to radiographically significant amounts of bone, whereas, as mentioned previously, human MSCs exposed to similar conditions did not substantially increase bone formation [108].

Stem cells are exposed not only to chemical stimuli and scaffold interactions, but also to physical forces that act on these cells during the engineering process and after transplantation. Limited studies have been performed; however, application of physical force to the cells prior to transplantation seems to modulate their differentiation into osteoblasts. When human adult stem cells are exposed to either constant or intermittent mechanical or sheer stress, increased levels of osteogenic gene expression and mineralized matrix formation are observed [49, 59, 74, 77, 98].

Adequate oxygenation is critical to the successful generation and grafting of stem-cell-derived bone. Prior to implantation, cells must be adequately oxygenated so that they can expand into multiple layers and migrate into the inner surfaces of the delivery scaffolds. Current culture systems cannot yet surpass the 150- to 200-μm limit of nutrient and oxygen penetration. For large defects in human long bones, for example, grafting tissues with thicknesses in the millimeter range would significantly decrease the time required for bone repair. To avoid cell necrosis, transplanted stem cell/scaffold constructs must be integrated rapidly into the recipient's vascular system. When growth factors such as vascular endothelial growth factor (VEGF), which stimulates vascularization, were made part of scaffolds, bone formation was found to be significantly enhanced [44, 66, 81].

Both adipose-derived and bone-marrow-derived stem cells can induce new blood vessel formation, because they synthesize physiologically significant amounts of angiogenic cytokines, including VEGF, placental growth factor (PlGF), hepatocyte growth factor, and transforming growth factor β (TGF-β) [54, 57, 96]. Moreover, as in wound sites, these cytokines increase in quantity when the cells are exposed to hypoxic conditions [57, 58, 96]. When adult stem cells are placed into models of hind limb ischemia, collateral perfusion is increased [58]. Stem cells therefore not only can differentiate into osteoblasts, but may also support vascularization of the new bone. Understanding how this response is regulated is critical not only to engineering bone, but also to the successful utilization of stem cells in generating avascular mesenchymal tissues such as cartilage. It must be remembered that too high a level of oxygen within cartilage can induce apoptosis [72].

## 1.5 Safety and Success

An important challenge for the tissue engineer is to assess the safety of human stem cells when they are used to form bone in vivo. Even in severely immunocompromised rodent models such as the NOD/SCID mouse, there appears to be at least a low-level immunological response to the MSCs, to the scaffolds onto which they are seeded, or both [122]. This response depends on differences in how the cells are isolated or expanded in vitro prior to transplantation. It has been argued that the safety of autologous stem cells used for tissue engineering of bone in preclinical studies should be an adequate indicator of human stem-cell safety. Indeed, human BM-MSCs not only are immunoprivileged but also can suppress immune function [2]. In the end, however, the answer to this question lies in the results of clinical trials yet to be undertaken.

Notwithstanding the many as yet unanswered questions, the use of human stem cells for bone repair has yielded encouraging initial results. Culture-enriched autologous BM-MSCs have been used to successfully treat refractory atrophic and hypotrophic nonunion fractures in a small phase I clinical trial in Spain [87]. Another case report from Germany describes how autologous ADSCs were used in combination with bone marrow to treat a nine-year-old girl who had sustained critical cranial defects as a result of trauma. Significant bone formation was demonstrated after only 3 months [67]. Previous attempts at using autologous and allograft bone alone had failed to heal these defects (Fig. 1.2C and D). As more clinical trials are conducted, bone derived from stem-cell grafts may make the challenges of autograft and allograft transplants a thing of the past.

## 1.6 Stem-Cell-Engineered Cartilage: Microenvironmental Factors Influence Stem-Cell Chondrogenesis

As with bone, embryonic and adult-derived stem cells can give rise to cartilage in vitro [27, 43, 56, 71, 93, 107, 116, 125, 126]. And similarly to bone, the ability of stem cells to form cartilage in vitro depends on both physical and chemical stimuli. These include growth and differentiation factors, cell–cell interactions, cell–matrix interactions, and inorganic chemical and physical factors such as oxygen tension and the three-dimensional organization of the cells. Unlike bone, however, the physical elements of the cartilage microenvironment appear to be more critical for stem cells to differentiate into chondrocytes than for stem cells to become osteoblasts.

Three-dimensional interactions between cells are required for chondrocyte differentiation and subsequent cartilage tissue formation. When stem cells are plated as a monolayer, virtually no chondrogenesis results, even with added growth factors such as TGF-β and BMP [10, 27, 42, 45]. However, if the cells can establish three-dimensional polarity when cultured as condensed cell pellets or seeded into semisolid matrices such as alginate or hydrogel, they express proteoglycans and collagen isoforms, and a cartilage matrix is formed [11, 27, 71, 125]. The oxygen level in the culture is a second, important physicochemical parameter that affects chondrogenesis. Reducing the oxygen level in the culture to that which characterizes the cartilage environment in vivo enhances cartilage formation by ADSCs [14, 120], decreases cell proliferation, and increases the secretion of the essential protein, type II collagen, and of chondroitin-4-sulfate [120].

Stem-cell differentiation toward a chondrogenic phenotype also depends on activation of the TGF-β/BMP cell-signaling pathways [11, 71, 103, 126]. Thus, human ADSCs that had been predifferentiated in the presence of TGF-β in an alginate construct, when implanted subcutaneously, produced significantly more cartilaginous matrix than cells not so treated [27]. Other signaling mechanisms involving the parathyroid hormone-related peptide (PTHrP) receptor, glucocorticoid receptor, hyaluronic acid, and sonic hedgehog pathways have also been found to stimulate stem-cell chondrogenesis [25, 27, 32, 103]. The relative importance of growth and differentiation factors and of the resulting signaling pathways is, however, model-dependent. Therefore, the same developmental challenges that must be overcome to generate bone also apply to stem-cell chondrogenesis [48].

Cartilage generation by stem cells also depends on the type of substrate or scaffold used. Conventional scaffolds are composed of collagen and proteoglycans or other hydrated organic molecules such as agarose, alginate, or hydrogels. Some articular cartilage defects, however, are repaired with the aid of constructs that contain calcium/phosphate salts [11, 32, 36, 80]. As shown in Fig. 1.3, a large articular cartilage defect in sheep was almost completely repaired when autologous BM-MSCs were loaded onto a β-tricalcium phosphate scaffold [36].

Understanding the properties of the scaffolds or matrices in which the cells are delivered is especially critical for cartilage formation in vitro and its implantation. The microenvironment within the scaffold must be such as to support stem-cell growth and differentiation. As discussed in other chapters of this volume, the scaffold material itself must have physical properties comparable to those of the host cartilage and last until enough new cartilage has been produced to replace or supplement the implanted scaffold. For the scaffold to be replaced, it must be biodegradable.

Gelatin and agarose-based scaffolds have been used most commonly. When stem cells are seeded into scaffolds with nonosteogenic matrices such as gelatin and agarose, the matrices undergo changes in stress and compression that correlate with increased cartilage matrix accumulation. [11] When ADSCs are loaded onto gelatin scaffolds, their equilibrium

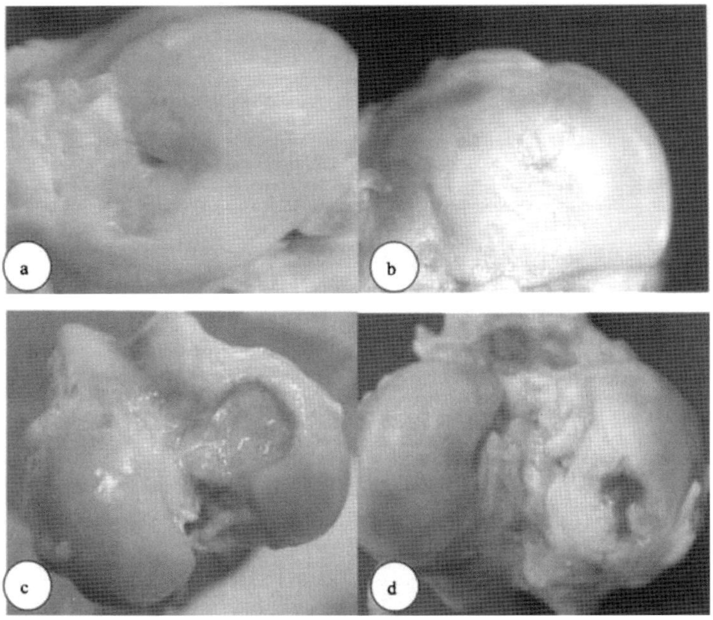


**Figure 1.3.** Bone marrow-derived mesenchymal stem cells (BM-MSCs) form new articular cartilage in vivo. At 12 weeks postoperation, the defects in the BM-MSC group were mostly repaired with tissue-engineered cartilage, resulting in a relatively smooth and consistent joint surface (A). At 24 weeks postoperation, the regenerated area was covered by smooth, consistent hyaline tissue that was indistinguishable from the surrounding normal cartilage (B). The defects in control group 1 were partially repaired with fibrous tissue, leaving some depression in the defect areas (C). In control group 2, a thin layer of red, irregular tissue surfacing the defects can be seen, and cracks on the surrounding normal cartilage are obvious (D). Reprinted from Guo et al. [36]. Copyright 2004, with permission from the European Association for Cranio-Maxillofacial Surgery.

compression and shear moduli increase after 28 days of culture, but this does not occur when they are loaded onto hydrogel-based matrices [11].

The efficacy of stem-cell chondrogenesis is model-dependent. Direct addition of stem cells to articular cartilage defects in rabbit and dog condyles generated tissue that was histologically comparable to native tissue [1]. However, both bone-marrow-derived and adipose-derived cartilage displayed less strength and elasticity than native cartilage. Direct addition of BM-MSCs in a caprine osteoarthritis model reduced cartilage loss and induced cartilage regeneration [80]. Cells derived from autologous rabbit bone marrow were able to regenerate a femoral condyle defect when loaded in a collagen gel [117]. Twenty-four weeks after transplantation, the reparative tissue from the BM-MSCs was stiffer and less compliant than the tissue derived from the empty defects, but it was less stiff and more compliant than normal cartilage [117]. Poly(lactic-co-glycolic acid) (PLGA) scaffolds supported cartilage formation by BM-MSCs transplanted into rabbit patellar defects [114]. Stem cells derived from allogenic rabbit adipose tissue, when delivered in a fibrin matrix, formed cartilage in an articular condyle defect that, on histological examination, appeared to have become integrated with the surrounding host tissue [82]. However, the new tissue became degraded after 12 weeks.

Gao and colleagues [32] recapitulated the host microenvironment and had greater long-term success. They utilized a two-layered matrix composed of a bottom layer of injectable calcium phosphate and a top layer of hyaluronan and found that by 12 weeks the defects had become filled with a stratified osteochondral tissue that was integrated into the surrounding tissue.

Alhadlaq and colleagues created a composite human articular condyle by predifferentiating BM-MSCs along the chondrogenic or osteogenic pathways and then loading the cells into photopolymerization gels [3]. The mold was made from a human condyle, photopolymerized, and then transplanted subcutaneously into immunocompromised mice. After 4 weeks, the resultant construct retained both the shape and the dimensions of the condyle and contained osteoid and cartilaginous matrix. More complex scaffolds can take advantage of stem-cell multipotentiality and may better stimulate the host environment, thereby providing appropriate niches for both bone and cartilage repair.

Spinal disc repair represents another field for the use of stem cells in cartilage tissue engineering. Cultures of ADSCs that also contain nucleus pulposus cells give rise to type II collagen and to aggrecan that is typical of nucleus pulposus cells [68]. In conventional spheroid cultures, adult MSCs express genes typical of intervertebral disc nucleus pulposus cells, including type II collagen, aggrecan, decorin, fibromodulin, and cartilage oligomeric matrix protein, with the levels of expression typical of disc cells rather than of hyaline articular cartilage [109]. In contrast, chondrogenically induced stem cells express type X collagen, an indicator of chondrocyte hypertrophy and eventual ossification [83]. Ossification in articular cartilage repair is necessary for tissue integration with the surrounding bone tissue, but in disc regeneration, ossification of the tissue is undesirable. However, if the surface properties of the substrates on which the stem cells are grown are altered, type X collagen gene expression in BM-MSCs can be inhibited [83]. Whether this also induces the expression of the desirable proteoglycan proteins remains to be determined.

A number of other stem-cell-dependent variables are likely to influence the efficacy of the stem cell/scaffold constructs, and studies to identify these variables are therefore warranted. For example, the site from which stem cells are harvested may influence their chondrogenic potential. Adult stem cells derived from bone marrow and those derived from adipose tissue appear to differ in their ability to form cartilage in vitro [42, 48]. The reasons for these differences and whether the observed differences are relevant in vivo remain to be determined but may be important in planning future therapies.

As yet, the mechanical properties of stem-cell-derived cartilage have not been characterized for most model systems. Tissue strength and viscoelastic, tribological, and anisotropic properties must be assessed to determine whether the new tissue can withstand in vivo stress loads. Secondly, appropriate studies are needed to establish the number of cells needed per scaffold for the formation of adequate amounts of cartilage, while avoiding necrosis or apoptosis. If too many cells are implanted into a wound, tear, or defect, the implant will not be sustained because of insufficient amounts of nutrients in the surrounding avascular, acellular matrix.

Stem cells from different body sites should be evaluated systematically for their chondrogenic potential [48]. The evaluation should take into consideration the relative ease of obtaining the stem cells, donor site morbidity, and the requirements for ex vivo expansion, as well as the quantitative and qualitative differences in the efficacy of the engineered tissue generated by the cells. Recent attempts to address these issues include reports [102, 103] that stem cells derived from synovium produced more cartilage than BM-MSCs, periosteal progenitors, skeletal muscle, and ADSCs from the same donor.

The therapeutic potential of cartilage synthesized by stem cells is illustrated by a report of two cases in which human BM-MSCs were successfully used to treat patellar articular cartilage defects, with the two individuals reporting that they had less joint pain after 1 and 2 years of follow-up [118]. Arthroscopy of the injured sites showed that they contained fibrocartilage [118].

## 1.7 Keeping Things Together: Stem-Cell-Engineered Ligament and Tendon

The use of stem cells to generate tissue-engineered ligament and tendon holds great promise. The cost of ligament repair alone exceeded five billion dollars in 2002 [89]. The current "gold standard" for repairing the most commonly injured ligament, the anterior cruciate ligament, is by implantation of autografts that consist of either patellar tendon or two hamstring tendons that are harvested at the time of surgery [115]. The rates of failure and recurrence of anterior cruciate ligament injury treated by these autograft methods, however, are still unacceptably high. As is the case for chondrocyte-loaded cartilage autografts, the supply of autologous tenocytes and ligament fibroblasts is limited, and their harvest often leads to donor-site morbidity.

The field of stem-cell-engineered tendons and ligaments is still in its infancy, even though the observation that BM-MSCs can differentiate into tendons and ligaments was made over a decade ago [20]. As is the case for cartilage and bone, the abundance of stem cells makes up for the limited availability of donor tissue and the high donor-site morbidity. Stem-cell-generated grafts, however, like ligament fibroblast- and tenocyte-seeded grafts, must be able to synthesize and remodel collagen, elastin, and other ECM proteins so that physiologically relevant levels of mechanical resistance and organization can be attained. Secondly, they must be delivered on a scaffold that is initially strong enough to endure cyclic stresses yet can undergo gradual degradation, thereby allowing the stem cells to differentiate and to secrete matrix proteins that can replace the scaffold. Finally, the new tissue must integrate with the host tissue so as to avoid recurrence of the injury.

Identifying optimal in vitro conditions that permit implantation has been challenging. Three factors seem essential: the absolute number of stem cells, the ratio of cells to collagen, and the ability of the cultured cells to synthesize the collagen in vitro prior to implantation [8, 9, 53]. Furthermore, as with ligament fibroblast-loaded constructs, exposure to appropriate cyclic strain is important to establish appropriate orientation and cross linking of matrix fibers within the new tissue [34].

To date, the characteristics that have been attained by stem-cell-engineered tendons and ligaments have fallen short of the desired outcome. In early studies, BM-MSCs initially seeded onto collagen scaffolds did not induce ligament regeneration, because the collagen scaffold did not stimulate the stem cells to produce adequate amounts of ligament matrix. In addition, the collagen fiber scaffolds did not support long-term anchoring of the grafts in vivo [115]. In a rabbit full-length, full-thickness tendon-defect model, the average maximum force and stress values of the BM-MSC-engineered collagen implants were approximately 30% that of normal patellar tendons [53]. The average repair stiffness and modulus values were 30% and 20%, respectively, of the values in normal patellar tendon [53]. Similarly, rabbit BM-MSCs loaded onto collagen gels and contracted onto sutures possessed only 25% of the maximum stress capacity of the normal tendon when implanted in a patellar tendon defect model. More disconcerting was the observation that bone formed in 28% of the patellar implant sites [7]. The results point to the obvious conclusion that the mesenchymal tissue engineer must continue efforts to identify the relevant mechanisms involved in tendon and ligament differentiation.

The recent utilization of silk-fiber-based delivery scaffolds with BM-MSCs has improved stem-cell-engineered ligaments [4, 115]. The silk fibers have superior mechanical properties and biodegrade within a more compatible timeframe. When woven into a six-cord rope configuration, the constructs display mechanical properties similar to the anterior cruciate ligament, and the constructs possess a greater surface area for cell attachment and ECM deposition [4, 115].

Integration of tendons and ligaments into the bone is critical for the long-term success of any engineered graft. Autografts and allografts used for ligament and tendon reconstruction have a poor record in this regard. Because of their ability to differentiate into multiple tissues, stem cells have the potential to generate the different tissues required for appropriate integration into the host tissue. When tendon autografts coated with fibrin glue were loaded with MSCs, cartilage cells covered a large area at the tendon–bone junction within 2 weeks [70]. By 8 weeks, a mature zone of cartilage blended from bone into the tendon grafts. At 8 weeks, the MSC-enhanced grafts had a significantly higher failure load and greater stiffness than the grafts loaded with fibrin glue.

For stem-cell-based therapeutics to be a success in clinical trials, research must be done to address key factors known to influence stem-cell efficacy. For example, autologous stem cells may be influenced by both the health status and the age of the patient. Bruder and colleagues have observed that the number of stem cells in bone marrow appears to decline with age [17]; however, whether this is true of adult stem cells from other body sites remains to be determined. Fewer stem cells are available as an individual ages, but their ability to proliferate remains the same [39, 110]. Therefore, for repairs in the elderly, more stem cells have to be harvested; alternatively, allogenic cells can be used.

Although the effects of aging on the ability of human stem cells to form tendon and ligament are unknown, an intradonor rabbit study utilizing BM-MSCs extracted from animals 1 and 4 years of age found no statistically significant differences in the mechanical properties of tendon regenerated by cells from the younger and the older animals. The stem cells from the older animals, however, exhibited reduced mechanical properties. Therefore, banking stem cells early in life for later use may lead to a better outcome. When the clinical and biomechanical factors involved in tendon and ligament differentiation are understood, tendons and ligaments grown from adipose or bone marrow cells are likely to become commonplace.

## 1.8 The Answers Are on the Horizon

Bone, cartilage, tendon, and ligament engineered from stem cells hold great promise to reduce suffering resulting from orthopedic injury and disease. With proper selection of stem cells and an appropriate supply of environmental signals, outcomes approaching 100% recovery may become possible. Indeed, mesenchymal tissue-engineered therapies that use novel combinations of scaffolds, stem cells, and differentiation factors are being reported almost monthly.

These novel approaches represent attempts to overcome the limitations of conventional stem-cell delivery systems. For example, replacing collagen with silk fibers generates porous silk fibroin scaffolds that are biodegradable and stronger than collagen scaffolds and that can support higher rates of human stem-cell differentiation than can conventional scaffolds [75, 76]. When BM-MSCs were loaded onto a biodegradable scaffold embedded with DNA that encodes an osteodifferentiation factor (BMP-4) and a proangiogenic factor (VEGF), greater amounts of properly vascularized bone were formed than when scaffolds containing only stem cells or one factor alone were used [44].

An exciting use of adult stem cells in mesenchymal tissue engineering is to take advantage of subtle differences found among cells from different body sites [95, 101]. Shi and colleagues recently described the isolation, characterization, and propagation of stem cells from different regions of adult human dental tissues that, when combined with appropriate scaffolds, developed into tissues resembling bone, dentin pulp, and cementum [105]. From an industry perspective, multiorigin stem-cell-engineered tissue poses significant intellectual property and regulatory hurdles for those who are brave enough to attempt to bring such tissue to the medical community. Ultimately, however, this approach may provide the regenerative capacity needed fully to restore or replace a damaged organ.

These considerations lead to what is perhaps an obvious conclusion: as tissue engineers identify and implement the essential multifactorial requirements for growing new or fixing old mesenchymal tissues, the full therapeutic potential of stem cells may ultimately be realized.

## References


1. Abdel-Hamid M, Hussein MR, Ahmad AF, Elgezawi EM (2005) Enhancement of the repair of meniscal wounds in the red-white zone (middle third) by the injection of bone marrow cells in canine animal model. Int J Exp Pathol 86:117–123.
2. Aggarwal S, Pittenger MF (2005) Human mesenchymal stem cells modulate allogenic immune cell responses. Blood 105:1815–1822.
3. Alhadlaq A, Elisseeff JH, Hong L, Williams CG, Caplan AI, Sharma B, Kopher RA, Tomkoria S, Lennon DP, Lopez A, Mao JJ (2004) Adult stem cell driven genesis of human-shaped articular condyle. Ann Biomed Eng 32:911–923.
4. Altman GH, Horan RL, Lu HH, Moreau J, Martin I, et al (2002) Silk matrix for tissue engineered anterior cruciate ligaments. Biomaterials 23:4131–4141.
5. Asakura A, Komaki M, Rudnicki M (2001) Muscle satellite cells are multipotential stem cells that exhibit myogenic, osteogenic, and adipogenic differentiation. Differentiation 68:245–253.
6. Aust L, Devlin B, Foster SJ, Halvorsen YD, Hicok K, du Laney T, Sen A, Willingmyre GD, Gimble JM (2004) Yield of human adipose-derived adult stem cells from liposuction aspirates. Cytotherapy 6:7–14.

7. Awad HA, Boivin GP, Dressler MR, Smith FN, Young RG, Butler DL (2003) Repair of patellar tendon injuries using a cell-collagen composite. J Orthop Res 21:420–431.
8. Awad HA, Butler DL, Boivin GP, Smith FN, Malaviya P, Huibregtse B, Caplan AI (1999) Autologous mesenchymal stem cell-mediated repair of tendon. Tissue Eng 5:267–277.
9. Awad HA, Butler DL, Harris MT, Ibrahim RE, Wu Y, Young RG, Kadiyala S, Boivin GP (2000) In vitro characterization of mesenchymal stem cell-seeded collagen scaffolds for tendon repair: effects of initial seeding density on contraction kinetics. J Biomed Mater Res 51:233–240.
10. Awad HA, Halvorsen YD, Gimble JM, Guilak F (2003) Effects of transforming growth factor β1 and dexamethasone on the growth and chondrogenic differentiation of adipose-derived stromal cells. Tissue Eng 9:1301–1312.
11. Awad HA, Wickham MQ, Leddy HA, Gimble JM, Guilak F (2004) Chondrogenic differentiation of adipose-derived adult stem cells in agarose, alginate, and gelatin scaffolds. Biomaterials 25:3211–3222.
12. Baksh D, Song L, Tuan RS (2004) Adult mesenchymal stem cells: characterization, differentiation, and application in cell and gene therapy. J Cell Mol Med 8:301–316.
13. Baum CM, Weissman IL, Tsukamoto AS, Buckle AM, Peault B (1992) Isolation of a candidate human hematopoietic stem-cell population. Proc Natl Acad Sci USA 89:2804–2808.
14. Betre H, Ong SR, Guilak F, Chilkoti A, Fermor B, Setton LA (2006) Chondrocytic differentiation of human adipose-derived adult stem cells in elastin-like polypeptide. Biomaterials 27:91–99.
15. Brittberg M, Lindahl A, Nilsson A, Ohlsson C, Isaksson O, Peterson L (1994) Treatment of deep cartilage defects in the knee with autologous chondrocyte transplantation. N Engl J Med 331:889–895.
16. Bruder SP, Jaiswal N (1996) The osteogenic potential of human mesenchymal stem cells is not diminished after one billion-fold expansion in vitro. Trans Orthop Res Soc 21:580.
17. Bruder S, Jaiswal N, Haynesworth S (1997) Growth kinetics, self-renewal and osteogenic potential of purified human mesenchymal stem cells during extensive subcultivation and following cryopreservation. J Cell Biochem 64:278–294.
18. Bruder SP, Kraus KH, Goldberg VM, Kadiyala S (1998) The effect of implants loaded with autologous mesenchymal stem cells on the healing of canine segmental bone defects. J Bone Joint Surg Am 80: 985–996.
19. Bruder S, Kurth A, Shea M, Hayes W, Jaiswal N, Kadiyala S (1998) Bone regeneration by implantation of purified, culture-expanded human mesenchymal stem cells. J Orthop Res 16:155–162.
20. Caplan AI (2005) Review: mesenchymal stem cells: cell-based reconstructive therapy in orthopedics. Tissue Eng 11:1198–1211.
21. Cheng SL, Yang JW, Rifas L, Zhang SF, Avioli LV (1994) Differentiation of human bone marrow osteogenic stromal cells in vitro: induction of the osteoblast phenotype by dexamethasone. Endocrinology 134:277–286.
22. Cho HH, Park HT, Kim YJ, Bae YC, Suh KT, Jung JS (2005) Induction of osteogenic differentiation of human mesenchymal stem cells by histone deacetylase inhibitors. J Cell Biochem 96:533–542.
23. Cowan CM, Shi YY, Aalami OO, Chou YF, Mari C, Thomas R, Quarto N, Contag CH, Wu B, Longaker MT (2004) Adipose-derived adult stromal cells heal critical-size mouse calvarial defects. Nat Biotechnol 22:560–567.
24. Ding C, Cicuttini F, Scott F, Cooley H, Jones G (2005) Association between age and knee structural change: a cross sectional MRI based study. Ann Rheum Dis 64:549–555.
25. Edwards PC, Ruggiero S, Fantasia J, Burakoff R, Moorji SM, Paric E, Razzano P, Grande DA, Mason JM (2005) Sonic hedgehog gene-enhanced tissue engineering for bone regeneration. Gene Ther 12:75–86.
26. Elisseeff J, Puleo C, Yang F, Sharma B (2005) Advances in skeletal tissue engineering with hydrogels. Orthod Craniofac Res 8:150–161.
27. Erickson GR, Gimble JM, Franklin DM, Rice HE, Awad H, Guilak F (2002) Chondrogenic potential of adipose tissue-derived stromal cells in vitro and in vivo. Biochem Biophys Res Commun 290:763–769.
28. Flachsmann R, Kim W, Broom N (2005) Vulnerability to rupture of the intact articular surface with respect to age and proximity to site of fibrillation: a dynamic and static-investigation. Connect Tissue Res 46:159–169.
29. Ford CE, Evans EP, Gardner RL (1975) Marker chromosome analysis of two mouse chimaeras. J Embryol Exp Morphol 33:447–457.
30. Friedenstein AJ (1976) Precursor cells of mechanocytes. Int Rev Cytol 47:327–355.
31. Gafni Y, Turgeman G, Liebergal M, Pelled G, Gazit Z, Gazit D (2004) Stem cells as vehicles for orthopedic gene therapy. Gene Ther 11:417–426.
32. Gao J, Dennis JE, Solchaga LA, Goldberg VM, Caplan AI (2002) Repair of osteochondral defect with tissue-engineered two-phase composite material of injectable calcium phosphate and hyaluronan sponge. Tissue Eng 8:827–837.
33. Gimble J, Guilak F (2003) Adipose-derived adult stem cells: isolation, characterization, and differentiation potential. Cytotherapy 5:362–369.
34. Goulet F (1997) In: Lanza RP, Langer R, Chick WL, eds. Principles of Engineering. 2nd edition. Chapter 50. 711–721 Academic Press, S Diego, CA.
35. Grigoriadis AE, Heersche JNM, Aubin JE (1988) Differentiation of muscle, fat, cartilage and bone from progenitor cells present in a bone-derived clonal cell population: effect of dexamethasone. J Cell Biol 106: 2139–2151.
36. Guo X, Wang C, Zhang Y, Xia R, Hu M, Duan C, Zhao Q, Dong L, Lu J, Qing Song Y (2004) Repair of large articular cartilage defects with implants of autologous mesenchymal stem cells seeded into beta-tricalcium phosphate in a sheep model. Tissue Eng 10:1818–1829.
37. Halvorsen YD, Franklin D, Bond AL, Hitt DC, Auchter C, Boskey AL, Paschalis EP, Wilkison WO, Gimble JM (2001) Extracellular matrix mineralization and osteoblast gene expression by human adipose tissue-derived stromal cells. Tissue Eng 7:729–741.

38. Hattori H, Sato M, Masuoka K, Ishihara M, Kikuchi T, Matsui T, Takase B, Ishizuka T, Kikuchi M, Fujikawa K, Ishihara M (2004) Osteogenic potential of human adipose tissue-derived stromal cells as an alternative stem cell source. Cells Tissue Organs 178:2–12.
39. Haynesworth SE, Reuben D, Caplan AI (1998) Cell-based tissue engineering therapies: the influence of whole body physiology. Adv Drug Deliv Rev 33:3–14.
40. Hicok KC, Du Laney TV, Zhou YS, Halvorsen YD, Hitt DC, Cooper LF, Gimble JM (2004) Human adipose-derived adult stem cells produce osteoid in vivo. Tissue Eng 10:371–380.
41. Hicok KC, Thomas T, Gori F, Rickard DJ, Spelsberg TC, Riggs BL (1998) Development and characterization of conditionally immortalized osteoblast precursor cell lines from human bone marrow stroma. J Bone Miner Res 13:205–217.
42. Huang JI, Kazmi N, Durbhakula MM, Hering TM, Yoo JU, Johnstone B (2005) Chondrogenic potential of progenitor cells derived from human bone marrow and adipose tissue: a patient-matched comparison. J Orthop Res (in press).
43. Huang JI, Zuk PA, Jones NF, Zhu M, Lorenz HP, Hedrick MH, Benhaim P (2004) Chondrogenic potential of multipotential cells from human adipose tissue. Plast Reconstr Surg 113:585–594.
44. Huang YC, Kaigler D, Rice KG, Krebsbach PH, Mooney DJ (2005) Combined angiogenic and osteogenic factor delivery enhances bone marrow stromal cell-driven bone regeneration. J Bone Miner Res 20: 848–857.
45. Hwang NS, Kim MS, Sampattavanich S, Baek JH, Zhang Z, Elisseeff J (2005) The effects of three dimensional culture and growth factors on the chondrogenic differentiation of murine embryonic stem cells. Stem Cells (in press).
46. Hwang WS, Roh SI, Lee BC, Kang SK, Kwon DK, Kim S, Kim SJ, Park SW, Kwon HS, Lee CK, Lee JB, Kim JM, Ahn C, Paek SH, Chang SS, Koo JJ, Yoon HS, Hwang JH, Hwang YY, Park YS, Oh SK, Kim HS, Park JH, Moon SY, Schatten G (2005) Patient-specific embryonic stem cells derived from human SCNT blastocysts. Science 308:1777–1783.
47. Illmensee K, Mintz B (1976) Totipotency and normal differentiation of single teratocarcinoma cells cloned by injection into blastocysts. Proc Natl Acad Sci USA 73:549–553.
48. Im GI, Shin YW, Lee KB (2005) Do adipose-derived mesenchymal stem cells have the same osteogenic and chondrogenic potential as bone marrow-derived cells? Osteoarthritis Cartilage 13:845–853.
49. Jagodzinski M, Drescher M, Zeichen J, Hankemeier S, Krettek C, Bosch U, van Griensven M (2004) Effects of cyclic longitudinal mechanical strain and dexamethasone on osteogenic differentiation of human bone marrow stromal cells. Eur Cell Mater 7:35–41; Discussion 41.
50. Jahoda CA, Whitehouse J, Reynolds AJ, Hole N (2003) Hair follicle dermal cells differentiate into adipogenic and osteogenic lineages. Exp Dermatol 12: 849–859.
51. Jay KE, Rouleau A, Underhill TM, Bhatia M (2004) Identification of a novel population of human cord blood cells with hematopoietic and chondrocytic potential. Cell Res 14:268–282.
52. Jiang Y, Jahagirdar BN, Reinhardt RL, Schwartz RE, Keene CD, Ortiz-Gonzalez XR, Reyes M, Lenvik T, Lund T, Blackstad M, Du J, Aldrich S, Lisberg A, Low WC, Largaespada DA, Verfaillie CM (2002) Pluripotency of mesenchymal stem cells derived from adult marrow. Nature 418:41–49.
53. Juncosa-Melvin N, Boivin GP, Galloway MT, Gooch C, West JR, Sklenka AM, Butler DL (2005) Effects of cell-to-collagen ratio in mesenchymal stem cell-seeded implants on tendon repair biomechanics and histology. Tissue Eng 11:448–457.
54. Kaigler D, Krebsbach PH, Polverini PJ, Mooney DJ (2003) Role of vascular endothelial growth factor in bone marrow stromal cell modulation of endothelial cells. Tissue Eng 9:95–103.
55. Kaplan D, Meyer K (1959) Ageing of human cartilage. Nature 183:1267–1268.
56. Kawaguchi J, Mee PJ, Smith AG (2005) Osteogenic and chondrogenic differentiation of embryonic stem cells in response to specific growth factors. Bone 36:758–769.
57. Kinnaird T, Stabile E, Burnett MS, Lee CW, Barr S, Fuchs S, Epstein SE (2004) Marrow-derived stromal cells express genes encoding a broad spectrum of arteriogenic cytokines and promote in vitro and in vivo arteriogenesis through paracrine mechanisms. Circ Res 94:678–685.
58. Kinnaird T, Stabile E, Burnett MS, Shou M, Lee CW, Barr S, Fuchs S, Epstein SE (2004) Local delivery of marrow-derived stromal cells augments collateral perfusion through paracrine mechanisms. Circulation 109:1543–1549.
59. Knippenberg M, Helder MN, Zandieh Doulabi B, Semeins CM, Wuisman P, Klein-Nulend J (2005) Adipose tissue-derived mesenchymal stem cells acquire bone cell-like responsiveness to fluid shear stress on osteogenic stimulation. Tissue Eng. (11–12): 1780–1788.
60. Krebsbach PH, Kuznetsov SA, Bianco P, Robey PG (1999) Bone marrow stromal cells: characterization and clinical application. Crit Rev Oral Biol Med 10: 165–181.
61. Krebsbach PH, Mankani MH, Satomura K, Kuznetsov SA, Robey PG (1998) Repair of craniotomy defects using bone marrow stromal cells. Transplantation 66:1272–1278.
62. Kume S, Kato S, Yamgishi S, Inagaki Y, Ueda S, Arima N, Okawa T, Kojiro M, Nagata K (2005) Advanced glycation end-products attenuate human mesenchymal stem cells and prevent cognate differentiation into adipose tissue, cartilage, and bone. J Bone Miner Res 20:1647–1658.
63. Kuznetsov SA, Krebsbach PH, Satomura K, Kerr J, Riminucci M, Benayahu D, Robey PG (1997) Single-colony derived strains of human marrow stromal fibroblasts form bone after transplantation in vivo. J Bone Miner Res 12:1335–1347.
64. Lawson MA, Barralet JE, Wang L, Shelton RM, Triffitt JT (2004) Adhesion and growth of bone marrow stromal cells on modified alginate hydrogels. Tissue Eng 10:1480–1491.
65. Leboy PS, Beresford JN, Devlin C, Owen ME (1991) Dexamethasone induction of osteoblast mRNAs in

rat marrow stromal cell cultures. J Cell Physiol 146: 370–378.
66. Lee KY, Peters MC, Anderson KW, Mooney DJ (2000) Controlled growth factor release from synthetic extracellular matrices. Nature 408:998–1000.
67. Lendeckel S, Jodicke A, Christophis P, Heidinger K, Wolff J, Fraser JK, Hedrick MH, Berthold L, Howaldt HP (2004) Autologous stem cells (adipose) and fibrin glue used to treat widespread traumatic calvarial defects: case report. J Craniomaxillofac Surg 32:370–373.
68. Li X, Lee JP, Balian G, Greg Anderson D (2005) Modulation of chondrocytic properties of fat-derived mesenchymal cells in co-cultures with nucleus pulposus. Connect Tissue Res 46:75–82.
69. Lian JB, Shalhoub V, Aslam F, Frenkel B, Green J, Hamrah M, Stein GS, Stein JL (1997) Species-specific glucocorticoid and 1,25-dihydroxyvitamin D responsiveness in mouse MC3T3-E1 osteoblasts: dexamethasone inhibits osteoblast differentiation and vitamin D down-regulates osteocalcin gene expression. Endocrinology 138:2117–2127.
70. Lim JK, Hui J, Li L, Thambyah A, Goh J, Lee EH (2004) Enhancement of tendon graft osteointegration using mesenchymal stem cells in a rabbit model of anterior cruciate ligament reconstruction. Arthroscopy 20:899–910.
71. Mackay AM, Beck SC, Murphy JM, Barry FP, Chichester CO, Pittenger MF (1998) Chondrogenic differentiation of cultured human mesenchymal stem cells from marrow. Tissue Eng 4:415–428.
72. Mansfield K, Pucci B, Adams CS, Shapiro IM (2003) Induction of apoptosis in skeletal tissues: phosphate-mediated chick chondrocyte apoptosis is calcium dependent. Calcif Tissue Int 73:161–172.
73. Martin JA, Buckwalter JA (2003) The role of chondrocyte senescence in the pathogenesis of osteoarthritis and in limiting cartilage repair. J Bone Joint Surg Am 85-A Suppl 2:106–110.
74. Mauney JR, Sjostorm S, Blumberg J, Horan R, O'Leary JP, Vunjak-Novakovic G, Volloch V, Kaplan DL (2004) Mechanical stimulation promotes osteogenic differentiation of human bone marrow stromal cells on 3-D partially demineralized bone scaffolds in vitro. Calcif Tissue Int 74:458–468.
75. Meinel L, Fajardo R, Hofmann S, Langer R, Chen J, Snyder B, Vunjak-Novakovic G, Kaplan D (2005) Silk implants for the healing of critical size bone defects. Bone 37:688–698.
76. Meinel L, Hofmann S, Karageorgiou V, Zichner L, Langer R, Kaplan D, Vunjak-Novakovic G (2004) Engineering cartilage-like tissue using human mesenchymal stem cells and silk protein scaffolds. Biotechnol Bioeng 88:379–391.
77. Meinel L, Karageorgiou V, Fajardo R, Snyder B, Shinde-Patil V, Zichner L, Kaplan D, Langer R, Vunjak-Novakovic G (2004) Bone tissue engineering using human mesenchymal stem cells: effects of scaffold material and medium flow. Ann Biomed Eng 32:112–122.
78. Mendes SC, Tibbe JM, Veenhof M, Bakker K, Both S, Platenburg PP, Oner FC, de Bruijn JD, van Blitterswijk CA (2002) Bone tissue-engineered implants using human bone marrow stromal cells: effect of culture conditions and donor age. Tissue Eng 8:911–920.
79. Miles JS, Eichelberger L (1964) Biochemical studies of human cartilage during the aging process. J Am Geriatr Soc 12:1–20.
80. Murphy JM, Fink DJ, Hunziker EB, Barry FP (2003) Stem cell therapy in a caprine model of osteoarthritis. Arthritis Rheum 48:3464–3474.
81. Murphy WL, Simmons CA, Kaigler D, Mooney DJ (2004) Bone regeneration via a mineral substrate and induced angiogenesis. J Dent Res 83:204–210.
82. Nathan S, Das De S, Thambyah A, Fen C, Goh J, Lee EH (2003) Cell-based therapy in the repair of osteochondral defects: a novel use for adipose tissue. Tissue Eng 9:733–744.
83. Nelea V, Luo L, Demers CN, Antoniou J, Petit A, Lerouge SR, Wertheimer M, Mwale F (2005) Selective inhibition of type X collagen expression in human mesenchymal stem cell differentiation on polymer substrates surface-modified by glow discharge plasma. J Biomed Mater Res 75:216–223.
84. O'Driscoll SW, Saris DB, Ito Y, Fitzimmons JS (2001) The chondrogenic potential of periosteum decreases with age. J Orthop Res 19:95–103.
85. Ohgushi H, Goldberg VM, Caplan AI (1989) Heterotopic osteogenesis in porous ceramics induced by marrow cells. J Orthop Res 7:568–578.
86. Oreffo RO, Bord S, Triffitt JT (1998) Skeletal progenitor cells and ageing human populations. Clin Sci (Lond) 94:549–555.
87. Orozco L, Rodriguez L, Torrico C, Douville J, Hock JM, Armstrong RD, Garcia J, Solano C (2005) Clinical feasibility study: the use of cultural enriched autologous bone marrow cells to treat refractory atrophic and hypotrophic nonunion fractures. www.aastrom.com/pdf/whitepaper_Barcelona_051205.pdf
88. Owen ME, Friedenstein AJ (1988) Stromal stem cells: Marrow-derived osteogenic precursors. In: Evered D, Harnett S, eds. Cell and Molecular Biology of Vertebrate Hard Tissues. Wiley, Chichester, UK, pp 42–60.
89. Pennisi E (2002) Tending tender tendons. Science 295:1011.
90. Pereira RF, Halford KW, O'Hara MD, Leeper DB, Sokolov BP, Pollard MD, Bagasra O, Prockop DJ (1995) Cultured adherent cells from marrow can serve as long-lasting precursor cells for bone, cartilage, and lung in irradiated mice. Proc Natl Acad Sci USA 92:4857–4861.
91. Petersen BE, Bowen WC, Patrene KD, Mars WM, Sullivan AK, Murase N, Boggs SS, Greenberger JS, Goff JP (1999) Bone marrow as a potential source of hepatic oval cells. Science 284:1168–1170.
92. Peterson B, Zhang J, Iglesias R, Kabo M, Hedrick M, Benhaim P, Lieberman JR (2005) Healing of critically sized femoral defects, using genetically modified mesenchymal stem cells from human adipose tissue. Tissue Eng 11:120–129.
93. Pittenger MF, Mackay AM, Beck SC, Jaiswal RK, Douglas R, Mosca JD, Moorman MA, Simonetti DW, Craig S, Marshak DR (1999) Multilineage potential of adult human mesenchymal stem cells. Science 284: 143–147.
94. Pittenger M, Vanguri P, Simonetti D, Young R (2002) Adult mesenchymal stem cells: potential for muscle and tendon regeneration and use in

gene therapy. J Musculoskelet Neuronal Interact 2:309–320.
95. Rahaman MN, Mao JJ (2005) Stem cell-based composite tissue constructs for regenerative medicine. Biotechnol Bioeng 91:261–284.
96. Rehman J, Traktuev D, Li J, Merfeld-Clauss S, Temm-Grove CJ, Bovenkerk JE, Pell CL, Johnstone BH, Considine RV, March KL (2004) Secretion of angiogenic and antiapoptotic factors by human adipose stromal cells. Circulation 109:1292–1298.
97. Reyes M, Lund T, Lenvik T, Aguiar D, Koodie L, Verfaillie CM (2001) Purification and ex vivo expansion of postnatal human marrow mesodermal progenitor cells. Blood 98:2615–2625.
98. Richards M, Goulet JA, Weiss JA, Waanders NA, Schaffler MB, Goldstein SA (1998) Bone regeneration and fracture healing. Experience with distraction osteogenesis model. Clin Orthop Relat Res (355 Suppl):S191–204.
99. Rickard DJ, Kassem M, Hefferan TE, Sarkar G, Spelsberg TC, Riggs BL (1996) Isolation and characterization of osteoblast precursor cells from human bone marrow. J Bone Miner Res 11:312–324.
100. Riggs BL, Melton LJ III (1986) Involutional osteoporosis. N Engl J Med 314:1676–1686.
101. Risbud MV, Shapiro IM (2005) Stem cells in craniofacial and dental tissue engineering. Orthod Craniofac Res 8:54–59.
102. Sakaguchi Y, Sekiya I, Yagishita K, Muneta T (2005) Comparison of human stem cells derived from various mesenchymal tissues: superiority of synovium as a cell source. Arthritis Rheum 52: 2521–2529.
103. Sekiya I, Vuoristo JT, Larson BL, Prockop DJ (2002) In vitro cartilage formation by human adult stem cells from bone marrow stroma defines the sequence of cellular and molecular events during chondrogenesis. Proc Natl Acad Sci USA 99:4397–4402.
104. Sen A, Lea-Currie YR, Sujkowska D, Franklin DM, Wilkison WO, Halvorsen YD, Gimble JM (2001) Adipogenic potential of human adipose derived stromal cells from multiple donors is heterogeneous. J Cell Biochem 81:312–319.
105. Shi S, Bartold PM, Miura M, Seo BM, Robey PG, Gronthos S (2005) The efficacy of mesenchymal stem cells to regenerate and repair dental structures. Orthod Craniofac Res 8:191–199.
106. Shigeno Y, Ashton BA (1995) Human bone-cell proliferation in vitro decreases with human donor age. J Bone Joint Surg Br 77:139–142.
107. Shirasawa S, Sekiya I, Sakaguchi Y, Yagishita K, Ichinose S, Muneta T (2005) In vitro chondrogenesis of human synovium-derived mesenchymal stem cells: Optimal condition and comparison with bone marrow-derived cells. J Cell Biochem Aug 8 [Epub ahead of print].
108. Srouji S, Livne E (2005) Bone marrow stem cells and biological scaffold for bone repair in aging and disease. Mech Ageing Dev 126:281–287.
109. Steck E, Bertram H, Abel R, Chen B, Winter A, Ritcher W (2005) Induction of intervertebral disc-like cells from adult mesenchymal stem cells. Stem Cells 23:403–411.
110. Tergeman G, Pittman DD, Muller R, Kurkalli BG, Zhou S, et al (2001) Engineered human mesenchymal stem cells: a novel platform for skeletal cell mediated gene therapy. J Gene Med 3:240–251.
111. Thomson JA, Itskovitz-Eldor J, Shapiro SS, Waknitz MA, Swiergiel JJ, Marshall VS, Jones JM (1998) Embryonic stem cell lines derived from human blastocysts. Science 282:1145–1147.
112. Toma JG, Akhavan M, Fernandes KJ, Barnabe-Heider F, Sadikot A, Kaplan DR, Miller FD (2001) Isolation of multipotent adult stem cells from the dermis of mammalian skin. Nat Cell Biol 3:778–784.
113. Tuli R, Nandi S, Li WJ, Tuli S, Huang X, Manner PA, Laquerriere P, Noth U, Hall DJ, Tuan RS (2004) Human mesenchymal progenitor cell-based tissue engineering of a single-unit osteochondral construct. Tissue Eng 10:1169–1179.
114. Uematsu K, Hattori K, Ishimoto Y, Yamauchi J, Habata T, Takakura Y, Ohgushi H, Fukuchi T, Sato M (2005) Cartilage regeneration using mesenchymal stem cells and a three-dimensional poly-lactic-glycolic acid (PLGA) scaffold. Biomaterials 26:4273–4279.
115. Vunjak-Navakovi G, Altman G, Horan R, Kaplan DL (2004) Tissue engineering of ligaments. Annu Rev Biomed Eng 6:131–156.
116. Wakitani S, Aoki H, Harada Y, Sonobe M, Morita Y, Mu Y, Tomita N, Nakamura Y, Takeda S, Watanabe TK, Tanigami A (2004) Embryonic stem cells form articular cartilage, not teratomas, in osteochondral defects of rat joints. Cell Transplant 13:331–336.
117. Wakitani S, Goto T, Pineda SJ, Young RG, Mansour JM, Caplan AI, Goldberg VM (1994) Mesenchymal cell-based repair of large, full-thickness defects of articular cartilage. J Bone Joint Surg Am 76: 579–592.
118. Wakitani S, Mitsuoka T, Nakamura N, Toritsuka Y, Nakamura Y, Horibe S (2004) Autologous bone marrow stromal cell transplantation for repair of full-thickness articular cartilage defects in human patellae: two case reports. Cell Transplant 13:595–600.
119. Wang CJ, Chan YS, Weng LH, Yuan LJ, Chen HS (2004) Comparison of autogenous and allogenous posterior cruciate ligament reconstructions of the knee. Injury 35:1279–1285.
120. Wang DW, Fermor B, Gimble JM, Awad HA, Guilak F (2005) Influence of oxygen on the proliferation and metabolism of adipose derived adult stem cells. J Cell Physiol 204:184–191.
121. Woodbury D, Schwarz EJ, Prockop DJ, Black IB (2000) Adult rat and human bone marrow stromal cells differentiate into neurons. J Neurosci Res 61: 364–370.
122. Xia Z, Ye H, Choong C, Ferguson DJ, Platt N, Cui Z, Triffitt JT (2004) Macrophagic response to human mesenchymal stem cell and poly(epsilon-caprolactone) implantation in nonobese diabetic/severe combined immunodeficient mice. J Biomed Mater Res A 71:538–548.
123. Yang M, Ma QJ, Dang GT, Ma K, Chen P, Zhou CY (2005) In vitro and in vivo induction of bone formation based on ex vivo gene therapy using rat adipose-derived adult stem cells expressing BMP-7. Cytotherapy 7:273–281.
124. Yoon Y, Wecker A, Heyd L, Park J, Tkebuchava T, Kusano K, Hanley A, Scadova H, Qin G, Cha D, Johnson KL, Aikawa R, Asahara T, Losordo DW

(2005) Clonally expanded novel multipotent stem cells from human bone marrow regenerate myocardium after myocardial infarction. J Clin Invest 115: 326–338.
125. Zuk PA, Zhu M, Mizuno H, Huang J, Futrell JW, Katz AJ, Benhaim P, Lorenz HP, Hedrick MH (2001) Multilineage cells from human adipose tissue: implications for cell-based therapies. Tissue Eng 7: 211–228.
126. zur Nieden NI, Kempka G, Rancourt DE, Ahr HJ (2005) Induction of chondro-, osteo- and adipogenesis in embryonic stem cells by bone morphogenetic protein-2: effect of cofactors on differentiating lineages. BMC Dev Biol 5:1.

# 2. Osteogenic Growth Factors and Cytokines and Their Role in Bone Repair

*Louis C. Gerstenfeld, Cory M. Edgar, Sanjeev Kakar, Kimberly A. Jacobsen, and Thomas A. Einhorn*

## 2.1 Introduction

Ontogenetic development is initiated at the time of fertilization and terminates with the differentiation, growth, and maturation of specialized tissues and organs. These developmental processes are characterized by molecular specialization that accompanies cellular differentiation and tissue morphogenesis. Most developmental processes terminate after birth or when animals reach sexual maturity, but some morphogenetic processes are reinitiated in response to injury in specific tissues. One such regenerative process is the repair of skeletal fractures and bone tissue after surgery, a process that recapitulates specific aspects of the initial developmental processes in the course of healing [58, 209]. Several aspects of the postnatal tissue environment of fracture healing, however, are unique and differ from what occurs in embryological and postnatal development. Understanding how cytokines and morphogens affect fracture or postsurgical healing is essential to the development of pharmacological and molecular approaches intended to enhance bone healing after surgery or traumatic injury, as well as to promote skeletal tissue engineering.

The first half of this review (Section 2.2) will focus on several groups of soluble protein factors that regulate postnatal bone repair: the tumor necrosis factor α (TNF-α) family, the transforming growth factor β (TGF-β) superfamily, angiogenic factors, and parathyroid hormone/parathyroid hormone-related peptide (PTH/PTHrP). Major emphasis has been directed to these molecules because their activities constitute current targets of pharmacological studies to promote or alter bone healing. Short reviews of the fibroblast growth factor (FGF) and Wnt families of factors are also presented in the context of their known functions in skeletal development and intended use as therapeutic agents. The second half of the review (sections 2.3–5) is focused on the anatomy and cell biology of bone healing, on what is known about the temporal and spatial expression of the various cytokines during bone healing, and how cytokines and morphogens may therapeutically modify the repair process.

## 2.2 Cytokines, Morphogens, and Growth Factors: The TNF-α Family

### 2.2.1 The TNF Family of Cytokines and Their Intracellular Functions

TNF was first identified in the early 1980s, and a large superfamily of related molecules has since been identified. So far, 18 members with

15% to 25% amino acid sequence homology and at least six cell-surface receptors have been described. The two members of this cytokine family that have been the most extensively characterized are TNF and Fas ligand (FasL). The ligands of this family are all predominantly type II transmembrane proteins. The receptors are all type I transmembrane proteins and are believed to aggregate upon interaction with their ligands. Although the extracellular side of the receptors is conserved and composed of cysteine repeats, the cytoplasmic domains of the receptors are different and mediate unique activities that lead to a multitude of biological responses through variations in their coupled signal transduction processes. These cytokines have been implicated in a wide variety of diseases, including tumorigenesis, septic shock, viral replication, bone resorption, rheumatoid arthritis, diabetes, and other inflammatory diseases [19, 121, 153, 187]. Recently, several therapeutic regimens have been approved that antagonize TNF-α activity to treat a variety of autoimmune diseases, including rheumatoid arthritis and Crohn's disease [163, 184]. Preliminary studies have also examined whether these approaches can be used to impede the loosening of orthopedic prostheses [37].

The TNF family members with the most homogeneity are TNF-α, TNF-β (LT-α), and LT-β. Both TNF-α ligands and TNF-β (LT-α) are homotrimers, whereas LT-β is a heterotrimer of $(LT\text{-}\alpha)_1(LT\text{-}\beta)_2$. There are three receptors in this family: TNFR1/p55/death receptor 1/DR1, TNFR2 (p75), and LT-β receptor. Both TNF ligands bind both TNF receptors, but LT-β/TNF-α trimers only bind to the LT receptor. FasL is a unique family member and is solely recognized by its receptor, FAS/Apo1/DR2 [211]. Most cells express TNF-α and its receptors, but the expression of TNF-β and its receptor appears to be restricted to T cells and natural killer cells. TNFR1 (p55) is constitutively expressed by almost all cells, but TNFR2 (p75) is strongly induced in immune and inflammatory responses. FasL and Fas are also expressed by many cells but show unique expression during many developmental processes, including the hypertrophy of chondrocytes [72, 174] and the regulation of immune cell differentiation [17, 23, 55, 192]. TNF-α and related cytokines either mediate programmed cell death (apoptosis) or facilitate cell survival and growth, primarily through the activation of the nuclear factor κB (NFκB) and c-Jun N-terminal kinase (JNK) transcription factors. The dichotomy of cellular responses to these cytokines resides in the receptors that are activated and the downstream signal transduction molecules that interact with these receptors. Signal transduction is mediated through a two-part system of docking proteins including MORT/FADD, TRADD, RIP, and CRADD, which bind to the death domain (DD) of the receptors, and the adaptor proteins that have been named TRAFs. Downstream from the coupled responses to TNFR1 and TNFR2 that mediate cell survival are the various mitogen-activated protein (MAP)-related kinases. Downstream from the apoptotic activation of TNFR1 and FAS is the activation of specific proteases (caspases) [19, 121, 153, 187]. There is a further bifurcation of the apoptotic cascade, with two separate pathways that can mediate apoptosis: an intrinsic (mitochondria-dependent low caspase 8) pathway and an extrinsic (mitochondria-independent high caspase 8) pathway [185]. To understand the complex regulatory functions within a tissue that are mediated through the actions of the TNF cytokine family, it is necessary to define the ligands and to specify the actions of specific receptors and the specific mechanisms of intracellular transduction within that tissue.

#### 2.2.1.1 TNF Cytokines as Arbitrators of the Tissue Microenvironment by Selective Promotion of Cell Death or Survival

The TNF family of cytokines plays a central role in the timing of the immune response, namely, when to terminate activation of the innate inflammatory response and initiate the acquired immune response, and when to terminate an innate or acquired response and initiate local tissue repair and regeneration. Thus both TNFR1 and Fas mediate activation-induced cell death in macrophages, T cells, and B cells [99, 111, 187]. The pathological manifestations of inappropriate control of the apoptotic processes in immune function are seen in mice that are deficient in TNFR1, Fas, and FAS/TNFR1. These animals exhibit more severe autoimmune disease and accelerated lymphoproliferation. These responses indicate that whereas Fas and TNFR1 receptors both activate the apoptotic cascade and carry out compensatory or redundant functions, each receptor mediates a unique set of biological responses [229]. Thus failure to initiate the programmed cell death of one or another population of immune cells that mediate the transition of the specific stages of an immune response leads to a variety of systemic autoimmune pathologies [204]. In essence, these cytokines act as the central arbitrators of a tissue's microenvironment during immune activation. They do so by promoting the survival of one population of cells while causing another to undergo apoptosis.

The TNF-α family of cytokines has been the primary focus of many immune function studies, but the death receptor family also plays a pivotal regulatory role in many developmental processes [43]. It is interesting that during postnatal tissue repair and regeneration these cytokines directly and indirectly regulate many nonimmune cell types downstream from an initial immune response [82]. The signaling functions by immune cell cytokines during postnatal tissue repair derive from functions carried out during embryogenesis. Alternatively, these cells may initiate postnatal repair or regenerative processes that replace mechanisms that functioned during embryological development. TNF-α thus functions within skeletal tissues either during the course of normal skeletal homeostasis or in response to tissue injury [158]. It does so by acting on both apoptotic and nonapoptotic events within mesenchymal cell types found in skeletal tissues. This includes specific types of mesenchymal precursors [78], osteogenic cells [1], and synovial fibroblasts [64, 137].

Recent studies have shown that activation of TNF-α and/or NFκB can affect tissue repair, response to injury, and arthritic pathology by specifically inducing the expression of morphogenetic factors of the TGF-α family [64]. It may also alter second signal activity of SMADs that mediate bone morphogenetic protein (BMP) signaling [21, 36, 57].

It is now well established that cartilage cells undergo apoptosis during normal endochondral development and during arthritic disease [3, 4, 5, 56, 72]. Currently, three members of the TNF family of cytokines have been implicated: Fas ligand (FasL), TNF-α, and TRAIL [6, 39, 83, 126]. Treatment of human articular chondrocytes with FasL in vitro causes apoptosis. Because the Fas system is present in growth-plate chondrocytes in vivo, it may play a role in chondrocyte apoptosis during endochondral development [6, 83]. In previous studies, cartilage cells within the fracture callus [224] have been shown to express Fas, and articular chondrocytes will undergo programmed cell death in response to TNF-α [69]. The relationship between the apoptotic process and the normal progression of endochondral development can be observed in pathological conditions such as rickets, as well as in the numerous genetically engineered defects that affect growth-cartilage development. The hallmark of almost all of these defects is either a foreshortening or an expansion of the growth plates. Two examples of factors causing an expansion of the growth plate are vitamin D deficiency in growing animals and the genetically engineered ablation of matrix metalloproteinase 9 (MMP-9) [210]. Ablation of the PTHrP gene, on the other hand, causes an osteochondrodysplasia, primarily manifested in an accelerated hypertrophy and removal of the chondrocytes. A phenomenon common to these very different pathologies of the endochondral process is that in all three the timing or rate of chondrocyte apoptosis has been altered. The consequence of an abnormally timed apoptosis is that the microenvironment of the endochondral tissue is altered by retention or loss of the chondrocytes. This is important because osteogenesis, vascular invasion, and marrow formation follow in sequence as the chondrogenic cells hypertrophy and undergo apoptosis [120]. Thus, in analogy with their role during the immune response, the death receptors and ligands during endochondral development promote the removal of one cell population (chondrocytes) and are permissive for osteogenic and marrow cell populations to move into the space previously occupied by the cartilage tissue.

#### 2.2.1.2 Role of the TNF-α Family of Cytokines in Bone Remodeling

As just discussed, embryologic development and postnatal growth are regulated by ontogenetic and systemic hormonal mechanisms.

Fracture and skeletal tissue healing after surgery, on the other hand, are initiated in response to regulatory mechanisms associated with inflammation and the innate immune response [16, 54]. Two discrete types of resorption take place during fracture repair. The first occurs at the end of the endochondral period, in the course of which mineralized cartilage is removed and primary bone is formed. TNF-α and its receptors remain largely unexpressed during the initial periods of endochondral differentiation, but are expressed as the cartilage cells hypertrophy and tissue resorption begins. During this same period, there is an increase in the concentration of RANKL and osteoprotegrin (OPG) (two members of the TNF-α superfamily) as well as macrophage colony-stimulating factor (M-CSF), all key regulatory factors in osteoclastogenesis [118]. However, other cytokines that are associated with bone remodeling, including interleukin 1α (IL-1α), IL-1β, and IL-6 [115], are not expressed. The other type of resorption occurs during secondary bone formation, which follows the endochondral phase. These events are comparable to the process of coupled remodeling seen in normal bone homeostasis. During this period, expression of IL-1 and IL-6 increases, whereas the levels of OPG, M-CSF, and RANKL decline.

These data suggest that the processes mediating endochondral resorption and the more prolonged phase of secondary bone remodeling differ and that the resorption of the mineralized cartilage is more dependent on the activities of M-CSF, OPG, and RANKL. In contrast, bone remodeling appears to depend on the levels of RANKL and to be coregulated by the activities of the cytokines IL-1, IL-6, and TNF-α found in bone marrow. Differences between bone and cartilage remodeling are apparent from studies of RANKL (TRANCE)-deficient mice and of mice whose RANKL expression was rescued by engineering RANKL expression in their lymphocytes. When RANKL was expressed by lymphocytes in the knockout mice, their osteopetrosis was overcome and osteoclast development was promoted. However, it was not possible to correct the chondrodysplasia of the epiphyseal and metaphyseal regions. The authors therefore concluded that cartilage and bone possess different mechanisms that induce RANKL expression [114]. In this context it is interesting to note the extreme differences in the avascular microenvironment of cartilage and bone. Indeed, the interactions of hematopoietic/lymphopoietic and osteogenic microenvironments in regulating bone remodeling are emerging as a major area of research, and changes in cytokines that alter lymphopoiesis affect both bone homeostasis and immune function [30, 105, 175, 206].

## 2.2.2 The Bone Morphogenetic Proteins (BMPs)

### 2.2.2.1 BMPs and Signaling

On the basis of their distinct structural characteristics, BMPs (with the exception of what has been named BMP-1) are members of the transforming growth factor β (TGF-β) superfamily. This family also includes activins, inhibins, and growth and differentiation factors (GDFs). BMP-1 belongs to the astacin family of metalloendopeptidases and exhibits BMP-like activity by proteolytically activating mixtures that contain the proforms of BMP.

The TGF-β superfamily of pre-proproteins displays extensive amino acid sequence homology across species and can carry out a wide diversity of biological functions. The proteins share a characteristic pattern of seven conserved cysteine residues within the carboxy-terminal mature region that are essential for the formation of cysteine knot domains. This tertiary protein structure is thought to be critical for receptor interaction [220]. The mature-region cysteines are also important in the formation of intermonomeric disulfide bonds necessary for the formation of physiologically functional dimers [119]. As in most secreted proteins, there are numerous potential N-linkage glycosylation sites located throughout the amino acid sequence. Most BMPs induce some level of glycosylation, which varies among species, with mouse BMP inducing the lowest and bovine BMP inducing the highest degree of glycosylation [183, 212].

As an example of a typical BMP structure, BMP-2 is translated as a 396-amino-acid pre-proprotein that contains a 19-amino-acid signal sequence for targeted secretion, a 263-amino-acid proregion, and a 114-amino-acid mature segment. Within the mature region of BMP-2, seven cysteines and one N-linked glycosylation recognition site are identifiable. The mature protein has a predicted mass of 14 kDa with an observed mass of 18 kDa, presumably due to glycosylation. The functional protein exists as a homodimer that is linked by two disulfide bridges. There is some speculation on the existence of heterodimeric complexes in some situations, although in normal physiological settings homodimeric complexes among the BMPs are most common [183, 217]. Considerable amino acid sequence similarity exists between species for the various family members. Approximately 16 BMPs have been characterized, with the majority demonstrating a high percentage of amino acid sequence homology among the different isotypes, in addition to a high level of amino acid conservation between species [119, 217].

BMPs initiate their signaling at the cell surface through interaction with two distinct serine/threonine kinase receptors: a type I receptor (50–55 kDa) and a type II receptor (more than 75 kDa) [220]. It appears that they weakly interact with certain members of the type II receptors independently of type I receptors, but in the presence of both receptors their binding affinity is increased dramatically [133]. Following receptor dimerization induced by BMP ligand binding, the type II receptor transphosphorylates the type I receptor, which subsequently transmits the BMP signal by activation of intracellular Smad (Sma and Mad) proteins. This activation is accomplished by the directed phosphorylation of specific serine or threonine residues within the Smad proteins. The structures of the two receptors are similar in that they contain N-glycosylated extracellular domains, a single membrane-spanning domain, and an intracellular serine/threonine kinase domain. The extracellular domains have several conserved cysteine residues believed to facilitate the formation of essential three-dimensional structures involved in BMP binding [59]. One distinction between the two receptor types is the presence of a glycine- and serine-rich domain (GS domain) found on the type I receptor within the intracellular N-terminal to the serine/threonine kinase domain. This region is important for the transmission of the BMP signal to intracellular second-messenger proteins by facilitating the receptors' ability to interact with Smad proteins. This was highlighted in an amino acid mutagenesis study linking Smad 7 activation to type I receptor phosphorylation in the GS domain by the type II receptor [134].

Currently, seven type I receptors, termed activin receptor-like kinases (ALKs) 1–7, have been identified in mammals. ALK-3 (BMP type IA) and ALK-6 (BMP type IB) receptors share an 85% amino acid sequence identity in the kinase domains, and both bind BMP-4, BMP-2, GDF-5, and BMP-7 [149]. Truncated forms of the ALK receptors are currently being used to examine the role of BMP signaling during the development of numerous types of tissues. On the other hand, there are only three BMP type II receptors that can interact with BMPs. The BMPR-II receptor seems to bind exclusively to BMPs, but the activin types IIA and IIB have affinities for specific BMPs (BMP-7, BMP-2 and GDF-5), in addition to their activin binding [220]. BMPR-II binds all BMPs weakly by itself, with a dramatic increase in the binding affinity following recruitment of the type I receptors.

BMP-2 ligand and receptor interactions have been carefully studied (160). During BMP-2 receptor activation, the BMP-2 protein contains two distinct domains that facilitate receptor interaction. The first is a large, high-affinity binding site (termed the "wrist epitope"), which interacts with the BMPR-IA. The second is a low-affinity binding site (termed the "knuckle epitope"), which interacts with BMPR-II [59]. The wrist epitopes from monomers (BMPs are dimeric structures) contribute to the binding of the BMPR-IA receptor, whereas the knuckle epitope from only one monomer binds to BMPR-IA. The juxtapositioning of these regions facilitates a close proximity of the receptors and initiation of intracellular signaling from inter-receptor type II phosphorylation to type I. Transphosphorylation eventually leads to the activation of Smad proteins and signal transmission to target downstream responsive genes [149].

Within the cell BMP signals are transduced by the Smad molecules. To date, eight Smad mammalian proteins have been isolated and characterized. Smad proteins are the direct downstream signaling molecules of BMPs and other TGF-β superfamily members and are activated directly by their serine/threonine kinase receptors. These proteins can be classified into three distinct groups based on their intracellular function. The receptor-regulated Smads (R-Smads) are the direct signal transducers from the BMP receptor complex following receptor transphosphorylation events. Smads 1, 5, and 8 interact with types I and II BMP receptors and are subsequently phosphorylated by the type I receptor within their COOH-terminus at the conserved SSXS motif [207]. They are then rapidly released from the receptor and subsequently interact with a common mediator Smad (co-Smad). Smad 4 is the only known co-Smad that signals in both the BMP and the TGF-β transduction pathways [207]. The R-Smad and the co-Smad proteins form active hetero-oligomeric complexes, which can then translocate to the nucleus and regulate the transcription of specific downstream genes. The nuclear localization of the Smad complexes is dependent on nuclear localization signals present on Smad 4. Consequently, this protein displays constant nuclear–cytoplasmic shuttling and is capable of autonomous nuclear import and export [218]. The third class of Smad proteins consists of the inhibitory Smads (I-Smads), Smad 6 and Smad 7, which exert their inhibitory effect by binding to the type I receptor and competing with the R-Smads for binding to the phosphorylated type I receptor.

All Smads share two conserved regions termed Mad homology domains 1 (MH1) and 2 (MH2). MH1 is found in the N-terminal portion of the protein, whereas MH2 is in the C-terminal portion, with a linker region of variable length and amino acid sequence separating the two domains [150]. The MH2 domain contains protein–protein interaction sequences and is important in R-Smad/co-Smad oligomerization. The MH1 domain seems to carry specific DNA-binding sequences necessary to act at the DNA level in the discrimination of gene regulation. However, a putative "Smad consensus sequence" has yet to be determined [107].

### 2.2.2.2 BMPs and Developmental Regulation

BMPs are considered one of the major groups of morphogenetic factors that mediate patterning and growth of many tissue types during embryogenesis and organogenesis. In the absence of specific BMPs, certain systems fail to develop, resulting in embryonic defects or lethality. The complete ablation of BMP-2 by homologous recombination resulted in embryonic lethality when bred to homozygosity [228]. These animals had distinct cardiac defects consistent with the expression patterning of BMP-2 in the extraembryonic mesoderm and promyocardium [228]. BMP-2 is expressed in a variety of embryonic nonskeletal epithelial and mesenchymal tissues known to play important roles in morphogenesis [139]. For example, during limb development high levels of transcripts were found in the ventral ectoderm and apical ectodermal ridges of the developing limb buds. In addition, BMP-2 expression was detectable in the developing heart, whisker follicles (ectodermal placodes), tooth buds (epithelial buds, dental papillae, and odontoblasts), and craniofacial mesenchyme [139]. Although other studies have validated the importance of BMP-2 during a wide array of mesodermal developmental processes, the protein also plays important roles in regulating the postnatal development of mesenchymal skeletal tissues [176]. In animals with a homozygous deletion of the mature coding region of the BMP-4 gene, development fails at an extremely early stage. The mice fail to develop the necessary primordial germ cells (PGCs) to form a functional embryo [123]. Lawson et al. have shown that BMP-4 promoter-driven LacZ expression in embryos prior to gastrulation results in BMP-4 expression in the extraembryonic ectoderm, followed by expression in the extraembryonic mesoderm [123]. These authors concluded that the initiation of the germ line in the mouse was dependent on secreted BMP-4 signals from the previously segregated, extraembryonic, trophectoderm lineage. This places BMP-4 function at one of the earliest stages of development [123]. However, BMPs do not appear to act individually but in a coordinated network. For example, the above-mentioned PGC cell generation is directed by more than just BMP-4. In fact, one study has demonstrated that BMP-2 is primarily expressed in the endoderm of mouse pregastrula and gastrula embryos and that the PGC generation in the mouse embryo is regulated not only by extraembryonic ectoderm-derived BMP-4 and BMP-8B, but also by endoderm-derived BMP-2 [223].

BMP-7 has been extensively studied. It is expressed later during mammalian development, but its function is redundant with that of other BMPs, since knockout animals survive through gestation. However, BMP-7-deficient mice die shortly after birth because their kidneys do not develop normally [86]. In situ hybridization analysis has shown that the absence of BMP-7 affects the expression of molecular markers of nephrogenesis, such as Pax-2 and Wnt-4, between 12.5 and 14.5 days postcoitum [138]. In addition, BMP-7-deficient mice have defects in the eye that appear to originate during lens development. Skeletal patterning defects affect the rib cage, the skull, and the hind limbs; this shows the wide influence BMPs have in mammalian development [138]. The importance of BMPs, however, is restricted neither to skeletal development nor to prenatal development. BMP-2 expression has been reported to be critical for both extraembryonic and embryonic development [101], with BMP-2 shown to be essential for cranial neural crest production. Without it, the skeletal and neural derivatives failed to develop. The importance of BMPs during development has been most extensively studied in *Xenopus*. If BMP-4 signaling is disrupted transgenically by expression of a dominant negative form of its receptor, the ventral mesoderm is converted to a dorsal mesoderm [197]. In situ hybridization in *Xenopus* showed that BMP-4 is expressed in a spatially and temporally restricted manner. Disruption of the pattern of BMP-4 expression by localized microinjections of rhBMP-4 severely disturbed embryonic development [49]. These experiments make it clear that BMP-4 regulates dorsal-ventral patterning in terms of both location and temporal expression. As a morphogen, BMP-4 modulates mesodermal patterning by establishing concentration gradients that cells detect during migration. Further evidence of the significant role active BMPs play in the control of differentiation comes from experiments that have examined the regulation and responsive expression of specific BMP antagonists, such as Noggin [44, 49]. The coordinated expression of BMP antagonists interferes with BMP function in somite and limb development [172, 190].

#### 2.2.2.3 BMP Function in Skeletal Repair

In 1965, Marshall R. Urist demonstrated that the implantation of demineralized bone at extraskeletal sites induces de novo formation of cartilage and bone [203]. This seminal observation led to investigations culminating in the extensive purification of the osteoinductive activity of demineralized bone matrix (DBM) and the sequencing and cloning of the individual BMPs [35, 166, 217]. The subsequent expression of BMPs in recombinant systems permitted their use in a variety of animal models, in particular to demonstrate their stimulating effects on the repair of fracture and skeletal defects [53, 67, 221]. Even though exogenous BMPs may enhance fracture healing, our understanding of their role in skeletal repair and regeneration remains incomplete.

Using reverse-transcriptase polymerase chain reaction (PCR) amplification, Nakase et al. were the first to demonstrate the temporal and spatial distribution of BMP-4 in fracture healing [157]. In an investigation using a monoclonal antibody against BMP-2 and BMP-4, Bostrom et al. delineated the expression of these BMPs over a 4-week period of fracture healing [26]. Recently, Cho et al. [38] have shown that specific members of the TGF-β superfamily, including the BMPs, may act in combination to promote the various stages of intramembranous and endochondral bone formation observed during fracture healing. Using ribonuclease protection analysis, this study demonstrated that BMP-2 has an early peak in expression on day 1 of fracture healing. This suggests that BMP-2 may be the most upstream mediator in the cascade of BMP expression. BMP-3 appeared to be preferentially associated with intramembranous bone formation, whereas BMP-4, -7, and -8 may function in osteoblast recruitment during both intramembranous and endochondral ossification. Taken together, these studies suggest that the coordinated expression of multiple BMPs and their receptors during fracture healing is important in both skeletal development and skeletal repair. However, the roles of specific BMPs during fracture healing need to be investigated.

### 2.2.3 Angiogenic Factors

Angiogenesis is the process by which new blood vessels are formed from pre-existent vessels. It is important for almost all embryological development and in wound healing, because the higher metabolic activities of cells within developing and healing tissues increase their nutrient and oxygen requirements [32].

Two classes of angiogenic factors and their receptors are associated with new vessel formation [60, 130]. These are the vascular endothelial growth factor (VEGF) [61] and the angiopoietin (Ang) [98] families.

VEGFs promote vascular permeability and stimulate mitogenesis in vascular endothelial cells. In conjunction with the angiopoietins (see below), VEGF stimulates endothelial-cell survival by inhibiting endothelial-cell apoptosis. The VEGFs are produced primarily in response to hypoxia-induced transcription factors (Hif 1$\alpha$ and Hif 2$\alpha$), which are expressed by many stromal and extracellular matrix (ECM)-producing cells in tissues with a high degree of vascularization. Vascular endothelial cells express most receptors for the various VEGF isoforms and are the primary responders to VEGF.

The angiopoietins, like the VEGFs, are expressed by stromal, mesenchymal, and smooth-muscle cells of larger vessels. Their receptors are expressed primarily on endothelial cells. Angiopoietins appear to be intimately involved in vessel remodeling and may play a particular role in wound-healing and tissue-repair situations where there are pre-existent vessels [171, 202]. The expression of Ang 2 is up-regulated by hypoxia and the associated Hif 1$\alpha$ factor, VEGF, angiotensin II, leptin, and estrogen. Ang 2 expression is down-regulated by basic fibroblast growth factor (bFGF). TNF-$\alpha$ also regulates Ang 2 expression, with up- or down-regulation dependent on the tissue type [75]. Ang 1 expression, although not extensively characterized as yet, appears to be up-regulated in response to hypoxia [167].

Unlike VEGF, angiopoietins are not mitogenic but promote cell survival by blocking apoptotic signals. Ang 1 also has strong chemoattractant properties for endothelial cells and promotes the adhesion of hematopoietic stem cells. Angiopoietins appear to stimulate both dissolution and migration of endothelial cells from pre-existent vessels and, in conjunction with VEGF promote cell survival and stabilize newly formed vessels in [98, 171]. Recent studies have shown Tie 2/angiopoietin signaling to regulate the hematopoietic stem-cell quiescence niche in the bone marrow niche. As a result, the hematopoietic cells are protected from myelosuppressive stresses [9].

#### 2.2.3.1 The VEGF Family and Receptors

The VEGF family of genes is currently known to comprise five related genes: VEGF A, B, C, and D and placental growth factor (PlGF). All of these have some sequence similarity to platelet-derived growth factor (PDGF). The VEGF proteins are roughly 45 kDa in size and exist as homodimers. Some of the VEGF isotypes bind to heparin. This enhances retention in the ECM and presentation to cellular receptors. Of the genomic subtypes, VEGF A is the most prevalent, based on tissue distribution and expression levels. Selective exon splicing leads to variants of VEGF A, of which six have been identified. They are denoted as VEGF 121, 145, 165, 183, 189, and 206, based on their amino acid lengths. Of these, VEGF 121 and 165 appear to be the most commonly expressed, whereas the 165, 189, and 206 variants maintain exons that encode the heparin-binding domains [201].

VEGFs have multiple receptors, including VEGFR1, also known as Flt-1, VEGFR2 (KDR or Flk1), and VEGFR3 (Flt-4). Each of these receptors is characterized by multiple IgG-like extracellular domains, and each is coupled to intracellular signaling networks through an intracellular domain that has tyrosine kinase activity. Two other more distantly related membrane receptors, neuropilin 1 and 2, also interact selectively with various VEGF molecules. VEGFR1 also exists in a soluble form that lacks the ability for intracellular signaling and antagonizes the less soluble form of VEGFR1. Each of the VEGF receptors differs in its interaction with the VEGF isotypes. VEGF A interacts with VEGFR1 and 2 and both neuropilins, VEGF B interacts with VEGFR1 and neuropilin 1, and VEGF C and D interact with VEGFR2 and 3, whereas PlGF only interacts with VEGFR1. These receptors also have the ability to signal utilizing a variety of intracellular pathways and can activate PLC, Ras, Shc, Nck PKC, and PI3 kinase.

#### 2.2.3.2 Angiopoietins and Tie Receptors

Three angiopoietins (Ang 1, 2, and 3/4) have been identified. They are made up of 498 amino acids and have a coil domain that is separated by a hinged region from a fibrinogen-like domain. Angiopoietins exist as multiple splicing variants and only interact with the Tie 2 receptor.

Angiopoietins bind to Tie 1 and 2 receptors, a tyrosine kinase with immunoglobulin and epidermal growth factor homology domains. Ligand binding induces receptor dimerization, which causes autophosphorylation of the receptor, thereby activating its kinase signaling [143]. Other studies have shown that the Tie 1 receptor is proteolytically modified when endothelial cells interact with VEGF. This suggests some coordination between the signaling events that are mediated by angiopoietins and VEGF. Interestingly, although the Tie receptors are tyrosine kinases, they do not signal through the MAP kinase system used by VEGF, but appear to recruit various phosphatases selectively, including SHP2, a factor that promotes cell migration by altering activities of focal adhesion kinases.

### 2.2.3.3 The Role of Angiogenic Factors in Bone Development

Angiogenesis is important during intramembranous and endochondral bone formation. Vascularization of the growth plate contributes to the coupling of chondrogenesis and osteogenesis. Chondrocyte apoptosis and osteoclast recruitment and activation are essential terminal stages of cartilage hypertrophy. The osteoclasts resorb the mineralized cartilage and thereby permit bone formation by osteoblasts. Morphological evidence suggests that chondrocyte apoptosis occurs readily following the invasion of endothelial cells [56, 90] and that chondrocyte death is induced by diffusible factors that arise either from the vasculature or from hematopoietic elements brought in during angiogenesis [71, 72]. The newly developing blood vessels in addition establish the conduit for the cells that form primary bone following resorption of the mineralized trabeculae of cartilage [59].

The interrelationship between blood-vessel formation and osteogenesis has been studied by various approaches aimed at inhibiting VEGF signaling. Because mice whose VEGF has been ablated die as embryos, studies have utilized inhibitors of VEGF signaling or select ablation strategies to assess the contribution of vessel formation to new bone formation. Administration of a soluble VEGF receptor 1-immunoadhesin, mFlt(1–3)-IgG, completely blocked new vessel formation in the growth plates of mice and impaired chondrocyte apoptosis and trabecular bone formation [66]. Studies by Gerber et al. (66) have identified VEGF as the key factor that regulates capillary invasion, growth-plate morphogenesis, and cartilage remodeling. In mice, systemic inhibition of VEGF during periods of rapid growth has led to inhibition of angiogenesis and to a decrease in the number of chondroclasts/osteoclasts/osteoblasts at the growth plates. Chondroclasts/osteoclasts belong to the monocyte cell lineage, express VEGFR, and migrate in response to VEGFR1-selective ligands. This indicates that VEGFR1 has a role in monocyte migration [15]. Because osteoblasts express both VEGF receptors and neuropilin 1 [80], the decrease in osteoblasts at the growth plates in anti-VEGF-treated mice reflects an impairment of VEGFR or neuropilin signaling. This in turn has impaired recruitment and/or differentiation of these cell types. Thus, VEGF contributes importantly not only to angiogenesis, but also to osteogenesis. In mice lacking the VEGF gene, the long bones demonstrate a disturbed vascular pattern at birth, consistent with reduced bone growth [140]. Osteoblast and hypertrophic chondrocyte development are also impaired [140].

VEGFs play an important role in regulating bone remodeling. They do so by attracting endothelial cells, osteoblasts, and osteoclasts [46, 147] and by autocrine regulation of chondrocyte function [33]. Local administration of VEGF also enhances osteoclast number [100]. Further linkage between VEGF and bone formation was recently described by studies in which hypoxia was shown to drive BMP expression through VEGF [27]. A number of recent studies have also shown that BMPs stimulate the expression of VEGF by osteoblasts and osteoblast-like cells [47, 222]. Finally, the tissue-specific regulation of VEGF expression during bone development seems to be dependent on the expression of Cbfa1/Runx2, known as the key transcriptional factor that regulates the commitment of mesenchymal cells to the skeletal-cell lineage [226]. Taken together, these findings provide a considerable body of evidence in support of the concept that VEGF mediates bone formation by direct stimulation of osteogenesis and indirectly by its effects on vascularization.

#### 2.2.3.4 The Role of Angiogenic Factors in Tissue Healing

Fracture healing and bone or tissue repair result in an up-regulation of blood flow, so that bone regeneration can occur within the callus or repair tissues [10, 52, 179]. The importance of vascularization during fracture repair was confirmed by studies showing that broad-spectrum angiogenic inhibitors completely prevented fracture healing, callus formation, and the formation of periosteal woven bone [84, 195]. In contrast, treatment of healing fractures with VEGF improved bone healing and led to more rapid mineralization of the callus and regaining of mechanical strength [195].

The role angiogenesis plays in osteogenesis following distraction rupture has been extensively studied with the aid of an artificially produced gap following osteotomy. When this technique is used, new bone forms primarily via an intramembranous mechanism with extensive revascularization of the regenerated bone. Within the marrow space, venous sinusoids are formed that parallel the newly grown trabeculae. Analysis of experimental models of osteogenesis following distraction has revealed an early intense vascular response, with the newly formed vessels maturing into sturdier vessels capable of withstanding the tensile forces that are generated in the distraction gap [129, 179].

As discussed above, angiogenesis appears to involve two separate pathways: a VEGF-dependent pathway and an angiopoietin-dependent pathway. Interestingly, both Ang 1 and Ang 2 have been identified in bone cells during development [89] and in bone cells that arise in osteogenesis following distraction and fracture healing [34, 127, 128]. Indeed, in studies of mice that had undergone distraction fracture, Ang 1, Ang 2, and their Tie receptors were expressed throughout healing at the same time that VEGF A and VEGF C were expressed. These regulators of angiogenesis were expressed throughout the chondrogenic phase of healing, reaching maximum levels during the late phases of endochondral remodeling and during bone formation. These studies, as well as fracture-healing studies, showed that during fracture healing Ang 2 was the factor with the highest expression. Unlike the VEGF family, which promotes new vessel formation by stimulating endothelial cell division, Ang 2 promotes destabilization and regression of blood vessels in the absence of VEGF A or bFGF [87, 135, 141]. Recent findings have suggested that Ang 2, along with VEGF, promotes new vessel formation by inducing remodeling of the capillary basal lamina and by stimulating endothelial-cell sprouting and migration [141]. This suggests that Ang 2 expression plays a role similar to that of VEGF in bone repair. By itself, Ang 2 inhibits blood-vessel formation, but in combination with VEGF it stimulates new vessel formation and plasticity in existing vessels.

These studies also pointed to collaborative interactions between VEGI (vascular endothelial growth inhibitor)-induced angiogenesis and the TNF-$\alpha$ family of regulators. Interaction of VEGI with death receptor 4 and the primary regulator of the progression of vascularization reflects a dual role: maintaining growth arrest of endothelial cells in G0/G1 interfaces, while at the same time inducing apoptosis in cells that enter the S phase [79, 225, 227]. Taken together, these results suggest that, after injury, vessels are dissociated into a pool of nondividing endothelial cells through the actions of Ang 2. They are then are held in this state through the actions of VEGI, which stimulates apoptosis of all cells that enter the S phase. When endochondral remodeling is initiated, VEGF levels rise, stimulating cell division and allowing endothelial cells to contribute to neoangiogenic processes. The concept of controlled cell regression and growth is also consistent with the role that angiopoietin is thought to play in blood-vessel formation [87].

### 2.2.4 Parathyroid Hormone (PTH)/ Parathyroid Hormone-Related Peptide (PTHrP) and PTHrP Signaling

#### 2.2.4.1 PTH Versus PTHrP: Endocrine Versus Paracrine Effects

PTH, a peptide, is an hormone that is synthesized by the parathyroid gland. The mature form of the peptide is 84 amino acids in length. PTHrP, on the other hand, is an autocrine/paracrine factor that was first discovered as the primary cause of malignant hypocalcemia in many cancers [28]. It is normally expressed during development and in many postnatal tissues, including cardiac, vascular, mammary, cartilage, and renal tissues, as well as a number of other epithelial surfaces. The mature form of PTHrP is 141 amino acids in length. Even though PTH and PTHrP bind to the same receptor, the two molecules share only a limited sequence homology along the first 34 amino acids of their amino terminal sequences and diverge considerably in their carboxyl domains.

The effects of the major calcitropic hormone PTH on skeletal cells are very important clinically, owing to the role played by the skeleton in mineral homeostasis. Both molecules have similar systemic effects on mineral metabolism, yet they differ in amino acid composition and physiological function. PTH and PTHrP share a common receptor (PTHR1) and, when in the circulation, are primarily targeted to the kidney and skeleton [62]. In the kidney, PTH and PTHrP bring about their calcitropic effects by stimulating calcium reabsorption and phosphate excretion in the distal end of the collecting tubules. They also regulate formation of the active vitamin $D_3$ metabolite, 1,25-dihydroxyvitamin $D_3$ by activating the enzyme that carries out the 1α-hydroxylation of 25-hydroxyvitamin $D_3$ in the proximal tubules. This leads to a rise in serum calcium and a lowering of phosphate level. The effects on the skeletal system are less well understood. PTH binds to the receptors of osteoblasts [177], which produce paracrine factors that induce increased activation and recruitment of osteoclasts.

PTH and PTHrP, like other peptide hormones, mediate their effects through interaction with a receptor. Two forms of this receptor are known, but the two peptides interact primarily with PTH1R. This receptor has seven transmembrane domains and is closely related to a subset of similar receptors that include the calcitonin and secretin receptors [65]. PTH and PTHrP bind almost identically to the receptor, which has both endocrine and autocrine/paracrine functions in the tissues in which it is expressed [2].

#### 2.2.4.2 PTH Receptor Signal Transduction and Nuclear Effects

Receptor activities are modulated through interaction with heterodimeric (α, β, γ) G proteins that activate or inhibit cyclase production of cAMP. The levels of cAMP then control the activity of protein kinase A (PKA), which serves as the cAMP intracellular second signal transducer [77]. The activation of the receptor by ligand binding also activates phospholipases Cβ through Gαq11. Activated phospholipase generates diacylglycerol and 1,4,5-inositol triphosphate (IP3). These two molecules activate both protein kinase C (PKC) and $Ca^{2+}$ release. Study of the "cross-talk" between the PKA-arm, PKC kinases [125], and $Ca^{2+}$ will likely sort out the many parallel and sometimes antagonistic functions of the PTH and PTHrP ligands in different target-cell populations [76].

At the nuclear level, both the PKA and the PKC families of kinases mediate their actions through the phosphorylation of members of the leucine zipper family of transcription factors [125]. These transcription factors, when phosphorylated, may activate or inhibit the transcription of specific genes [113, 132] and may be classified into two broad groups: the cAMP response element-binding protein family (CREBs) and members of the AP-1 family. The CREBs include the CREB, CREM, and ATF classes of factors; the primary members of the AP-1 family include fos, jun, and fra [8, 77, 125, 131, 173, 205]. In general, the actions of PKA are mediated through the phosphorylation of members of the CREB family, while PKC appears to act on members of the AP-1 family. However, phosphorylation may not be restricted to one type of kinase or individual factors. The factors are active when dimeric. Members of both families can undergo specific heterodimerization with one another [125]. Heterodimerization gives rise to a diversity of specific transcription factors. As a result, genes may be expressed or silenced in a tissue-specific fashion in response to common second signals [77]. Similarly targeted changes or ablation of these transcriptional regulators give rise to specific skeletal tissue phenotypes.

Extensive data have been accumulated to suggest that the leucine zipper family of transcription factors plays a major role in the regulation of gene expression and development in the skeleton. Studies in which both the *c-fos* and the *v-fos* genes were virally introduced have shown that *fos* expression generated osteosarcomas [74, 180, 181]. *C-fos* knockout mice develop osteochondrodysplasia, overproduce hypertrophic cartilage, and cannot replace bone. This condition in some ways looks like osteopetrosis [213]. Other studies have also shown increases in *c-fos* proto-oncogene in bone from patients with fibrous dysplasia in whom bone formation is overexpressed or bone forms ectopically [31]. Studies examining different members of the basic leucine zipper protein family have demonstrated that both ATF-2 and hXBP are expressed in skeletal tissues [40, 173]. Ablation of ATF-2 leads to a defect in endochondral ossification with a histopathology similar to human hypochondroplasia [173].

#### 2.2.4.3 The Role of PTHrP in Endochondral Development

The discovery of PTHrP as the primary factor in malignant hypocalcemia constituted a major advance in understanding the systemic effects of many malignancies. However, the subsequent characterization of PTHrP as an essential autocrine/paracrine factor in skeletal-tissue development was equally important. Initial animal studies demonstrated that PTHrP mRNA was fully expressed in perichondral cells and in the chondrocytes found in the proliferating zones of endochondral growth plates. The PTHrP receptor was expressed progressively more fully as endochondral chondrocytes matured toward their terminal hypertrophic state. Missense expression studies of PTHrP in developing avian embryo growth plates and studies in transgenic animals with the targeted ablation of PTHrP have demonstrated a complex negative feedback loop that involves Indian hedgehog (Ihh) regulation of the progression of chondrocyte development during endochondral bone formation [102, 122, 208]. These studies demonstrate that Ihh positively regulates PTHrP expression, as a result of which cells are maintained in an undifferentiated state, with Ihh promoting proliferation and thereby expanding the population. The increased output of PTHrP causes Ihh expression to be down-regulated by expanding the pool of proliferating immature chondrocytes. Ihh-ablated transgenic mice have no detectable PTHrP in their growth plates, which are made up mainly of hypertrophic chondrocytes. Conversely, animals lacking PTHrP have very small zones of proliferating chondrocytes and exhibit a premature transition to cellular hypertrophy and mineralization [136, 194]. Interestingly, if the Ihh-ablated mice are engineered to have a constitutively active PTH1R receptor, premature chondrocyte hypertrophy is prevented, but proliferation of the chondrocytes in the growth zones is still diminished. These findings suggest molecular mechanisms that are regulated by Ihh and not activated by PTHrP control cell proliferation [103].

A further understanding of the developmental role of PTHrP has been gained from studies in two different human chondrodysplasias. One type of chondrodysplasia is a nonlethal, autosomal dominant disorder that was first identified by Jansen in 1934 [95]. It has subsequently been characterized at a molecular level as a constitutively activating mutation in the PTH1R receptor [186]. The patients are characterized by short limbs caused by severe abnormalities in their growth plates and associated hypocalcemia. The second type of chondrodysplasia was identified by Blomstrand et al. [22]. It is a prenatal lethal chondrodysplasia characterized by abnormal bone ossification and shortened limbs. In these fetuses, the limbs show very advanced endochondral development, and the disease is characterized by an autosomal recessive pattern of inheritance. Molecular analysis of these patients suggested the disease is due to an inactivating mutation in the PTH1R receptor [97].

#### 2.2.4.4 PTH as a Therapeutic for Osteoporosis and Augmentation of Fracture Healing

Numerous recent studies have focused on the systemic effects and potential therapeutic applications of PTH [48, 159, 169]. The continuous infusion of PTH into mammals induces catabolic events, increases bone remodeling, and leads to a loss of skeletal bone mass. On the other hand, intermittent dosing seems to have anabolic effects and results in increased bone mass [116, 151, 199]. Clinical trials utilizing the 1–34 PTH peptide have increased bone mineral density and reduced the risk of vertebral and nonvertebral fractures in postmenopausal women. Intermittent treatment also has improved the bone mass in osteoporotic men [85].

The recent approval of PTH(1–34) as an anabolic treatment for osteoporosis has been a major impetus to the use of PTH in bone healing. PTH administration has enhanced early fracture healing in parathyroidectomized rats [63], with PTH doses ranging from 10 to 200 μg/kg having significantly improved the mechanical and histological aspects of normal fracture repair in the rat [7, 88, 155]. PTH analogs have also been shown to reverse the inhibition of bone healing in ovariectomized rats [112] and in corticosteroid-treated rabbits [25]. PTH(1–34) is reported to increase bone ingrowth and pullout strength in porous metallic implants [193].

One drawback of the rat studies is that the hormone doses were much higher than would be tolerated in humans. To evaluate the clinical potential of PTH for fracture healing, patient-appropriate doses of recombinant human parathyroid hormone [PTH(1–34); teriparatide; Forteo™] were used in a well-established rat model. As early as day 21 of this study, calluses from the group treated with 30 μg/kg of PTH showed significant increases over controls in terms of torsional strength, stiffness, bone mineral content (BMC), bone mineral density (BMD) and cartilage volume. By day 35, both the 5-μg and the 30-μg/kg PTH-treated groups showed significant increases in BMC, BMD, and total osseous tissue volume; the experimental groups also showed significant decreases in void space and cartilage volume. At day 35, torsional strength was also significantly increased in the group treated with 30 μg of PTH. Even after 84 days, the group that had received 30 μg of PTH for 21 days, with treatment discontinued thereafter, exhibited increases in torsional strength and BMD over comparable control values. Thus, daily systemic administration of a low dose of PTH(1–34) enhanced fracture healing and induced an anabolic effect throughout the entire remodeling phase.

## 2.2.5 Other Growth Factors Within Skeletal Tissues

### 2.2.5.1 Fibroblast Growth Factor (FGF)

FGFs were originally isolated as oncogenes and shown to stimulate cell proliferation [196]. The current family of closely structurally related proteins is encoded by at least 22 genes. Four distinct FGF receptors each a unique gene product, mediate activity through tyrosine kinase activity. Each receptor appears to be activated by all members of the FGF ligand family. The ligands in general have heparin-binding activity and, when complexed with heparin, have improved activity. FGF ligands regulate a wide variety of cellular functions and can act as mitogens, chemoattractants, and mediators of cellular differentiation. FGF receptor activity appears to directly regulate the expression of a number of different proteins, including metalloproteinases and morphogens [142, 164, 165].

The roles played by FGF in skeletal development have been elucidated by identifying autosomal dominant mutations that constitutively activate the FGF receptors [191, 219]. Mutations in the receptors lead two types of disorders. One, in FGFR3, affects axial long-bone development and leads to the dwarfing chondrodysplasia syndromes. These include hypochondroplasia [18], achondroplasia [191], and thanatophoric dysplasia [178]. The second group of mutations, in FGFR2, causes a variety of craniosynostosis syndromes, including the Apert syndrome [215] and the Crouzon syndrome [93]. To date, changes in growth due to inactivating mutations in individual FGF ligands have not been identified. This suggests that the developmental functions of the FGF ligands involve collaboration among various molecules.

FGF family signaling pathways play multiple and essential roles in the early stages of skeletal patterning and in the recruitment and ultimate apoptosis of mesenchymal cells. They also seem to participate in the control of endochondral growth in the axial skeleton and of cranial bone growth at suture lines. During early limb-bud development, FGF signaling plays a role in mesenchymal epithelia [144]. As a result, FGF-10 is produced and acts on the FGF receptor 2b in the apical ectodermal ridge. Cells in the latter then express FGF-8, which signals back to FGFR1c in the limb mesoderm.

The role of FGF signaling in endochondral growth has been made apparent by activating mutations in FGFR3. However, the exact effect of the signaling pathways involved in endochondral development and the downstream FGF signaling on chondrocytes and osteoblasts is less well understood. As with BMPs, multiple forms of the FGFs are expressed in the perichondrium. FGFR1 is expressed in prehypertrophic and hypertrophic zones and FGFR3 by proliferating chondrocytes. More recently, the actions of FGF signaling have been shown to depend on the stage of chondrocyte differentiation and the nature of the individual ligands. Thus, specific receptors are expressed and interact with specific ligands in chondrocytes only at specific stages of differentiation. Studies of limb cell cultures have indicated that FGF signaling interacts with both the Ihh/PTHrP and the BMP signaling systems in a complex network. FGF signaling seems to accelerate both the onset and the pace of hypertrophic differentiation, in actions that are antagonistic to those of BMPs, and to regulate chondrocyte Ihh expression and hypertrophic differentiation. BMP, on the other hand, seems to rescue the remaining proliferating and hypertrophic chondrocytes in achondroplastic mice. This has led to the conclusion that the interaction of BMP and FGF in the growth cartilage regulates the rate of chondrocyte differentiation and proliferation [148].

The intracellular effects of FGFs are mediated by two signaling pathways: the mitogen-activated protein kinase/ERK kinase 1 (MEK1) pathway and the Janus kinase-signal transducer and activator of transcription (JAK-STAT) pathway [152, 165, 182]. The JAK-STAT signaling pathway mediates the ability of FGF signaling to inhibit chondrocyte proliferation and enhances hypertrophic chondrocyte apoptosis, whereas the MEK1 pathway mediates FGF inhibition of hypertrophic differentiation.

A number of studies have examined whether FGF has utility in promoting bone formation. Systemic low doses of basic FGF (FGF-2) stimulate endosteal and endochondral bone formation, but depress periosteal bone formation in growing rats [145, 156]. Local administration of acidic FGF (FGF-1) increases new bone formation and bone density, whereas systemic FGF-1 appears to restore bone microarchitecture and prevent bone loss associated with estrogen withdrawal [50]. Both FGF-1 and FGF-2 appear immediately at injury sites after fracture. FGF-2 was shown to improve bone healing in a study that induced a large segmental defect and in another with a metaphyseal fracture. In a 32-week study of beagle dogs, a single dose of bFGF injected into the fracture sites resulted in increased callus area and BMC and significant recovery in strength by week 16. Thus, FGF has therapeutic potential to enhance bone healing after surgery or injury.

### 2.2.5.2 Wnts (Wingless)

Wnts are 39- to 46-kDa cysteine-rich, secreted glycoproteins that are closely associated with both the cell surface and the ECM [161, 214]. Wnts are considered one of the major morphogenetic gene families responsible for appropriate embryonic development [146]. Genetic studies first performed in *Drosophila* have defined the function of this gene family. In *Drosophila*, the wingless gene is required for normal patterning in the adult and larval body segments [13, 14, 162]. The lack of this wingless gene results in the deletion of the posterior region of each body segment [14, 162]. Ectopic gene expression in *Xenopus* and gene knockout models in mice have since led to further understanding of the crucial role that Wnts play in organ development, segmentation, CNS patterning, cell fate and growth, limb development, and organization of asymmetric cell divisions [11, 45, 168, 216]. To date, approximately 100 Wnt genes have been identified in species ranging from *Caenorhabditis elegans* to humans [216].

Once the Wnt proteins are secreted, they bind to two families of cell-surface receptors, the Frizzled (Fzd) receptors and the low-density lipoprotein (LDL) receptor-related proteins (LRPs). The Fzd receptor generally consists of an extracellular cysteine-rich domain (CRD) that binds the specific Wnt protein. This receptor also consists of a seven membrane spanning domain on the cytoplasmic tail towards the carboxy-terminus of the protein. In contrast, the LRP-5 and -6 receptors have a single transmembrane domain [200]. A variety of secreted proteins, such as Frizzled-related proteins (sFRPs), Wnt inhibitory factor 1 (WIF1), and Cereberus, have been shown to be moderators of extracellular Wnt signaling. The Dickkopf (DKK) protein also exerts regulatory action by directly binding to the LRPs, thereby blocking signal transduction [11].

When the ligand becomes bound to the Fzd receptor, three signaling pathways are activated: the Wnt/β-catenin (canonical) pathway, the Wnt/$Ca^{2+}$ pathway, and the Wnt/polarity pathway. The latter two are defined as noncanonical [11]. It is of interest that a given Wnt protein can activate more than one signaling cascade. The canonical pathway involves stabilization of β-catenin, followed by translocation to the nucleus where transcription genes are activated via the TCF/LEF1 family of transcription factors [24].

The noncanonical signaling is not as well understood, though studies in *Drosophila* and *C. elegans* are being continued. The Wnt/$Ca^{2+}$ pathway is thought to induce an increase in intracellular $Ca^{2+}$ and activation of PKC, but further signaling steps have not yet been identified. Genetic studies in *Drosophila* indicate that the c-Jun N-terminal kinase (JNK) pathway is involved in the Wnt/polarity pathway, which in turn regulates cell polarity by controlling cytoskeletal organization, utilizing at some stage the disheveled (Dsh) scaffold protein [11]. The exact mechanism by which the LRP-5 and -6 coreceptors function is not understood, but they are essential for appropriate signaling. Loss of function of Arrow, the *Drosophila* analog to the vertebrate LRP receptor, mimics the wingless mutation that was first observed in the early 1980s and therefore provides evidence for the synergism between these receptors [200].

The roles Wnt signaling plays during skeletal development and postnatal bone repair were recognized as a result of mutations in humans. One is the autosomal recessive disorder osteoporosis pseudoglioma, characterized by low bone mass, frequent deformations and fractures, and defects in eye vascularization, all of which are linked to mutations in LRP-5 [73]. Children with osteoporosis pseudoglioma have normal endochondral growth and bone turnover, but their trabecular bone volume is significantly decreased [106]. Furthermore, gain-of-function experiments in humans and in mouse models have shown that organisms with an activated LRP-5 mutation exhibit a high bone mass [12].

Because the canonical signal transduction pathway is fairly well known, Wnt 3a was studied in transgenic mice. Previous studies had shown that Wnt 3a acts in the apical ectodermal ridge of the limb bud to keep cells in an undifferentiated and proliferative state [108, 110]. When Wnt 3a expression was analyzed in a murine knockout model, severe skeletal defects were observed [91, 198]. Studies of the direct effects of Wnt 3 on mesenchymal stem cells (MSCs) demonstrated that exogenous addition of Wnt 3 to murine MSCs inhibited osteogenic differentiation and decreased matrix mineralization; however, the suppression of osteogenesis can be fully reversed when Wnt 3a is removed.

The noncanonical effects of Wnt signaling have been examined through studies of Wnt 5. In contrast to the inhibitory effects of Wnt 3a, Wnt 5 appeared to promote osteogenic differentiation of the MSCs. These findings suggest that canonical Wnt signaling functions to maintain an undifferentiated, proliferative MSC population, whereas the noncanonical Wnts stimulate osteogenic differentiation [24]. Interestingly, ectopic expression of Wnt 5a delayed chondrocyte maturation and collagen type X expression, processes involved in cartilage formation [81, 109].

Many questions remain on the functional role of the Wnts, their receptors, their intracellular signaling, and their possible interaction with other morphogenic factors, such as the TGF-β family.

## 2.3 Origins of Postnatal Skeletal Stem Cells, Cytokines, and Morphogenetic Signals During Bone Repair

Bone is unique in that after fracture or surgery, it can regenerate the original structure and biomechanical competency of the damaged tissue. Bone repair involves four stages that overlap and cause the various tissue types to interact, as shown in Fig. 2.1. The fracture line in the bone determines the spatial relationships of the morphogenetic fields during tissue regeneration. This is evidenced by the development of two circular centers of cartilage (ECB) that form symmetrically with respect to the fracture line and taper proximally and distally along the

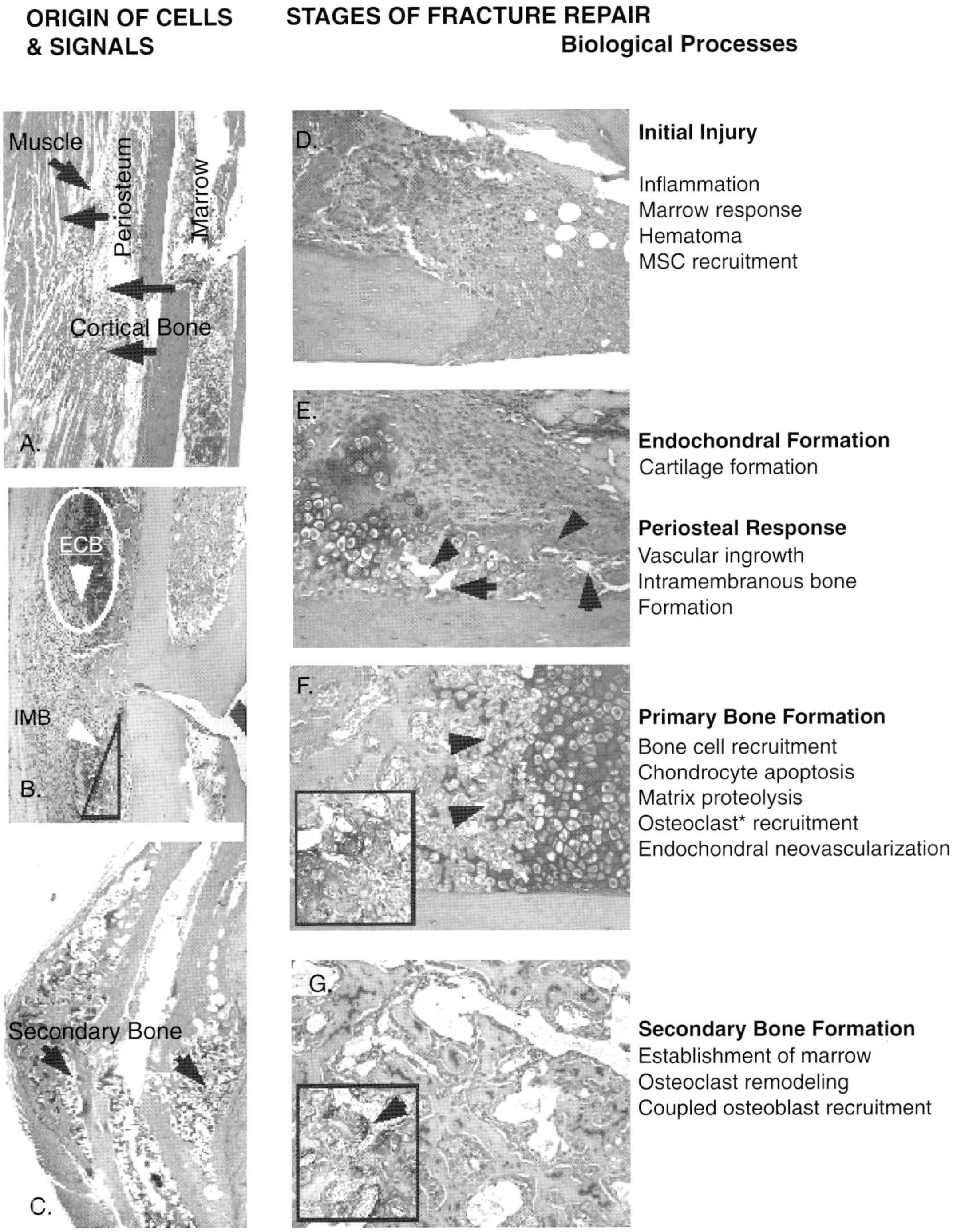
ORIGIN OF CELLS & SIGNALS
STAGES OF FRACTURE REPAIR
Biological Processes
Muscle
Periosteum
Marrow
Cortical Bone
A.
ECB
IMB
B.
Secondary Bone
C.
D.
Initial Injury
Inflammation
Marrow response
Hematoma
MSC recruitment
E.
Endochondral Formation
Cartilage formation
Periosteal Response
Vascular ingrowth
Intramembranous bone
Formation
F.
Primary Bone Formation
Bone cell recruitment
Chondrocyte apoptosis
Matrix proteolysis
Osteoclast* recruitment
Endochondral neovascularization
G.
Secondary Bone Formation
Establishment of marrow
Osteoclast remodeling
Coupled osteoblast recruitment

bone cortices (see the middle microphotograph in Fig. 2.1A, B, and E). At the same time, a crescent-shaped region of intramembranous bone formation appears at the proximal and distal ends of the area of periosteal response and tapers inward toward the fracture line deep in the cartilage ring. Thus, endochondral and intramembranous bone formation both contribute to bone healing.

During bone repair, cell interactions are initiated between the external soft tissues that surround the injured bone, the underlying cortical bone and marrow, and the developing endochondral and intramembranous bone tissues (Fig. 2.1A). The origin of the MSCs that contribute to bone repair and the identity of the cells that initiate morphogenetic signals are still unresolved. Figure 2.1 shows potential sources of cells and signals that lead to the construction of these developmental fields.

MSCs involved in fracture repair may originate in the periosteum, the surrounding tissues, or both (Fig. 2.1A). The periosteum appears to be the primary source of MSCs that then give rise to the intramembranous bone that forms in the callus [154]. If the periosteum is removed, callus development is diminished [29], because periosteal cells robustly produce BMPs during the initial phases of fracture healing [26]. These observations suggest that morphogens recruit stem cells locally and induce them to differentiate.

MSCs may also originate in the surrounding muscle or marrow space. Data to support a muscle origin come from studies showing that demineralized bone powder or purified BMPs, when implanted or injected into muscle tissue, induce bone formation [92, 96, 203]. Other studies have shown that a variety of premyogenic cell lines can differentiate into chondrogenic or osteogenic cells when treated with BMPs [41, 70, 104]. Marrow stroma also can differentiate into osteoblasts and chondrocytes [20, 94, 188, 189]. Once recruited, their numbers increase as a result of other morphogenetic factors. It is important to identify the source of the stem cells, because they make up much of the callus tissue and may make up as much as 30% of the original volume of the uninjured long bone

Vascular tissues grow into the developing callus as new periosteal bone develops and progresses toward the fracture line from the proximal and distal edges. The interaction of the vascular elements and the initiation and propagation of the periosteal response thus appear to be the primary driving mechanisms that facilitate intramembranous bone formation. Perivascular mesenchymal cells in blood-vessel walls may also contribute to this process [27]. Figure 2.2 summarizes the mesenchymal lineage and types of morphogens that are involved in lineage selection, expansion, survival, and programmed cell removal.

**Figure 2.1.** Anatomic characterization of fracture repair. Left panels (A-C) show an overview of the morphogenetic fields of tissue development and the proximate tissue interactions. (A) Histological section of the fracture site immediately postfracture. Potential tissue origins of mesenchymal stem cells (MSCs) and morphogenetic signals are denoted by the arrows and denoted in the figure. (B) Histological section of the fracture site at 7 days postfracture. The two types of bone-formation processes are denoted as endochondral bone (ECB) and intramembraneous bone (IMB) formation. The two proceed in a symmetrical manner around the fracture site. (C) Histological section of the fracture site at 28 days postfracture. Secondary bone formation and coupled remodeling predominate in the late stage of bone repair. Right panels (D-G) show a summary of the multiple stages of fracture healing. Histological sections are presented for each stage, and the various processes associated with each stage are summarized. All histological specimens are from sagittal sections of mouse tibia transverse fractures and were stained with safranin O and fast green; micrographic images are at 200× magnification. (D) Section for the initial injury was taken from the fracture site 24 hours postinjury. (E) Section depicting the initial periosteal response and endochondral formation is from 7 days postinjury. Arrows denote vascular ingrowth from the peripheral areas of the periosteum. (F) Section depicting the period of primary bone formation is from 14 days postinjury. Arrows denote neovascular growth areas in the underlying new bone. Inset depicts images of an osteoclast (*chondroclast) resorbing an area of calcified cartilage. (G) Sections depicting the period of secondary bone formation are from 21 days postinjury. Callus sites. Inset depicts 400× images of an osteoclast resorbing an area of primary bone. Reproduced with permission from Gerstenfeld LC, Cullinane DM, Barnes GL, et al. Fracture healing as a post-natal developmental process: molecular, spatial, and temporal aspects of its regulation. J Cell Biochem. 2003 Apr 1;88(5):873–84. Copyright © 2003 Wiley-Liss, Inc., A Wiley Company.

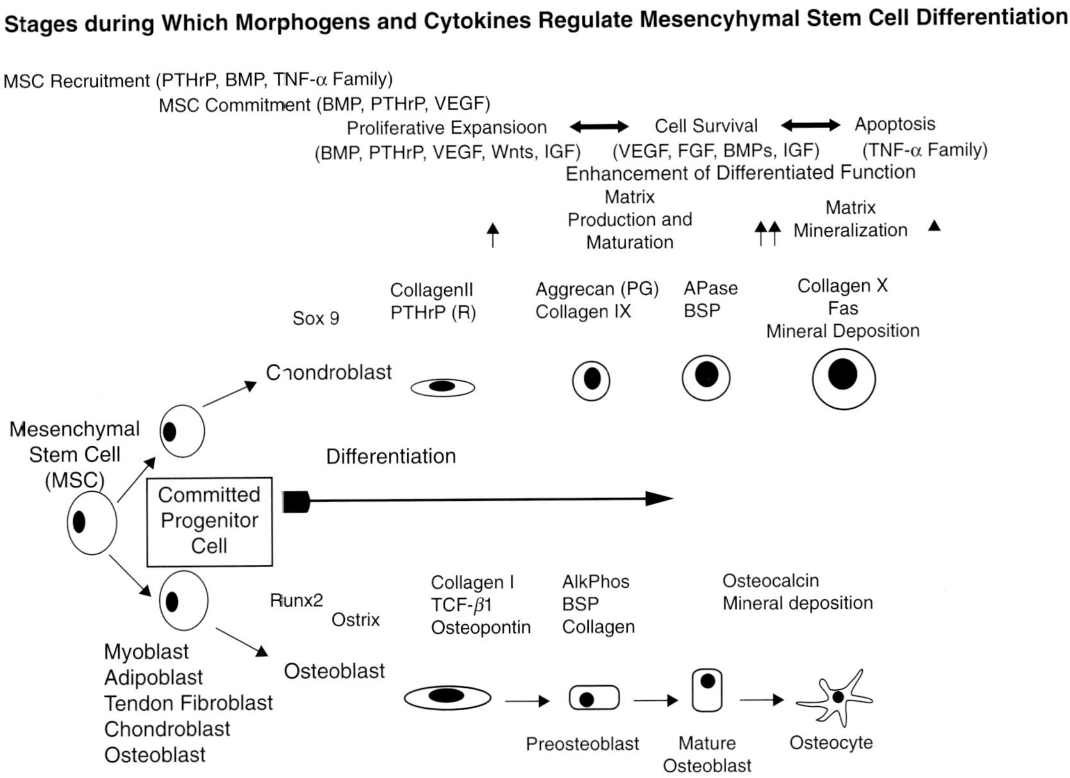


**Figure 2.2.** Schematic summary of the lineage progression of mesenchymal stem-cell (MSC) differentiation. Upper panel: Multiple stages of the life cycle of an MSC. The morphogenetic regulators of each stage are in parentheses. Lower panel: The separate stages of each of the major anabolic skeletal cell lineages are indicated with known markers that define each stage of their lineage progression. PTHrP, parathyroid hormone-related peptide; BMP, bone morphogenetic protein; TNF, tumor necrosis factor; VEGF, vascular endothelial growth factor; IGF, insulin-like growth factor; FGF, fibroblast growth factor; PG, large proteoglycan; BSP, bone sialoprotein; TGF, transforming growth factor.

Restoration of the original anatomic geometry of the tissue is an important aspect of bone repair. For this to occur there must be some relationship between the original structure of the tissue and the gradients of the morphogens that promote the developmental process and the characteristics of the injury. One obvious functional role must be attributed to the signals that initiate and establish the symmetry of bone repair around the fracture line. These signals may be thought of as arising from the marrow or from the injured cortical bone matrix. In this connection, how the injury influences tissue responses may have considerable relevance, because the inflammatory signals spread out from the point of origin of the injury [16, 51, 54]. Data that support the role inflammatory cytokines play in the initiation of skeletal tissue repair come from studies showing that in the absence of TNF-α signaling in receptor-null animals, the callus does not develop symmetrically around the fracture line. The absence of TNF-α signaling also leads to a delay in intramembranous and endochondral bone formation. Thus, TNF-α signaling facilitates the repair process, perhaps by stimulating MSC recruitment or differentiation [68].

The structural geometry of the callus may also depend on the muscular anatomy or vascularization of the tissue and on the biomechanical environment at the site of injury. The latter seems to be particularly important. When bending and shear loading were introduced at an osteotomy site, osteogenesis was favored over chondrogenesis [42]. Other studies have similarly shown that mechanical instability leads to persistence of cartilage tissue at the fracture site. This involves up-regulation of molecular signals such as Ihh that regulate chondrogenesis [124]. How morphogenetic fields are established and how biomechanical factors direct tissue differentiation and the geometry of the regenerative process are questions of considerable importance, because the answers may identify the signal molecules and relate them to the origins of MSCs. Defining how the morphogenetic fields are established also has clinical importance, since the therapeutic responses to bioactive factors may depend on whether they are correctly directed to the morphogenetic field.

## 2.4 Bone Repair Is Dependent upon Multiple Cellular and Molecular Signals

The cellular and molecular processes that govern bone repair after injury have many features that are similar to what occurs in a growth plate during embryonic and postnatal skeletal development. As reviewed earlier, fracture healing involves several stages and is mediated by very different biological processes. Figure 2.2 presents the stages and progression of MSC differentiation into cartilage and bone as the skeleton is formed.

Figure 2.2 also shows the stages at which various morphogens and cytokines become active and regulate MSC and skeletal-cell differentiation. In addition the figure lists the specific transcription factors (Runx 2, Osterix, and Sox 9 [117]) involved in lineage commitment and identifies stage-specific markers for the two skeletal-cell lineages. Of particular interest is the fact that some factors act at several stages during skeletal-cell lineage progression. For example, BMPs not only are associated with MSC lineage commitment but also are involved in cellular expansion. In contrast, the scope of morphogens such as VEGF and the Wnts appears to be more restricted, with their predominant effect on proliferative lineage expansion or survival. Members of the TNF-$\alpha$ family, which are part of the immune response to injury, regulate the initial stages of MSC recruitment and cell survival during the inflammatory stage and re-emerge at the end of the MSC cycle to control apoptosis during tissue remodeling. Finally, factors such as the FGFs control the rate or timing of entry and exit of committed cells during their period of proliferative expansion.

The functional contribution of specific cytokines and morphogens during fracture healing is presented in Fig. 2.3. These factors are expressed during different phases of fracture healing and therefore may vary in the roles they play during healing. For example, TGF-$\beta$2, TGF-$\beta$3, and GDF-5 show peak mRNA expression during chondrogenic differentiation and as the endochondral phases develop. This suggests that the two factors are functionally restricted to the periods in which chondrogenesis takes place.

Understanding the temporal pattern and molecular nature of the factors as they are expressed during bone healing can allow targeting and modification of their actions to lead to better fracture healing. Knowing the spatial nature of the morphogenetic fields during the temporal processes of fracture healing has clinical importance because the therapeutic responses to bioactive factors may be influenced by the moment in time when they contact the correct morphogenetic field. Such knowledge will help to develop therapeutic agents to treat osteoporosis and can equally well be applied to the development of therapeutic agents that promote bone formation. Table 2.1 lists the biological processes and approaches that can be modified in coupled bone remodeling, either to impede bone loss or to promote bone regeneration. The table also lists approaches that could enhance the rate or quality of bone healing.

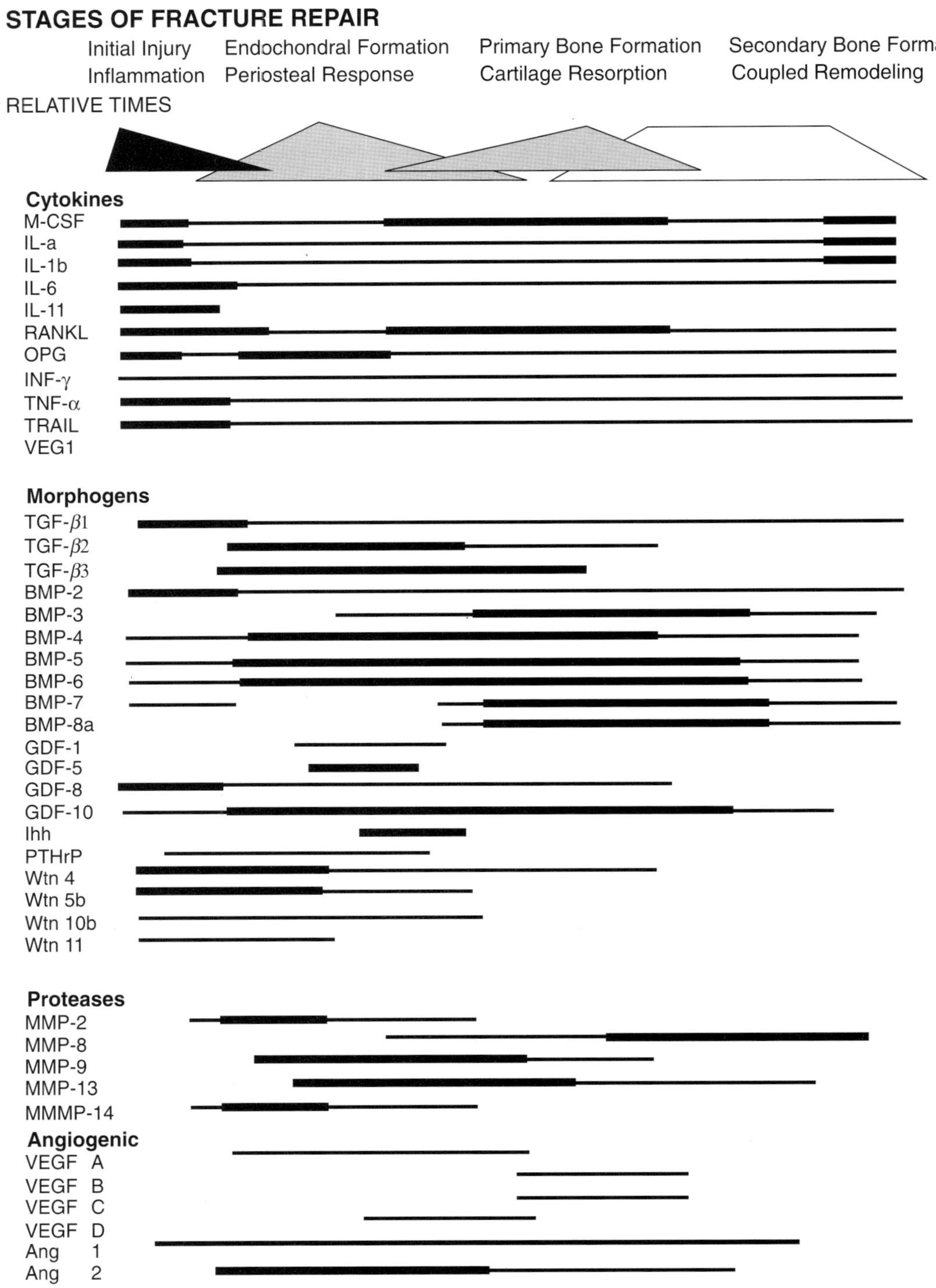
STAGES OF FRACTURE REPAIR
Initial Injury
Inflammation
Endochondral Formation
Periosteal Response
Primary Bone Formation
Cartilage Resorption
Secondary Bone Formation
Coupled Remodeling
RELATIVE TIMES
Cytokines
M-CSF
IL-a
IL-1b
IL-6
IL-11
RANKL
OPG
INF-γ
TNF-α
TRAIL
VEG1
Morphogens
TGF-β1
TGF-β2
TGF-β3
BMP-2
BMP-3
BMP-4
BMP-5
BMP-6
BMP-7
BMP-8a
GDF-1
GDF-5
GDF-8
GDF-10
Ihh
PTHrP
Wtn 4
Wtn 5b
Wtn 10b
Wtn 11
Proteases
MMP-2
MMP-8
MMP-9
MMP-13
MMMP-14
Angiogenic
VEGF A
VEGF B
VEGF C
VEGF D
Ang 1
Ang 2

**Figure 2.3.** Schematic summary of the stages of fracture repair and their associated molecular processes. The relative temporal aspects of each of the stages of the fracture healing process are denoted by basic geometric shapes that also connote the relative intensity of the molecular processes that define each of the stages. The relative levels of expression of various mRNAs that have been examined in our laboratories are denoted by three line widths. The levels of expression are in percent over baseline for each and are not comparable for the various mRNAs. Data for expression levels for the proinflammatory cytokines and the extracellullar matrix (ECM) mRNAs are from Kon et al., 2001 [118]; data for TGF-α family members are from Cho et al., 2002 [38]; data for proteases and angiogenic factors from are from Lehmann et al., 2002 [127]; and data for Cox2 are from Gerstenfeld et al., 2002 [70]. Data pertaining to Ihh and iNOs expression are unpublished. M-CSF, macrophage colony-stimulating factor; IL, interleukin; RANKL, RANK ligand; OPG, osteoprotegrin; INF, interferon; TNF, tumor necrosis factor; VEG1, xxx; TGF, transforming growth factor; BMP, bone morphogenetic protein; GDF, growth and differentiation factor; Ihh, Indian hedgehog; PTHrP, parathyroid hormone-related peptide; MMP, matrix metalloproteinase; VEGF, vascular endothelial growth factor; Ang, angiopoietin. Reproduced with permission from Gerstenfeld LC, Cullinane DM, Barnes GL, et al. Fracture healing as a post-natal developmental process: molecular, spatial, and temporal aspects of its regulation. J Cell Biochem. 2003 Apr 1;88(5):873–84. Copyright © 2003 Wiley-Liss, Inc., A Wiley Company.

**Table 2.1.** Comparison of strategies in the development of therapeutic agents to treat osteoporosis versus fracture and bone repair

| | | | |
|---|---|---|---|
| **A. Stages of fracture repair and strategies to enhance fracture repair** | | | |
| Initial injury | Endochondral formation | Primary bone formation | Secondary bone formation |
| Inflammation | Periosteal response | Cartilage resorption | Coupled remodeling |
| Factors that promote stem-cell recruitment (PTH, BMPs) | Increase ratio of bone/ cartilage differentiation (FGFs, Wnts, PTH) | Factors that change rates of endochondral remodeling (TNF family) | Factors that enhance coupled bone formation (TNF family) |
| **B. Stages of coupled remodeling and strategies to enhance bone mass** | | | |
| Activation | Diminish numbers of osteoclasts (TNF family) | | |
| Resorption | Diminish osteoclast activity/increase rate of osteoclast turnover (TNF family) | | |
| Formation | Increase osteoblast numbers/osteoblast activity (BMPs, PTH, Wnts) | | |

## 2.5 Future Perspectives on Therapeutic Uses of Morphogenetic Factors

Reduction of the morbidity associated with some 5% to 10% of fractures and improvement of healing after osteotomies, arthrodeses, spinal fusions, and other reconstructive orthopedic procedures depend on better understanding of the biology of fracture and bone healing (170). As discussed above, multiple morphogenetic factors regulate normal skeletal development, but it is not clear how they function in postnatal healing. Many factors act cooperatively or even antagonistically at different stages of bone development. Single use of individual factors has had mixed success in promoting bone healing. Regaining biomechanical competency more quickly is even more complicated than promoting stem-cell differentiation. Biomechanical competency involves many factors, including the restoration of the material properties of the tissue and of appropriate skeletal-tissue geometry. At the same time, it will be necessary to define appropriate modalities for using repair-promoting factors and to identify when, where, and how long the factors should be applied. Because many factors, once they activate receptors, utilize overlapping signal-transduction pathways to mediate intracellular effects, signal pathways need to be identified in the hope of making optimal use of the small-molecule pharmaceuticals that are being developed.

# References


1. Abbas S, Zhang YH, Clohisy JC, Abu-Amer Y (2003) Tumor necrosis factor-alpha inhibits pre-osteoblast differentiation through its type-1 receptor. Cytokine 22:33–41.
2. Abou-Samra AB, Juppner H, Force T, Freeman MW, Kong XF, Schipani E, Urena P, Richards J, Bonventre JV, Potts JT Jr, Kronenberg HM, Segre GV (1992) Expression cloning of a common receptor for parathyroid hormone and parathyroid hormone-related peptide from rat osteoblast-like cells: a single receptor stimulates intracellular accumulation of both cAMP and inositol trisphosphates and increases in trocellular free calcium. Proc Natl Acad Sci USA 89:2732–2736.
3. Adams CS, Shapiro IM (2002) The fate of the terminally differentiated chondrocyte: evidence for microenvironmental regulation of chondrocyte apoptosis. Crit Rev Oral Biol Med 13:465–473.
4. Aigner T, Kim H (2002) Apoptosis and cellular vitality: issues in osteoarthritic cartilage degeneration. Arthritis Rheum 46:1986–1996.
5. Aizawa T, Kokubun S, Tanaka Y (1997) Apoptosis and proliferation of growth plate chondrocytes in rabbits. J Bone Joint Surg Br 79B:483–486.
6. Aizawa T, Kon T, Einhorn TA, Gerstenfeld LC (2001) Induction of apoptosis in chondrocytes by tumor necrosis factor-alpha. J Orthop Res 19:785–796.
7. Andreassen TT, Ejersted C, Oxlund H (1999) Intermittent parathyroid hormone (1–34) treatment increases callus formation and mechanical strength of healing rat fractures. J Bone Miner Res 14:960–968.
8. Angel P, Imagawa M, Chiu R, Stein B, Imbra RJ, Rahmsdorf HJ, Jonat C, Herrlich P, Karin M (1987) Phorbol ester-inducible genes contain a common cis element recognized by a TPA-modulated trans-acting factor. Cell 49:729–739.
9. Arai F, Hirao A, Ohmura M, Sato H, Matsuoka S, Takubo K, Ito K, Koh GY, Suda T (2004) Tie2/angiopoietin-1 signaling regulates hematopoietic stem cell quiescence in the bone marrow niche. Cell 118:149–161.
10. Aronson J, Harp JH, Walker CW, Dalrymple GV (1990) Blood flow, bone formation and mineralization during distraction osteogenesis. Trans Orthop Res Soc 15:589–594.
11. Attisano L, Labbe E (2004) TGF-beta and Wnt pathway cross-talk. Cancer Metastasis Rev 23:53–61.
12. Babij P, Zhao W, Small C, Kharode Y, Yaworsky PJ, Bouxsein ML, Reddy PS, Bodine PV, Robinson JA, Bhat B, Marzolf J, Moran RA, Bex F (2003) High bone mass in mice expressing a mutant LRP5 gene. J Bone Miner Res 18:960–974.
13. Babu P (1977) Early developmental subdivisions of the wing disk in Drosophila. Mol Gen Genet 151: 289–294.
14. Baker NE (1988) Localization of transcripts from the wingless gene in whole Drosophila embryos. Development 103:289–298.
15. Barleon B, Sozzani S, Zhou D, Weich HA, Mantovani A, Marme D (1996) Migration of human monocytes in response to vascular endothelial growth factor (VEGF) is mediated via the VEGF receptor flt-1. Blood 87:3336–3343.
16. Barnes GL, Kostenuik PJ, Gerstenfeld LC, Einhorn TA (1999) Growth factor regulation of fracture repair. J Bone Miner Res 14:1805–1815.
17. Bazzoni F, Beutler B (1996) The tumor necrosis factor ligand and receptor family. N Engl J Med 334: 1717–1725.
18. Bellus GA, McIntosh I, Smith EA, Aylsworth AS, Kaitila I, Horton WA, Greenhaw GA, Hecht JT, Francomano CA (1995) A recurrent mutation in the tyrosine kinase domain of fibroblast growth factor receptor 3 causes hypochondroplasia. Nat Genet 10:357–359.
19. Bhardwaj A, Aggarwal BB (2003) Receptor-mediated choreography of life and death. J Clin Immunol 23:317–332.
20. Bianco P, Riminucci M, Gronthos S, Robey PG (2001) Bone marrow stromal stem cells: nature, biology, and potential applications. Stem Cells 19:180–192.
21. Bitzer M, von Gersdorff G, Liang D, Dominguez-Rosales A, Beg AA, Rojkind M, Bottinger EP (2000) A mechanism of suppression of TGF-beta/SMAD signaling by NFkappa B/RelA. Genes Dev 14:187–197.
22. Blomstrand S, Claesson I, Save-Soderbergh J (1985) A case of lethal congenital dwarfism with accelerated skeletal maturation. Pediatr Radiol 15:141–143.
23. Bluethmann H (1998) Physiological, immunological, and pathological functions of tumor necrosis factor (TNF) revealed by TNF receptor-deficient mice. In: Durum S, Muegge K, eds. Cytokine Knockouts. Totowa: Humana Press, pp 69–87.
24. Boland GM, Perkins G, Hall DJ, Tuan RS (2004) Wnt 3a promotes proliferation and suppresses osteogenic differentiation of adult human mesenchymal stem cells. J Cell Biochem 93:1210–1230.
25. Bostrom MP, Gamradt SC, Asnis P, Vickery BH, Hill E, Avnur Z, Waters RV (2000) Parathyroid hormone-related protein protein analog RS-66271 is an effective therapy for impaired bone healing in rabbits on corticosteroid therapy. Bone 2:437–442.
26. Bostrom MP, Lane JM, Berberian WS, Missri AA, Tomin E, Weiland A, Doty SB, Glaser D, Rosen VM (1995) Immunolocalization and expression of bone morphogenetic protein 2 and 4 in fracture healing. J Orthop Res 13:357–367.
27. Bouletreau PJ, Warren SM, Spector JA, Peled ZM, Gerrets RP, Greenwald JA, Longaker MT (2002) Hypoxia and VEGF up-regulate BMP-2 mRNA and protein expression in microvascular endothelial cells: implications for fracture healing. Plast Reconstr Surg 109:2384–2397.
28. Broadus AE, Stewart AF (1994) Parathyroid hormone-related protein: structure, processing, and physiological actions. In: Bilezikian JP, Levine MA, Marcus R, eds. The Parathyroids: Basic and Clinical Concepts. New York: Raven Press, pp 259–294.
29. Buckwalter JA, Einhorn TA, Marsh LJ (2001) Bone and joint healing. In: Bucholz RW, Heckman JD, eds. Rockwood and Green's Fractures in Adults. Philadelphia: Lippincott, Williams, and Wilkins, pp 245–271.
30. Calvi LM, Adams GB, Weibrecht KW, Weber JM, Olson DP, Knight MC, Martin RP, Schipani E, Divieti P, Bringhurst FR, Milner LA, Kronenberg HM, Scadden DT (2003) Osteoblastic cells regulate the hematopoietic stem cell niche. Nature 425:841–846.

31. Candeliere GA, Glorieux FH, Prud'homme J, St-Arnaud R (1995) Increased expression of the c-fos proto-ocogene in bone from patients with fibrous dysplasia. N Engl J Med 332:1546–1551.
32. Carano RA, Filvaroff EH (2003) Angiogenesis and bone repair. Drug Discov Today 8:980–989.
33. Carlevaro MF, Cermelli S, Cancedda R, Descalzi Cancedda F (2000) Vascular endothelial growth factor (VEGF) in cartilage neovascularization and chondrocyte differentiation: auto-paracrine role during endochondral bone formation. J Cell Sci 113:59–69.
34. Carvalho RS, Einhorn TA, Lehmann W, Edgar C, Al-Yamani A, Apazidis A, Pacicca D, Clemens TL, Gerstenfeld LC (2004) The role of angiogenesis in a murine tibial model of distraction osteogenesis. Bone 34:849–861.
35. Celeste AJ, Iannazzi JA, Taylor RC, Hewick RM, Rosen V, Wang EA, Wozney JM (1990) Identification of transforming growth factor beta family members present in bone-inductive protein purified from bovine bone. Proc Natl Acad Sci 87:9843–9847.
36. Chen S, Guttridge DC, Tang E, Shi S, Guan K, Wang CY (2001) Suppression of tumor necrosis factor-mediated apoptosis by nuclear factor kappaB-independent bone morphogenetic protein/Smad signaling. J Biol Chem 276:39259–39263.
37. Childs LM, Paschalis EP, Xing L, Dougall WC, Anderson D, Boskey AL, Puzas JE, Rosier RN, O'Keefe RJ, Boyce BF, Schwarz EM (2002) In vivo RANK signaling blockade using the receptor activator of NF-kappaB:Fc effectively prevents and ameliorates wear debris-induced osteolysis via osteoclast depletion without inhibiting osteogenesis. J Bone Miner Res 17:192–199.
38. Cho T-J, Gerstenfeld LC, Einhorn TA (2002) Differential temporal expression of members of the transforming growth factor beta superfamily during murine fracture healing. J Bone Miner Res 17:513–520.
39. Cho T-J, Lehmann W, Edgar C, Sadeghi C, Hou A, Einhorn TA, Gerstenfeld LC (2003) Tumor necrosis factor alpha activation of the apoptotic cascade in murine articular chondrocytes is associated with the induction of metalloproteinases and specific pro-resorptive factors. Arthritis Rheum 48:2845–2854.
40. Clauss IM, Gravallese EM, Darling JM (1993) In situ hybridization studies suggest a role for the basic region-leucine zipper protein hXBP-1 in exocrine gland and skeletal development during mouse embryogenesis. Dev Dyn 197:146–156.
41. Constantinides PG, Taylor SM, Jones PA (1978) Phenotypic conversion of cultured mouse embryo cells by aza pyrimidine nucleosides. Dev Biol 66:57–71.
42. Cullinane DM, Fredrick A, Eisenberg SR, Pacicca D, Elman MV, Lee C, Salisbury K, Gerstenfeld LC, Einhorn TA (2002) Induction of a neoarthrosis by precisely controlled motion in an experimental mid-femoral defect. J Orthop Res 20:579–586.
43. Danial NN, Korsmeyer SJ (2004) Cell death: critical control points. Cell 116:205–219.
44. De Robertis EM, Larrain J, Oelgeschlager M, Wessely O (2000) The establishment of Spemann's organizer and patterning of the vertebrate embryo. Nat Rev Genet 1:171–181.
45. Dealy CN, Roth A, Ferrari D, Brown AM, Kosher RA (1993) Wnt-5a and Wnt-7a are expressed in the developing chick limb bud in a manner suggesting roles in pattern formation along the proximo distal and dorsoventral axes. Mech Dev 43:175–186.
46. Deckers MM, Karperien M, van der Bent C, Yamashita T, Papapoulos SE, Lowik CW (2000) Expression of vascular endothelial growth factors and their receptors during osteoblast differentiation. Endocrinology 141:1667–1674.
47. Deckers MM, van Bezooijen RL, van der Horst G, Hoogendam J, van Der Bent C, Papapoulos SE, Lowik CW (2002) Bone morphogenetic proteins stimulate angiogenesis through osteoblast-derived vascular endothelial growth factor A. Endocrinology 143: 1545–1553.
48. Dempster DW, Cosman F, Kurland ES, Zhou H, Nieves J, Woelfert L, Shane E, Plavetic K, Muller R, Bilezikian J, Lindsay R (2001) Effects of daily treatment with parathyroid hormone on bone microarchitecture and turnover in patients with osteoporosis: a paired biopsy study. J Bone Miner Res 16:1846–1853.
49. Dosch R, Gawantka V, Delius H, Blumenstock C, Niehrs C (1997) BMP-4 acts as a morphogen in dorsoventral mesoderm patterning in Xenopus. Development 124:2325–2334.
50. Dunstan CR, Boyce R, Boyce BF, Garrett IR, Izbicka E, Burgess WH, Mundy GR (1999) Systemic administration of acidic fibroblast growth factor (FGF-1) prevents bone loss and increases new bone formation in ovariectomized rats. J Bone Miner Res 14:953–959.
51. Einhorn TA (1995) Enhancement of fracture healing. J Bone Joint Surg 77:940–956.
52. Einhorn T, Lee C (2001) Bone regeneration: new findings and potential clinical applications. J Am Acad Orthop Surg 9:157–165.
53. Einhorn TA, Majeska RJ, Mohaideen A, Kagel EM, Bouxsein ML, Turek TJ, Wozney JM (2003) A single percutaneous injection of recombinant human bone morphogenetic protein-2 accelerates fracture repair. J Bone Joint Surg Am 85-A:1425–1435.
54. Einhorn TA, Majeska RJ, Rush EB, Levine PM, Horowitz MC (1995) The expression of cytokine activity by fracture callus. J Bone Miner Res 10: 1272–1281.
55. Eugster H-P, Muller M, LeHir M, Ryffel B (1998) Immunodeficiency of tumor necrosis factor and lymophotoxin-alpha double deficient mice. In: Durum S, Muegge K, editors. Cytokine Knockouts. Totowa: Humana Press, pp 103–118.
56. Farnum CE, Willsman NJ (1989) Cellular turnover at the chondro-osseous junction of growth plate cartilage: analysis by serial sections at the light microscopical level. J Orthop Res 7:654–666.
57. Feng JQ, Xing L, Zhang JH, Zhao M, Horn D, Chan J, Boyce BF, Harris SE, Mundy GR, Chen D (2003) NFkappaB specifically activates BMP-2 gene expression in growth plate chondrocytes in vivo and in a chondrocyte cell line in vitro. J Biol Chem 278:29130–29135.
58. Ferguson C, Alpern E, Miclau T, Helms JA (1999) Does adult fracture repair recapitulate embryonic skeletal formation? Mech Dev 87:57–66.

59. Ferrara N (1999) Role of vascular endothelial growth factor in the regulation of angiogenesis. Kidney Int 56:794–814.
60. Ferrara N (2000) VEGF: an update on biological and therapeutic aspects. Curr Opin Biotechnol 11: 617–624.
61. Ferrara N (2001) Role of vascular endothelial growth factor in regulation of physiological angiogenesis. Am J Physiol Cell Physiol 280:C1358–C1366.
62. Fitzpatrick LA, Coleman DT, Bilezikian JP (1992) The target tissue actions of parathyroid hormone. In: Coe FL, Favus MJ, eds. Disorders of Bone and Mineral Metabolism. New York: Raven Press, pp 123–147.
63. Fukuhara H, Mizuno K (1989) The influence of parathyroid hormone on the process of fracture healing. Nippon Seikeigeka Gakkai Zasshi 63:100–115.
64. Fukui N, Zhu Y, Maloney WJ, Clohisy J, Sandell LJ (2003) Stimulation of BMP-2 expression by proinflammatory cytokines IL-1 and TNF-alpha in normal and osteoarthritic chondrocytes. J Bone Joint Surg Am 85A:59–66.
65. Gardella TJ, Jüppner H (2001) Molecular properties of the PTH/PTHrP receptor. Trends Endocrinol Metab 12:210–217.
66. Gerber HP, Vu TH, Ryan AM, Kowalski J, Werb Z, Ferrara N (1999) VEGF couples hypertrophic cartilage remodeling, ossification and angiogenesis during endochondral bone formation. Nat Med 5:623–628.
67. Gerhart TN, Kirker-Head CA, Kriz MJ, Holtrop ME, Hennig GE, Hipp J, Schelling SH, Wang E (1993) Healing segmental femoral defects in sheep using recombinant human bone morphogenetic protein. Clin Orthop 293:317–326.
68. Gerstenfeld LC, Cho T-J, Kon T, Aizawa T, Cruceta J, Graves BD, Einhorn TA (2001) Impaired intramembranous bone formation during bone repair in the absence of tumor necrosis factor-alpha signaling. Cells Tissues Organs 169:285–294.
69. Gerstenfeld LC, Cho T-J, Kon T, Aizawa T, Tsay A, Fitch J, Barnes GL, Graves DT, Einhorn TA (2003) Impaired fracture healing in the absence of TNF-alpha signaling: the role of TNF-alpha in endochondral cartilage resorption. J Bone Miner Res 18:1584–1592.
70. Gerstenfeld LC, Cruceta J, Shea CM, Sampath K, Barnes GL, Einhorn TA (2002) Chondrocytes provide morphogenic signals that selectively induce osteogenic differentiation of mesenchymal stem cells. J Bone Miner Res 17:221–230.
71. Gerstenfeld LC, Shapiro FD (1996) Expression of bone-specific genes by hypertrophic chondrocytes: implication of the complex functions of the hypertrophic chondrocyte during endochondral bone development. J Cell Biochem 62:1–9.
72. Gibson GJ, Kohler WJ, Schaffler MB (1995) Chondrocyte apoptosis in endochondral ossification of chick sterna. Dev Dyn 203:468–476.
73. Gong Y, Slee RB, Fukai N, Rawadi G, Roman-Roman S, Reginato AM, Wang H, Cundy T, Glorieux FH, Lev D, Zacharin M, Oexle K, Marcelino J, Suwairi W, Heeger S, Sabatakos G, Apte S, Adkins WN, Allgrove J, Arslan-Kirchner M, Batch JA, Beighton P, Black GC, Boles RG, Boon LM, Borrone C, Brunner HG, Carle GF, Dallapiccola B, De Paepe A, Floege B, Halfhide ML, Hall B, Hennekam RC, Hirose T, Jans A, Juppner H, Kim CA, Keppler-Noreuil K, Kohlschuetter A, LaCombe D, Lambert M, Lemyre E, Letteboer T, Peltonen L, Ramesar RS, Romanengo M, Somer H, Steichen-Gersdorf E, Steinmann B, Sullivan B, Superti-Furga A, Swoboda W, van den Boogaard MJ, Van Hul W, Vikkula M, Votruba M, Zabel B, Garcia T, Baron R, Olsen BR, Warman ML, for the Osteoporosis-Pseudoglioma Syndrome Collaborative Group (2001) LDL receptor-related protein 5 (LRP5) affects bone accrual and eye development. Cell 107:513–523.
74. Goralczyk R, Closs EI, Ruther U, Wagner EF, Strauss PG, Erfle V, Schmidt J (1990) Characterization of fos-induced osteogenic tumours and tumour-derived murine cell lines. Differentiation 44:122–131.
75. Gravallese EM, Pettit AR, Lee R, Madore R, Manning C, Tsay A, Gaspar J, Goldring MB, Goldring SR, Oettgen P (2003) Angiopoietin-1 is expressed in the synovium of patients with rheumatoid arthritis and is induced by tumor necrosis factor alpha. Ann Rheum Dis 62:100–107.
76. Guo J, Chung UI, Kondo H, Bringhurst FR, Kronenberg HM (2002) The PTH/PTHrP receptor can delay chondrocyte hypertrophy in vivo without activating phospholipase C. Dev Cell 3:183–194.
77. Habener JF (1990) Cyclic AMP response element binding proteins: a cornucopia of transcription factors. Mol Endocrinol 4:1087–1094.
78. Harbour ME, Gregory JW, Jenkins HR, Evans BA (2000) Proliferative response of different human osteoblast-like cell models to proinflammatory cytokines. Pediatr Res 48:163–168.
79. Haridas V, Shrivastava A, Su J, Yu GL, Ni J, Liu D, Chen SF, Ni Y, Ruben SM, Gentz R, Aggarwal BB (1999) VEGI, a new member of the TNF family activates nuclear factor-kappa B and c-Jun N-terminal kinase and modulates cell growth. Oncogene 18:6496–6504.
80. Harper J, Gerstenfeld LC, Klagsbrun M (2001) Neuropilin-1 expression in osteogenic cells: down-regulation during differentiation of osteoblasts into osteocytes. J Cell Biochem 81:82–92.
81. Hartmann C, Tabin CJ (2000) Dual roles of Wnt signaling during chondrogenesis in the chicken limb. Development 127:3141–3159.
82. Harty M, Neff AW, King MW, Mescher AL (2003) Regeneration or scarring: an immunologic perspective. Dev Dyn 226:268–279.
83. Hashimoto S, Setareh M, Ochs R, Lotz M (1997) Fas/Fas ligand expression and induction of apoptosis in chondrocytes. Arthritis Rheum 40:1749–1755.
84. Hausman MR, Schaffler MB, Majeska RJ (2001) Prevention of fracture healing in rats by an inhibitor of angiogenesis. Bone 29:560–564.
85. Hock JM, Gera I (1992) Effects of continuous and intermittent administration and inhibition of resorption on the anabolic response of bone to parathyroid hormone. J Bone Miner Res 7:65–72.
86. Hofmann C, Luo G, Balling R, Karsenty G (1996) Analysis of limb patterning in BMP-7-deficient mice. Dev Genet 19:43–50.
87. Holash J, Wiegand SJ, Yancopoulos GD (1999) New model of tumor angiogenesis: dynamic balance between vessel regression and growth mediated

by angiopoietins and VEGF. Oncogene 18:5356–5362.
88. Holzer G, Majeska RJ, Lundy MW, Hartke JR, Einhorn TA (1999) Parathyroid hormone enhances fracture healing. Clin Orthop 366:258–263.
89. Horner A, Bord S, Kelsall AW, Coleman N, Compston JE (2001) Tie2 ligands angiopoietin-1 and angiopoietin-2 are coexpressed with vascular endothelial cell growth factor in growing human bone. Bone 28:65–71.
90. Hunziker EB, Schenk RK, Cruz-Orive LM (1987) Quantitation of chondrocyte performance in growth-plate cartilage during longitudinal bone growth. J Bone Joint Surg Am 69:162–173.
91. Ikeya M, Takada S (2001) Wnt-3a is required for somite specification along the anteroposterior axis of the mouse embryo and for regulation of cdx-1 expression. Mech Dev 103:27–33.
92. Iwata H, Sakano S, Itoh T, Bauer TW (2002) Demineralized bone matrix and native bone morphogenetic protein in orthopaedic surgery. Clin Orthop 395:99–109.
93. Jabs EW, Li X, Scott AF, Meyers G, Chen W, Eccles M, Mao JI, Charnas LR, Jackson CE, Jaye M (1994) Jackson-Weiss and Crouzon syndromes are allelic with mutations in fibroblast growth factor receptor 2. Nat Genet 8:275–279.
94. Jaiswal N, Haynesworth SE, Caplan AI, Bruder SP (1997) Osteogenic differentiation of purified, culture-expanded human mesenchymal stem cells in vitro. J Cell Biochem 64:295–312.
95. Jansen M (1934) Über atypische Chondrodystophie (Achondroplasie) und über eine noch nicht beschriebene angeborene Wachstumsstörung des Knochensystems: Metaphysäre Dysostosis. Zeitschr Orthop Chir 61:253–286.
96. Jingushi S, Urabe K, Okazaki K, Hirata G, Sakai A, Ikenoue T, Iwamoto Y (2002) Intramuscular bone induction by human recombinant bone morphogenetic protein-2 with beta-tricalcium phosphate as a carrier: in vivo bone banking for muscle-pedicle autograft. J Orthop Sci 7:490–494.
97. Jobert AS, Zhang P, Couvineau A, Bonaventure J, Roume J, Le Merrer M, Silve C (1998) Absence of functional receptors for parathyroid hormone and parathyroid hormone-related peptide in Blomstrand chondrodysplasia. J Clin Invest 102:34–40.
98. Jones PF (2003) Not just angiogenesis—wider roles for the angiopoietins. J Pathol 201:515–527.
99. Ju ST, Panka DJ, Cui H, Ettinger R, el-Khatib M, Sherr DH, Stanger BZ, Marshak-Rothstein A (1995) Fas (CD95)/FasL interactions required for programmed cell death after T-cell activation. Nature 373:444–448.
100. Kaku M, Kohno S, Kawata T, Fujita I, Tokimasa C, Tsutsui K, Tanne K (2001) Effects of vascular endothelial growth factor on osteoclast induction during tooth movement in mice. J Dent Res 80:1880–1883.
101. Kanzler B, Foreman RK, Labosky PA, Mallo M (2000) BMP signaling is essential for development of skeletogenic and neurogenic cranial neural crest. Development 127:1095–1104.
102. Karaplis AC, Luz A, Glowacki J, Bronson RT, Tybulewicz VL, Kronenberg HM, Mulligan RC (1994) Lethal skeletal dysplasia from targeted disruption of parathyroid hormone-related peptide gene. Genes Dev 8:277–289.
103. Karp SJ, Schipani E, St-Jacques B, Hunzelman J, Kronenberg H, McMahon AP (2000) Indian hedgehog coordinates endochondral bone growth and morphogenesis via parathyroid hormone related-protein-dependent and -independent pathways. Development 127:543–548.
104. Katagiri T, Yamaguchi A, Komaki M, Abe E, Takahashi N, Ikeda T, Rosen V, Wozney JM, Fujisawa-Sehara A, Suda T (1994) Bone morphogenetic protein-2 converts the differentiation pathway of C2C12 myoblasts into the osteoblast lineage. J Cell Biol 127:1755–1766.
105. Katavic V, Lukic IK, Kovacic N, Grcevic D, Lorenzo JA, Marusic A (2003) Increased bone mass is a part of the generalized lymphoproliferative disorder phenotype in the mouse. J Immunol 170:1540–1547.
106. Kato M, Patel MS, Levasseur R, Lobov I, Chang BH, Glass DA 2nd, Hartmann C, Li L, Hwang TH, Brayton CF, Lang RA, Karsenty G, Chan L (2002) Cbfal-independent decrease in osteoblast proliferation, osteopenia, and persistent embryonic eye vascularization in mice deficient in Lrp5, a Wnt coreceptor. J Cell Biol 157:303–314.
107. Kawabata M, Imamura T, Miyazono K (1998) Signal transduction by bone morphogenetic proteins. Cytokine Growth Factor Rev 9:49–61.
108. Kawakami Y, Capdevila J, Buscher D, Itoh T, Rodriguez Esteban C, Izpisua Belmonte JC (2001) WNT signals control FGF-dependent limb initiation and AER induction in the chick embryo. Cell 104:891–900.
109. Kawakami Y, Wada N, Nishimatsu SI, Ishikawa T, Noji S, Nohno T (1999) Involvement of Wnt-5a in chondrogenic pattern formation in the chick limb bud. Dev Growth Differ 41:29–40.
110. Kengaku M, Capdevila J, Rodriguez-Esteban C, De La Pena J, Johnson RL, Belmonte JC, Tabin CJ (1998) Distinct WNT pathways regulating AER formation and dorsoventral polarity in the chick limb bud. Science 280:1274–1277.
111. Kiener PA, Davis PM, Starling GC, Mehlin C, Klebanoff SJ, Ledbetter JA, Liles WC (1997) Differential induction of apoptosis by Fas-Fas ligand interactions in human monocytes and macrophages. J Exp Med 185:1511–1516.
112. Kim HW, Jahng JS (1999) Effect of intermittent administration of parathyroid hormone on fracture healing in ovariectomized rats. Iowa Orthop J 19:71–77.
113. Kim KS, Tinti C, Song B, Cubells JF, Joh TH (1994) Cyclic AMP-dependent protein kinase regulates basal and cyclic AMP-stimulated but not phorbol ester-stimulated transcription of the tyrosine hydroxylase gene. J Neurochem 63:834–842.
114. Kim N, Odgren PR, Kim DK, Marks SC Jr, Choi Y (2000) Diverse roles of the tumor necrosis factor family member TRANCE in skeletal physiology revealed by TRANCE deficiency and partial rescue by a lymphocyte-expressed TRANCE transgene. Proc Natl Acad Sci USA 97:10905–10910.
115. Kimble RB, Bain S, Pacifici R (1997) The functional block of TNF but not IL-6 prevents bone loss in ovariectomized mice. J Bone Miner Res 12:935–941.

116. Kimmel DB, Bozzato RP, Kronis KA, Coble T, Sindrey D, Kwong P, Recker RR (1993) The effect of recombinant human (1–84) or synthetic human (1–34) parathyroid hormone on the skeleton of adult osteopenic ovariectomized rats. Endocrinology 132:1577–1584.
117. Kobayashi T, Kronenberg H (2005) Minireview: transcriptional regulation in development of bone. Endocrinology 146:1012–1017.
118. Kon T, Cho T-J, Aizawa T, Yamazaki M, Nooh N, Graves D, Gerstenfeld LC, Einhorn TA (2001) Expression of osteoprotegerin, receptor activator of NF-kappaB ligand (osteoprotegerin ligand) and related proinflammatory cytokines during fracture healing. J Bone Miner Res 16:1004–1014.
119. Kostiainen K (1996) Recombinant bone morphogenetic proteins: cloning and molecular structures (Chapter 4). In: Lindholm TS, ed. Bone Morphogenetic Proteins: Biology, Biochemistry, and Reconstructive Surgery. Austin: RG Landes Company and Academic Press, pp 39–46.
120. Kronenberg HM (2003) Developmental regulation of the growth plate. Nature 423:332–336.
121. Kucharczak J, Simmons MJ, Fan Y, Gelinas C (2003) To be, or not to be: NFkappaB is the answer—role of Rel/NFkappaB in the regulation of apoptosis. Oncogene 22:8961–8982.
122. Lanske B, Karaplis AC, Lee K, Luz A, Vortkamp A, Pirro A, Karperien M, Defize LH, Ho C, Mulligan RC, Abou-Samra AB, Juppner H, Segre GV, Kronenberg HM (1996) PTH/PTHrP receptor in early development and Indian hedgehog-regulated bone growth. Science 273:663–666.
123. Lawson KA, Dunn NR, Roelen BA, Zeinstra LM, Davis AM, Wright CV, Korving JP, Hogan BL (1999) BMP4 is required for the generation of primordial germ cells in the mouse embryo. Genes Dev 13: 66–68.
124. Le AX, Miclau T, Hu D, Helms JA (2001) Molecular aspects of healing in stabilized and non-stabilized fractures. J Orthop Res 19:78–84.
125. Lee KAW, Masson N (1993) Transcriptional regulations by CREB and its relatives. Biochem Biophys Acta 1174:221–233.
126. Lee SW, Lee HJ, Chung WT, Choi SM, Rhyu SH, Kim DK, Kim KT, Kim JY, Kim JM, Yoo YH (2004) TRAIL induces apoptosis of chondrocytes and influences the pathogenesis of experimentally induced rat osteoarthritis. Arthritis Rheum 50:534–542.
127. Lehmann W, Edgar CM, Wang K, Cho TJ, Barnes GL, Kakar S, Graves DT, Rueger JM, Gerstenfeld LC, Einhorn TA (2005) Tumor necrosis factor alpha (TNF-alpha) coordinately regulates the expression of specific matrix metalloproteinases (MMPS) and angiogenic factors during fracture healing. Bone 36:300–310.
128. Lewinson D, Maor G, Rozen N, Rabinovich I, Stahl S, Rachmiel A (2001) Expression of vascular antigens by bone cells during bone regeneration in a membranous bone distraction system. Histochem Cell Biol 116:381–388.
129. Li G, Simpson AH, Kenwright J, Triffitt JT (1999) Effect of lengthening rate on angiogenesis during distraction osteogenesis. J Orthop Res 17:362–367.
130. Liekens S, De Clercq E, Neyts J (2001) Angiogenesis: regulators and clinical applications. Biochem Pharmacol 61:253–270.
131. Liou H-C, Boothby MR, Finn PW, Davidon R, Nabavi N, Zeleznik-Le NJ, Ting JP, Glimcher LH (1990) A new member of the leucine zipper class of proteins that binds to the HLA DR alpha promoter. Science 247:1581–1584.
132. Liou H-C, Boothby MR, Glimcher LH (1988) Distinct cloned class II MHC DNA binding proteins recognize the X transcription element. Science 242:69–71.
133. Liu F, Ventura F, Doody J, Massague J (1995) Human type II receptor for bone morphogenic proteins (BMPs): extension of the two-kinase receptor model to the BMPs. Mol Cell Biol 15:3479–3486.
134. Liu X, Nagarajan RP, Vale W, Chen Y (2002) Phosphorylation regulation of the interaction between Smad7 and activin type I receptor. Fed Euro Biochem Soc Lett 519:93–98.
135. Lobov IB, Brooks PC, Lang RA (2002) Angiopoietin-2 displays VEGF-dependent modulation of capillary structure and endothelial cell survival in vivo. Proc Natl Acad Sci USA 99:11205–11210.
136. Long F, Zhang XM, Karp S, Yang Y, McMahon AP (2001) Genetic manipulation of hedgehog signaling in the endochondral skeleton reveals a direct role in the regulation of chondrocyte proliferation. Development 128:5099–5108.
137. Lories RJ, Derese I, Ceuppens JL, Luyten FP (2003) Bone morphogenetic proteins 2 and 6, expressed in arthritic synovium, are regulated by proinflammatory cytokines and differentially modulate fibroblast-like synoviocyte apoptosis. Arthritis Rheum 48:2807–2818.
138. Luo G, Hofmann C, Bronckers AL, Sohocki M, Bradley A, Karsenty G (1995) BMP-7 is an inducer of nephrogenesis, and is also required for eye development and skeletal patterning. Genes Dev 9:2808–2820.
139. Lyons KM, Pelton RW, Hogan BL (1990) Organogenesis and pattern formation in the mouse: RNA distribution patterns suggest a role for bone morphogenetic protein-2A (BMP-2A). Development 109:833–844.
140. Maes C, Carmeliet P, Moermans K, Stockmans I, Smets N, Collen D, Bouillon R, Carmeliet G (2002) Impaired angiogenesis and endochondral bone formation in mice lacking the vascular endothelial growth factor isoforms VEGF164 and VEGF188. Mech Dev 111:61–73.
141. Maisonpierre PC, Suri C, Jones PF, Bartunkova S, Wiegand SJ, Radziejewski C, Compton D, McClain J, Aldrich TH, Papadopoulos N, Daly TJ, Davis S, Sato TN, Yancopoulos GD (1997) Angiopoietin-2, a natural antagonist for Tie 2 that disrupts in vivo angiogenesis. Science 277:55–60.
142. Marie PJ (2003) Fibroblast growth factor signaling controlling osteoblast differentiation. Gene 316:23–32.
143. Marron MB, Hughes DP, Edge MD, Forder CL, Brindle NP (2000) Evidence for heterotypic interaction between the receptor tyrosine kinases TIE-1 and TIE-2. J Biol Chem 275:39741–39746.
144. Martin GR (1998) The roles of FGFs in the early development of vertebrate limbs. Genes Dev 12:1571–1586.

145. Mayahara H, Ito T, Nagai H, Miyajima H, Tsukuda R, Taketomi S, Mizoguchi J, Kato K (1993) In vivo stimulation of endosteal bone formation by basic fibroblast growth factor in rats. Growth Factors 9:73–80.
146. Maye P, Zheng J, Li L, Wu D (2004) Multiple mechanisms for Wnt11-mediated repression of the canonical Wnt signaling pathway. J Biol Chem 279:24659–24665.
147. Mayr-Wohlfart U, Waltenberger J, Hausser H, Kessler S, Gunther KP, Dehio C, Puhl W, Brenner RE (2002) Vascular endothelial growth factor stimulates chemotactic migration of primary human osteoblasts. Bone 30:472–477.
148. Minina E, Kreschel C, Naski MC, Ornitz DM, Vortkamp A (2002) Interaction of FGF, Ihh/Pthlh, and BMP signaling integrates chondrocyte proliferation and hypertrophic differentiation. Dev Cell 3:439–449.
149. Miyazono K (2002) Bone morphogenetic protein receptors and actions (Chapter 51). In: Bilezikian JP, Raisz LG, Rodan GA, eds. Principles of Bone Biology. 2nd edition. San Diego: Academic Press, pp 929–942.
150. Miyazono Y (1999) Signal transduction by bone morphogenetic protein receptors: functional roles of Smad proteins. Bone 25:91–93.
151. Mosekilde L, Sogaard CH, Danielsen CC, Torring O (1991) The anabolic effects of human parathyroid hormone (hPTH) on rat vertebral body mass are also reflected in the quality of bone, assessed by biomechanical testing: a comparison study between hPTH-(1–34) and hPTH-(1–84). Endocrinology 129:421–428.
152. Murakami S, Balmes G, McKinney S, Zhang Z, Givol D, de Crombrugghe B (2004) Constitutive activation of MEK1 in chondrocytes causes Stat1-independent achondroplasia-like dwarfism and rescues the Fgfr3-deficient mouse phenotype. Genes Dev 18:290–305.
153. Nagata S (1997) Apoptosis by death factor. Cell 88:356–365.
154. Nakahara H, Bruder SP, Haynesworth SE, Holecek JJ, Baber MA, Goldberg VM, Caplan AI (1990) Bone and cartilage formation in diffusion chambers by subcultured cells derived from the periosteum. Bone 11:181–188.
155. Nakajima A, Shimoji N, Shiomi K, Shimizu S, Moriya H, Einhorn TA, Yamazaki M (2002) Mechanisms for the enhancement of fracture healing in rats treated with intermittent low-dose human parathyroid hormone (1–34). J Bone Miner Res 17:2038–2047.
156. Nakamura T, Hanada K, Tamura M, Shibanushi T, Nigi H, Tagawa M, Fukumoto S, Matsumoto T (1995) Stimulation of endosteal bone formation by systemic injections of recombinant basic fibroblast growth factor in rats. Endocrinology 136:1276–1284.
157. Nakase T, Nomura S, Yoshikawa H, Hashimoto J, Hirota S, Kitamura Y, Oikawa S, Ono K, Takaoka K (1994) Transient and localized expression of bone morphogenetic protein 4 messenger RNA during fracture healing. J Bone Miner Res 9:651–659.
158. Nanes MS (2003) Tumor necrosis factor-alpha: molecular and cellular mechanisms in skeletal pathology. Gene 321:1–15.
159. Neer RM, Arnaud CD, Zanchetta JR, Prince R, Gaich GA, Reginster JY, Hodsman AB, Eriksen EF, Ish-Shalom S, Genant HK, Wang O, Mitlak BH (2001) Effect of parathyroid hormone (1–34) on fractures and bone mineral density in postmenopausal women with osteoporosis. N Engl J Med 344:1434–1441.
160. Nickel J, Dreyer MK, Kirsch T, Sebald W (2001) The crystal structure of the BMP-2:BMPR-IA complex and the generation of BMP-2 antagonists. J Bone Joint Surg Am 83A(Suppl 1):S7–S14.
161. Nishioka K, Dennis JE, Gao J, Goldberg VM, Caplan AI (2005) Sustained Wnt protein expression in chondral constructs from mesenchymal stem cells. J Cell Physiol 203:6–14.
162. Nusslein-Volhard C, Wieschaus E (1980) Mutations affecting segment number and polarity in Drosophila. Nature 287:795–801.
163. Olsen NJ, Stein CM (2004) New drugs for rheumatoid arthritis. N Engl J Med 350:2167–2179.
164. Ornitz DM, Itoh N (2001) Fibroblast growth factors. Genome Biol 2:REVIEWS3005.
165. Ornitz DM, Marie PJ (2002) FGF signaling pathways in endochondral and intramembranous bone development and human genetic disease. Genes Dev 16:1446–1465.
166. Ozkaynak E, Rueger DC, Drier EA, Corbett C, Ridge RJ, Sampath TK, Oppermann H (1990) OP-1 cDNA encodes an osteogenic protein in the TFG-beta family. EMBO J 9:2085–2093.
167. Park YS, Kim NH, Jo I (2003) Hypoxia and vascular endothelial growth factor acutely up-regulate angiopoietin-1 and Tie2 mRNA in bovine retinal pericytes. Microvasc Res 65:125–131.
168. Parr BA, McMahon AP (1995) Dorsalizing signal Wnt-7a required for normal polarity of D-V and A-P axes of mouse limb. Nature 374:350–353.
169. Podbesek R, Edouard C, Meunier PJ, Parsons JA, Reeve J, Stevenson RW, Zanelli JM (1983) Effects of two treatment regimes with synthetic human parathyroid hormone fragment on bone formation and the tissue balance of trabecular bone in greyhounds. Endocrinology 112:1000–1006.
170. Praemer AP, Furner S, Rice DP (1992) Musculoskeletal Conditions in the United States. Park Ridge: American Academy of Orthopaedic Surgeons.
171. Ramsauer M, D'Amore PA (2002) Getting Tie(2)d up in angiogenesis. J Clin Invest 110:1615–1617.
172. Reddi AH (2001) Bone morphogenetic proteins: from basic science to clinical applications. J Bone Joint Surg Am 83A:S1–S6.
173. Reimold AM, Grusby MJ, Kosaras B, Fries JW, Mori R, Maniwa S, Clauss IM, Collins T, Sidman RL, Glimcher MJ, Glimcher LH (1996) Chondrodysplasia and neurological abnormalities in ATF-2-deficient mice. Nature 379:262–265.
174. Roach HI, Erenpreisa J, Aigner T (1995) Osteogenic differentiation of hypertrophic chondrocytes involves asymmetric cell division. J Cell Biol 131: 483–494.
175. Roodman GD (2002) Role of the bone marrow microenvironment in multiple myeloma. J Bone Miner Res 17:1921–1925.
176. Rosen V, Thies RS (1992) The BMP proteins in bone formation and repair. Trends Genet 8:97–102.

177. Rouleau MF, Mitchell J, Goltzman D (1990) Characterization of the major parathyroid hormone target cell in the endosteal metaphysis of rat long bones. J Bone Miner Res 5:1043–1053.
178. Rousseau F, Saugier P, Le Merrer M, Munnich A, Delezoide AL, Maroteaux P, Bonaventure J, Narcy F, Sanak M (1995) Stop codon FGFR3 mutations in thanatophoric dwarfism type 1. Nat Genet 10:11–12.
179. Rowe NM, Mehara BJ, Luchs JS, Dudziak ME, Steinbrech DS, Illei PB, Fernandez GJ, Gittes GK, Longaker MT (1999) Angiogenesis during mandibular distraction osteogenesis. Ann Plast Surg 42:470–475.
180. Ruther U, Garber C, Komitowski D, Muller R, Wagner EF (1987) Deregulated c-fos expression interferes with normal bone development in transgenic mice. Nature 325:412–416.
181. Ruther U, Komitowski D, Schubert FR, Wagner EF (1989) C-fos expression induced bone tumors in transgenic mice. Oncogene 4:861–865.
182. Sahni M, Raz R, Coffin JD, Levy D, Basilico C (2001) STAT1 mediates the increased apoptosis and reduced chondrocyte proliferation in mice overexpressing FGF2. Development 128:2119–2129.
183. Sampath TK, Coughlin JE, Whetstone RM, Banach D, Corbett C, Ridge RJ, Ozkaynak E, Oppermann H, Rueger DC (1990) Bovine osteogenic protein is composed of dimers of OP-1 and BMP-2A, two members of the transforming growth factor-beta superfamily. J Biol Chem 265:13198–13205.
184. Sandborn WJ (2004) How future tumor necrosis factor antagonists and other compounds will meet the remaining challenges in Crohn's disease. Rev Gastroenterol Disord 4:S25–S33.
185. Scaffidi C, Fulda S, Srinivasan A, Friesen C, Li F, Tomaselli KJ, Debatin KM, Krammer PH, Peter ME (1998) Two CD95 (APO-1/Fas) signaling pathways. EMBO J 17:1675–1687.
186. Schipani E, Kruse K, Juppner H (1995) A constitutively active mutant PTH-PTHrP receptor in Jansen-type metaphyseal chondrodysplasia. Science 268:98–100.
187. Schultz DR, Harrington WJ Jr (2003) Apoptosis: programmed cell death at a molecular level. Semin Arthritis Rheum 32:345–369.
188. Sekiya I, Larson BL, Smith JR, Pochampally R, Cui JG, Prockop DJ (2002) Expansion of human adult stem cells from bone marrow stroma: conditions that maximize the yields of early progenitors and evaluate their quality. Stem Cells 20:530–541.
189. Sekiya I, Vuoristo JT, Larson BL, Prockop DJ (2002) In vitro cartilage formation by human adult stem cells from bone marrow stroma defines the sequence of cellular and molecular events during chondrogenesis. Proc Natl Acad Sci USA 99:4397–4402.
190. Sela-Donenfeld D, Kalcheim C (2002) Localized BMP4-noggin interactions generate the dynamic patterning of noggin expression in somites. Dev Biol 246:311–328.
191. Shiang R, Thompson LM, Zhu YZ, Church DM, Fielder TJ, Bocian M, Winokur ST, Wasmuth JJ (1994) Mutations in the transmembrane domain of FGFR3 cause the most common genetic form of dwarfism, achondroplasia. Cell 78:335–342.
192. Shinbrot E, Moore M (1998) Cooperation between the TNF receptors demonstrated by TNF receptor knockout mice. In: Durum S, Muegge K, eds. Cytokine Knockouts. Totowa: Humana Press, pp 89–101.
193. Skripitz R, Andreassen TT, Aspenberg P (2000) Parathyroid hormone (1–34) increases the density of rat cancellous bone in a bone chamber. A dose-response study. J Bone Joint Surg Br 82:138–141.
194. St-Jacques B, Hammerschmidt M, McMahon AP (1999) Indian hedgehog signaling regulates proliferation and differentiation of chondrocytes and is essential for bone formation. Genes Dev 13:2072–2086.
195. Street J, Bao M, deGuzman L, Bunting S, Peale FV Jr, Ferrara N, Steinmetz H, Hoeffel J, Cleland JL, Daugherty A, van Bruggen N, Redmond HP, Carano RA, Filvaroff EH (2002) Vascular endothelial growth factor stimulates bone repair by promoting angiogenesis and bone turnover. Proc Natl Acad Sci USA 99:9656–9661.
196. Sullivan R, Klagsbrun M (1985) Purification of cartilage-derived growth factor by heparin affinity chromatography. J Biol Chem 260:2399–2403.
197. Suzuki A, Thies RS, Yamaji N, Song JJ, Wozney JM, Murakami K, Ueno N (1994) A truncated bone morphogenetic protein receptor affects dorsal-ventral patterning in the early Xenopus embryo. Proc Natl Acad Sci USA 91:10255–10259.
198. Takada S, Stark KL, Shea MJ, Vassileva G, McMahon JA, McMahon AP (1994) Wnt-3a regulates somite and tailbud formation in the mouse embryo. Genes Dev 8:174–189.
199. Tam CS, Heersche JN, Murray TM, Parsons JA (1982) Parathyroid hormone stimulates the bone apposition rate independently of its resorptive action: differential effects of intermittent and continuous administration. Endocrinology 110:506–512.
200. Tamai K, Zeng X, Liu C, Zhang X, Harada Y, Chang Z, He X (2004) A mechanism for Wnt coreceptor activation. Mol Cell 13:149–156.
201. Thurston G, Gale NW (2004) Vascular endothelial growth factor and other signaling pathways in developmental and pathologic angiogenesis. Int J Hematol 80:7–20.
202. Uemura A, Ogawa M, Hirashima M, Fujiwara T, Koyama S, Takagi H, Honda Y, Wiegand SJ, Yancopoulos GD, Nishikawa S (2002) Recombinant angiopoietin-1 restores higher-order architecture of growing blood vessels in mice in the absence of mural cells. J Clin Invest 110:1619–1628.
203. Urist MR (1965) Bone: formation by autoinduction. Science 150:893–899.
204. Vaux DL, Flavell RA (2000) Apoptosis genes and autoimmunity. Curr Opin Immunol 12:719–724.
205. Verma IM, Mulligan R, Beauset A (1988) Gene Transfer in Animals. UCLA Symposia on Molecular and Cellular Biology. New York: Alan R. Liss.
206. Visnjic D, Kalajzic Z, Rowe D, Katavic V, Lorenzo J, Aguila HL (2004) Hematopoiesis is severely altered in mice with an induced osteoblast deficiency. Blood 103:3258–3264.
207. Von Bubnoff A, Cho K (2001) Intracellular BMP signaling regulation in vertebrates: pathway or network? Dev Biol 239:1–14.
208. Vortkamp A, Lee K, Lanske B, Segre GV, Kronenberg HM, Tabin CJ (1996) Regulation of rate of cartilage differentiation by Indian hedgehog and PTH-related protein. Science 273:613–622.

209. Vortkamp A, Pathi S, Peretti GM, Caruso EM, Zaleske DJ, Tabin CJ (1998) Recapitulation of signals regulating embryonic bone formation during postnatal growth and in fracture repair. Mech Dev 71:65–76.
210. Vu TH, Shipley JM, Bergers G, Berger JE, Helms JA, Hanahan D, Shapiro SD, Senior RM, Werb Z (1998) MMP-9/gelatinase B is a key regulator of growth plate angiogenesis and apoptosis of hypertrophic chondrocytes. Cell 93:411–422.
211. Wallach D, Varfolomeev EE, Malinin NL, Goltsev YV, Kovalenko AV, Boldin MP (1999) Tumor necrosis factor receptor and fas signaling mechanisms. Annu Rev Immunol 17:331–367.
212. Wang EA, Rosen V, Cordes P, Hewick RM, Kriz MJ, Luxenberg DP, Sibley BS, Wozney JM (1988) Purification and characterization of other distinct bone-inducing factors. Proc Natl Acad Sci USA 85:9484–9488.
213. Wang Z-Q, Ovitt C, Grigoriadis AE, Mohle-Steinlein U, Ruther U, Wagner EF (1992) Bone and haematopoietic defects in mice lacking c-fos. Nature 360: 741–745.
214. Westendorf JJ, Kahler RA, Schroeder TM (2004) Wnt signaling in osteoblasts and bone diseases. Gene 341:19–39.
215. Wilkie AO, Slaney SF, Oldridge M, Poole MD, Ashworth GJ, Hockley AD, Hayward RD, David DJ, Pulleyn LJ, Rutland P, Malcolm S, Winter RM, Reardon W(1995) Apert syndrome results from localized mutations of FGFR2 and is allelic with Crouzon syndrome. Nat Genet 9:165–172.
216. Wodarz A, Nusse R (1998) Mechanisms of Wnt signaling in development. Annu Rev Cell Dev Biol 14: 59–88.
217. Wozney JM (1989) Bone morphogenetic proteins. Prog Growth Factor Res 1:267–280.
218. Xiao Z, Latek R, Lodish HF (2003) An extended bipartite nuclear localization signal in Smad4 is required for its nuclear import and transcriptional activity. Oncogene 22:1057–1069.
219. Xu X, Weinstein M, Li C, Deng C (1999) Fibroblast growth factor receptors (FGFRs) and their roles in limb development. Cell Tissue Res 296:33–43.
220. Yamashita H, Ten Dijke P, Heldin CH, Miyazono K (1996) Bone morphogenetic protein receptors. Bone 19:569–574.
221. Yasko AW, Lane JM, Fellinger EJ, Rosen V, Wozney JM, Wang EA (1992) The healing of segmental bone defects induced by recombinant human bone morphogenetic protein (rhBMP-2). J Bone Joint Surg 74A:659–671.
222. Yeh LC, Lee JC (1999) Osteogenic protein-1 increases gene expression of vascular endothelial growth factor in primary cultures of fetal rat calvaria cells. Mol Cell Endocrinol 153:113–124.
223. Ying Y, Zhao GQ (2001) Cooperation of endoderm-derived BMP2 and extraembryonic ectoderm-derived BMP4 in primordial germ cell generation in the mouse. Dev Biol 232:484–492.
224. Young-In-Lee F, Choi YW, Behrens FF, DeFouw DO, Einhorn TA (1998) Programmed removal of chondrocytes during endochondral fracture healing. J Orthop Res 6:144–149.
225. Yu J, Tian S, Metheny-Barlow L, Chew LJ, Hayes AJ, Pan H, Yu GL, Li LY (2001) Modulation of endothelial cell growth arrest and apoptosis by vascular endothelial growth inhibitor. Circ Res 89: 1161–1167.
226. Zelzer E, Glotzer DJ, Hartmann C, Thomas D, Fukai N, Soker S, Olsen BR (2001) Tissue specific regulation of VEGF expression during bone development requires Cbfa1/Runx2. Mech Dev 106:97–106.
227. Zhai Y, Ni J, Jiang GW, Lu J, Xing L, Lincoln C, Carter KC, Janat F, Kozak D, Xu S, Rojas L, Aggarwal BB, Ruben S, Li LY, Gentz R, Yu GL (1999) VEGI, a novel cytokine of the tumor necrosis factor family, is an angiogenesis inhibitor that suppresses the growth of colon carcinomas in vivo. FASEB J 13:181–189.
228. Zhang H, Bradley A (1996) Mice deficient for BMP2 are nonviable and have defects in amnion/chorion and cardiac development. Development 122:2977–2986.
229. Zhou T, Edwards CK 3rd, Yang P, Wang Z, Bluethmann H, Mountz JD (1996) Greatly accelerated lymphadenopathy and autoimmune disease in lpr mice lacking tumor necrosis factor receptor I. J Immunol 156:2661–2665.

# 3. Bone Allograft Safety and Performance

*Calin S. Moucha, Regis L. Renard, Ankur Gandhi, Sheldon S. Lin, and Rocky S. Tuan*

## 3.1 Introduction

Bone allograft transplantation is a common practice; in the United States 650,000 procedures were performed in 1999, a 186% increase from 1990 [3]. This increase can be attributed to morbidities associated with bone autografts [6, 18, 30, 35, 59], the increased availability of bone allografts, and the expansion of these applications [9, 16, 21, 22, 29, 31, 42, 66]. A variety of musculoskeletal allografts are available for different reconstructive applications. Bone allograft is an alternative to autograft because it has osteoconductive properties, acts as a scaffold for bone growth, and induces bone formation by providing osteogenic factors, in addition to mesenchymal precursor cells, osteoblasts, and osteocytes. Although these properties are advantageous, the potential for the transmission of infectious diseases remains a great concern [1, 2, 4, 10, 12, 24, 26, 27, 32, 38, 49, 53]. Because of the biological origin of bone allografts, the clinician must be educated about the effects of tissue preparation and processing on the immunogenic, osteoinductive, osteoconductive, and structural properties of allografts in order to make appropriate clinical decisions. This chapter discusses the safety of bone allografts and the effects of donor selection, harvesting, processing, and implantation on the performance of bone allograft in reconstructive surgery.

## 3.2 An Overview of Musculoskeletal Graft Harvesting and Processing

In the United States, the Food and Drug Administration (FDA) currently regulates organ and tissue transplants with mandated donor and tissue screening protocols for human immunodeficiency virus (HIV) types 1 and 2, hepatitis B virus (HBV), and hepatitis C virus (HCV) (Table 3.1). The FDA also requires documentation to accompany the donor graft to provide a medical history that precludes any recent infections or patient "social" habits, such as drug abuse, which would increase the risk of allograft infection. In addition, the American Association of Tissue Banks (AATB), a nonprofit organization, provides industry guidelines and recommendations for its accredited members beyond those of the FDA, which include testing for human T-lymphocytic virus (HTLV) types 1 and 2 and syphilis [67] (Table 3.1). However, there are no uniform industry standards for tissue processing, and not all tissue banks are AATB-accredited. Medical conditions contraindicated by the FDA and AATB for tissue and organ donation include benign tumors near the allograft excision sites, malignant tumors, autoimmune or inflammatory diseases, severe endocrine/metabolic disease, and collagen diseases [22, 29, 36, 42, 62, 63]. Additional contraindications for donations include deaths resulting from trauma with large resuscitation volumes, with or without blood or blood products, and deaths resulting from poisoning or related to toxic overdoses [67].

Upon identification and screening of an acceptable donor, appropriate consent must be obtained from the donor or nearest relative prior to tissue and/or organ procurement. Musculoskeletal allografts may be obtained from living donors, multiorgan donors, and cadavers. Harvesting of a musculoskeletal allograft from a living donor (such as a femoral head allograft harvested from a total hip replacement) is performed in a sterile operating room, as is harvesting from a multiorgan donor. Cadaveric musculoskeletal tissues must be procured within 24 hours of death, with the time interval between death and refrigeration not to exceed twelve hours. Harvesting of a musculoskeletal graft from a cadaver is performed in an approved, aseptic environment. Musculoskeletal allografts can be categorized as (1) bone with soft-tissue attachments (such as a bone-patellar tendon-bone allograft), (2) bone devoid of soft-tissue attachments (such as a femoral head), or (3) an isolated soft-tissue allograft (such as a meniscus). After the tissue is harvested, the donor serum and allograft are cultured for microbial contamination. The allograft is then cleaned, soaked in an antiseptic solution such as BioCleanse (Regeneration Technologies, Alachua, FL), and irrigated with or without pressurized lavage or by ultrasonic/mechanical cleansing techniques. The allograft is then frozen and may be terminally sterilized (described below). In some cases, freezing is replaced by cryopreservation techniques to retain cell viability and possible osteogenic ability.

Freezing cannot substitute for sterilization and at best may only prevent bacteria, fungi, spores, or viruses from growing. As a result, some tissue banks perform terminal bactericidal and virucidal sterilization that includes heating, gamma-irradiation, chemical sterilization, and lyophilization. These procedures further reduce the risk of infection and allogenic response by musculoskeletal tissues. Some tissue banks routinely "pasteurize" or autoclave allografts [36, 42]; the resulting increase in temperature eliminates the biological activity of the cells, but may decrease the strength of the grafts as a result of the denaturation of structural proteins [7]. In addition, heat sterilization may not inactivate bacterial spores [27]. Gamma-irradiation at the level of 1.5 to 2.5 megarads or above [8, 13, 36, 42, 45, 62] is believed to inactivate bacterial contaminants and HCV, but not HIV [20, 45, 47]. Gamma-irradiation, moreover, weakens musculoskeletal allografts [14, 44]. Lyophilization, i.e., freeze-drying, is a process by which water is removed from the tissue to the point where cellular activity is no longer supported. This process involves partially freezing the tissues to allow sublimation of water, followed by further drying with the aid of other techniques. As a result, HIV and HCV are inactivated and the risk of transmission is minimized in the infected blood products and bone marrow [53]. This technique may, however, reduce the strength of the musculoskeletal allografts [14, 44]. With proper storage, freeze-dried allografts retain biological activity for several years.

Chemical sterilization with proprietary solutions or ethylene oxide has also been used for terminal sterilization. Adverse reactions, such as moderate inflammation from residual ethylene oxide in the allograft, have been reported [8, 54, 60, 62]. Proprietary solutions may contain particular bactericidal, virucidal, and fungicidal agents, but there is no industry-wide standard for their usage.

Allogenic bone can be machined and separated into cortical, corticocancellous, and cancellous preparations. Cortical and corticocancellous allografts are used for structural support and have limited osteoconductive capability, with no osteoinductive properties. Cortical and corticocancellous bone grafts undergo slow resorption in the host secondary to limited vascular invasion; this decreases the structural properties of the graft. The cortical/corticocancellous allograft is incorporated by the host through creeping substitution in conjunction with slow bone remodeling. These grafts are available in several forms: morsellized "bone chips," short segments of diaphyseal rings from femora or tibiae, iliac crest bone wedges, cortical struts, and whole bones en bloc, such as a fibula. Large areas of nonincorporated necrotic bone often remain in a patient for years after implantation. Cancellous allografts provide limited structural support and osteoconductivity that can be enhanced with demineralization. In the course of bone remodeling, cancellous allografts

**Table 3.1.** Graft donor infectious pathogens screened

| |
|---|
| Mandated by FDA |
| Human immunodeficiency virus (HIV) types 1 and 2 |
| Hepatitis B virus (HBV) |
| Hepatitis C virus (HCV) |
| Additional AATB screening |
| Human T-lymphocytic virus (HTLV) types 1 and 2 |
| Syphilis |

are resorbed more quickly than cortical grafts and are typically available as small, porous, spongy blocks that are used to fill segmental bone defects.

After terminal sterilization, bone allografts can be demineralized to make osteoinductive biological molecules, such as bone morphogenetic proteins (BMPs), more readily available to augment new bone formation [17, 37, 50]. The demineralization process is thought to destroy the antigenic surface of the bone graft, which reduces the host immune response. Like gamma-irradiation and lyophilization, the demineralization process weakens musculoskeletal allografts [14, 44]. Thus, choosing an appropriate allograft becomes critical when the primary requirement is structural augmentation.

Quality control of tissue banks is maintained through documentation and periodic audits of stored allografts. Some tissue banks routinely test stored tissues as new laboratory methods become available [12]. These periodic audits increase the chance of detection of potential cases of HIV transmission and/or epidemiological exposures to other previously undetected infections.

## 3.3 Infection from Musculoskeletal Transplants

Musculoskeletal transplantation is a safe, comprehensively regulated practice with a low incidence of infections, especially in light of its substantial usage in reconstructive procedures. However, the risk of potentially fatal complications from infectious transmission does exist. The literature describes many cases of contamination with HIV [2, 32, 53], HCV [1, 12], *Clostridium* species [4, 27, 38], and other bacteria [4, 24, 26, 65], and viruses in transplants procured from acceptable donors.

HIV infection is one of the most serious risks associated with allograft transplantation. There is currently no cure or vaccine for this lifelong, disabling disease. With proper donor screening and HIV antibody and antigen testing, the estimated risk of HIV transmission in musculoskeletal transplantation is 1 in 150,471 and can be reduced to 1 in 1.67 million with lymph node testing, serology, and checking for complications associated with grafts from the same donor [10]. The risk of infection following allograft transplantation is comparable to the risk of HIV infection from screened whole red blood cell transfusion; it is thought to be between 1 in 250,000 and 1 in 2,000,000 [10]. Between 1988 and 1992, four cases of HIV transmission were reported resulting from procedures that utilized fresh-frozen bone allografts in 1984 and 1985, which were traced to two donors [2, 53]. These investigations were initiated after the allograft recipients, whose only risk for HIV was transplantation, were found to be positive for HIV several years later. Other infected allograft recipients were then identified through analysis of banked tissue. The donors of these tissues were screened for HIV and tested negative. It is believed that the infection occurred during an early stage when HIV antibodies were not yet detectable. In another case, a fresh-frozen bone allograft was implanted that had been subjected to extensive intramedullary reaming prior to implantation and did not test positive, yet became the source of the HIV infection. Conceivably, the removal of blood and bone marrow from the allograft prior to implantation cleared infectious cells from the tissue and thus led to a negative test result [53]. To date, there have been no reports of HIV transmission from musculoskeletal allografts obtained from seronegative donors that were subjected to freeze-drying or other terminal sterilization methods [8, 32, 47, 48, 53, 62, 63, 64]. Since then, tests have been developed for other markers of HIV, including the p24 antigen assay and the use of the polymerase chain reaction (PCR) [20, 32, 47, 48, 53, 62, 63].

Hepatitis C is a chronic hepatic disease that for several years after infection may exhibit no clinical signs or symptoms, yet ultimately lead to severe morbidity and mortality. There is no cure or vaccine for hepatitis C. Nine cases of HCV transmission by musculoskeletal allografts from three donors were reported in the United States between 1995 and 2003 [1, 12]. These donors had negative medical and social histories and initially tested negative for HCV when subjected to an anti-HCV immunoassay. In these cases, previously undetected HCV was identified from retrospective testing of tissue and sera with newer anti-HCV immunoassays and PCR analysis. After the donor tissues had been identified, a protocol was initiated to inform and test all recipients of tissues or organs from these donors. Interestingly, this study reported that when the high-risk seroconverted individual was excluded, all recipients of minimally processed allografts seroconverted for HCV. However, recipients of irradiated tissue that had been freeze-dried, frozen or cryopreserved did not test positive for HCV infection [12]. In 2002, the Centers for Disease Control (CDC) reported four cases of HCV transmission that resulted from a screened donor of bone-patellar-bone and tendon allografts [1]. The CDC investigation was prompted by the fact that acute hepatitis C was diagnosed 6 weeks after a recipient received a bone-patellar-bone allograft. Further testing with an anti-HCV immunoassay showed that the donor serum was negative for the HCV protein, but PCR analysis showed a positive result for HCV mRNA. Testing of the other recipients of the infected allografts revealed no cases of HCV transmission if the bone allografts had undergone gamma irradiation [1, 12]. Gamma-irradiation of musculoskeletal allografts would therefore appear to reduce the risk of HCV transmission from infected tissues.

Studies of bacterial infection or contamination of musculoskeletal allografts have shown that most of the allograft contamination is due to *Staphylococcus* and other mixed skin flora [4, 8, 24, 26, 33, 65]. Kainer et al. [27] identified 14 cases of infection by *Clostridium* species that they traced to nine donors. The time between death and tissue procurement in two of the nine donors exceeded industry standards. The 14 infected patients had received nine frozen bone-patellar tendon-bone allografts, four fresh femoral condyles, and one meniscus graft. All of the processed allograft tissues from the 14 identified cases came from one tissue bank, and the unprocessed donor tissues originated from seven other tissue banks. The tissue banks that provided the allografts to the recipients had procured the tissues using aseptic techniques that included decontamination by suspension in a proprietary antibiotic solution, but did they did not employ terminal sterilization. However, when terminal sterilization was performed, whether by gamma-irradiation or by low temperature, or if chemical sterilization had been employed at other tissue banks, the resulting allografts from five of the nine identified donors did not induce infection. Even though the overall rate of *Clostridium* infection was less than 0.5% among recipients of allografts from the tissue bank that reported *Clostridium* infection, this rate was still significantly higher than the rate among recipients of allografts from the tissue banks with no *Clostridium* infection [27].

An additional means of reducing the risk of contamination involves harvesting the tissues in an operating room with sterile techniques [26, 65]. The degree of bacterial contamination

**Table 3.2.** Factors influencing allograft performance

| Factor | Implication |
|---|---|
| Graft donor age | Osteoinductive potential is greater from donors aged 42 years and younger. Mechanical properties of allograft bone are inversely proportional to donor age after the fifth decade. |
| Presence of osteoporosis or osteopenia | Osteoporotic and osteopenic bone have decreased mechanical properties. According to histologic appearance, the incidence of osteoporosis is higher in donors after the fifth decade of life. |
| Graft anatomic origin | Fibular strut grafts are stronger than femoral ring or tibial grafts. Iliac crest grafts from the anterior iliac spine are stronger than those from the posterior iliac spine. |
| Tissue processing | Gamma-irradiation of $\geq$3.0 megarads (virucidal levels) reduces mechanical properties. Lyophilization can also weaken allografts. Pasteurization may also decrease the mechanical strength of allograft bone. |

of a musculoskeletal allograft is a direct function of the time that elapses between death and refrigeration [38, 65]. Tissues obtained from living donors have lower rates of bacterial contamination than tissues harvested from cadavers at autopsy [26, 65]. Musculoskeletal allografts from donors who suffered multiple trauma, with or without resuscitation, had higher rates of bacterial contamination than allografts from organ donors [65]. These observations are best explained by the fact that as postmortem time increases, the risk of infection by intestinal flora such as *Clostridium* and *Escherichia* species also increases [27]. This is particularly true for spore-forming bacteria such as *Clostridium* that are capable of long dormancy. As with surgical infection rates, the rate of allograft contamination is directly proportional to the number of persons present in the operating room during procurement [65]. The order in which tissues are harvested also affects the rate of bacterial contamination; the rate is higher in specimens from the hemipelvis than in specimens from the femur or tibia [26], probably because the hemipelvis is typically the last large structural bone to be harvested. Prolonged handling of the skin also increases the risk of contamination [26]. The risk of contamination can be reduced by antiseptic soaking, irrigation, and terminal sterilization [24].

## 3.4 Donor Selection Factors Affecting Musculoskeletal Allograft Performance

All potential allograft transplant donors are screened for a variety of factors, including but not limited to sex, age, cause of death, and past medical and social history; the results of serological tests for medical diseases; and, most importantly, the presence of bacterial and viral pathogens. The most commonly reported exclusion factors for tissue donors include a medical history of infection at the excision sites, benign or malignant tumors at the excision sites, autoimmune diseases, severe endocrine/metabolic diseases, collagen diseases, and infection by HIV, HCV, and/or HBV. Age, sex, medical history, and the type of bone harvested from screened tissue donors have been evaluated for their effects on osteoinductive potential and structural support of the allografts (Table 3.2).

Increased donor age may be inversely related to the osteoinductive potential of bone allografts. Using an in vivo nude murine model, Schwartz et al. [50] reported that an increase in donor age decreased the osteoinductivity of the demineralized, freeze-dried bone allograft (DFDBA). Areas of new bone formation, new cortical bone function, and new bone-marrow production were smaller in allografts obtained from older donors (>50 years) than in allografts obtained from younger donors (<29 years). Osteoconductivity was not affected by donor gender. Lohman et al. [37] confirmed the age-dependent effect by noting that allograft osteoinductive potential was significantly greater for donors under 42 than for donors over 70 years of age.

Several studies have shown that the mechanical properties of bone decline with age. Burnstein et al., using cadaveric human specimens, observed a highly significant negative correlation between age and femoral yield stress, ultimate stress, elastic modulus, and ultimate strain [11]. Smith et al. [55] observed a negative correlation between the tensile stress of bone and age in vivo. McCalden et al. demonstrated that there is an inverse relationship between the mechanical properties of cortical bone and age, and theorized that the decrease in bone strength is the result of an age-dependent increase in bone porosity [39].

Allografts from donors with osteoporosis or osteopenia, conditions that are not contraindicated for bone transplant donation, may have less strength and stiffness [17]. Dickenson et al. [15] reported a significant decrease in the modulus of elasticity, the ultimate tensile strength, and the amount of plastic and elastic energy absorbed in osteoporotic bone in comparison with nonosteoporotic bone in vitro. They also theorized that the decrease in strength and stiffness in osteoporotic bone grafts was due to greater porosity. In vivo, Lill et al. observed a significant reduction in the bending stiffness of intact osteoporotic tibiae in comparison with normal tibiae, as well as delayed fracture healing in osteoporotic bone [34].

Histologic evaluation of bone allografts has shown that osteoporosis and osteopenia affect bone allograft performance [43, 51, 58]. Histo-

**Table 3.3.** Effects of graft tissue processing on allograft mechanical performance

| Processing technique | Study | Observations |
|---|---|---|
| Lyophilization | Brantigan et al. 1993 [9] | Fresh frozen cancellous bone is 219% stronger than lyophilized cancellous bone. |
| | Simonian et al. 1994 [54] | Lyophilization significantly decreases screw pullout strength. |
| | Kang and Kim 1995 [28] | In vivo lyophilized graft had decreases of 30.1% in bending strength and 41.3% in compressive strength. |
| | Thoren and Aspenberg 1995 [60] | Lyophilization decreased mechanical stiffness by 19%, yield by 16%, and energy to failure by 31%. |
| | Nather et al. 2004 [41] | Lyophilized allografts significantly weaker than deep-frozen grafts. |
| Gamma-irradiation | Anderson et al. 1992 [5] | Failure stress and elastic moduli of cancellous bone significantly decreased after 6.0 megarads but not after 2.5 megarads. |
| | Rasmussen et al. 1994 [45] | 12% decrease in stiffness and 26% decrease in maximum force after 4.0 megarads. |
| | Zhang et al. 1994 [68] | No significant difference in mechanical properties of iliac crest wedge grafts after 2.0 to 2.5 megarads. |
| | Fideler et al. 1995 [19] | Mechanical properties of fresh-frozen bone-patella-bone graft reduced by 15% after 2.0 megarads, with further reduction of 46% after 4.0 megarads. |
| | Hamer et al. 1996 [23] | Dose-dependent decreases of up to 46% in mechanical strength after irradiation. |
| | Currey et al. 1997 [13] | Virucidal irradiation levels decreased bending strength by 52% to 67%, work to fracture by 74% to 96%, and impact energy by 37% to 75%. |
| Pasteurization | Borcher et al. 1995 [7] | Boiling and autoclaving decreased allograft strength by 26% and 58%, respectively. Freezing did not compromise allograft strength. |
| Ethylene oxide | Wittenberg et al. 1990 [66] | Ethylene oxide had no significant effect on immediate compression strength of grafts. |

logical evaluation was performed on 27% of the 1,146 osteoarthritic femoral heads donated by patients undergoing elective total hip arthroplasty. More than 30% of the samples exhibited osteopenia on radiographic examination. Marked, generalized osteopenia with thinning of the cortical and cancellous bone was found in 3% of the samples [43]. The increased incidence of osteopenia clearly affects the bone quality, but the effects of metabolic and inflammatory diseases noted in some specimens are not known [43]. Siddiqui et al. [51] observed that 12% of 40 allografts from screened donors in their fifties had osteoporosis. They suggested that these allografts would not be suitable in cases where graft strength is required.

Bone allografts can be used in either orthotopic or heterotopic transplantations. In orthotopic transplantation, cortical bone allografts are placed in an anatomically appropriate site, as in an area of large bone loss. In heterotopic transplantation, bone allografts are placed in an anatomically abnormal location, such as a fibular strut allograft used adjunctively during a vertebral fusion [57]. In general, cortical bone allografts are stronger than cancellous bone allografts. Cortical bone graft strength varies according to anatomical location, with fibular struts being stronger than femoral rings, which in turn are stronger than tricortical iliac bone crest [21, 46, 52, 66]. Additionally, iliac bone grafts harvested close to the anterior superior iliac spine are stronger than those harvested near the posterior iliac spine [31]. Various combinations of cortical and cancellous bone allografts can augment reconstructive procedures, but terminal sterilization, though recommended, may reduce their strength.

## 3.5 Effects of Processing on Biomechanical Properties of Musculoskeletal Allografts

Gamma-irradiation and lyophilization (freeze-drying), two commonly used techniques for terminal sterilization of allografts, lead to weakening of grafts (Table 3.3). Gamma-irradiation of at least 3.0 megarads is required to inactivate viruses, whereas 1.5 to 2.5 megarads can inactivate bacteria [20, 45, 47]. Radiation dose weakens the biomechanical properties of musculoskeletal allografts. Gamma-irradiation at virucidal levels significantly increases fatigue while decreasing failure strength, failure energy, and stiffness [5, 13, 19, 23, 45, 54, 68]. Lyophilization reduces screw pullout strength and the maximum limits of strength, torque, and torsional stiffness and diminishes absorption energy [28, 41, 54]. Lyophilized allografts need to be rehydrated before transplantation; the quality of rehydration can also affect the mechanical parameters [28, 41, 61]. Terminal sterilization with ethylene oxide does not significantly weaken screw pullout strength [54]. Musculoskeletal tissues subjected to boiling or autoclaving exhibit significant reductions in strength, but freezing does not reduce their strength [7].

## 3.6 Conclusions

Musculoskeletal allografts are an alternative to autografts without the associated morbidities [6, 18, 30, 35, 59]. Allografts are widely available in a variety of preparations, and their transplantation is a safe, comprehensively regulated practice with a low incidence of HIV [2, 32, 53], HCV, and bacterial infection. The risk of infection is further decreased when musculoskeletal allografts are obtained from AATB-accredited tissue banks that practice comprehensive donor screening and tighter tissue procurement and employ more testing than required by the FDA [67]. The use of tissue banks that perform PCR analysis and/or histomorphometric testing of donor tissues further minimizes the risk of viral or bacterial transmission [47, 48, 51, 58]. The risk of HIV infection from fresh-frozen, non-terminally-sterilized musculoskeletal allografts is comparable to that from blood transfusion [10]. Terminal sterilization by gamma-irradiation or lyophilization of musculoskeletal tissues can further diminish HIV and HCV infection rates, but at the cost of a decrease in the mechanical properties of the allograft. Until more tissue-screening tests become available, active surveillance and audit of stored nonimplanted allografts will provide further assurance for the quality control of musculoskeletal transplants.

Using fresh, fibular strut or femoral ring cortical bone allografts from younger, nonosteoporotic donors permits the surgeon to maximize the structural integrity of reconstructive procedures [11, 15, 17, 39, 46, 52, 55]. The osteoinductive ability of bone allografts can be maximized by selecting demineralized, freeze-dried grafts from younger donors [37, 50].

In the future, new methods or modifications of existing tissue-processing techniques may be developed to maximize the osteoinductive and osteoconductive properties of bone allografts. Possible approaches to augment bone allograft performance may include the use of bone morphogenetic proteins (BMPs) or other local or systemic mediators of growth and inflammation. For example, the use of structural cortical bone allografts with osteoconductive and structural capabilities could add to the osteoinductive ability of the graft [25, 40, 56]. Although promising, the use of BMPs in conjunction with allografts needs further study.

## References

1. Hepatitis C virus transmission from an antibody-negative organ and tissue donor-United States, 2000–2002 (2003) MMWR 53:273–276.
2. Transmission of HIV through bone transplantation: case report and public health recommendations (1988) MMWR 37:597–599.
3. United States Census Bureau, Statistical Abstract of the United States (2001) No. 168 Organ Transplants and Grafts, 1990 to 2000.
4. Update: allograft-associated bacterial infections-United States (2002) MMWR 51:207–210.
5. Anderson MJ, Keyak JH, Skinner HB (1992) Compressive mechanical properties of human cancellous bone after gamma irradiation. J Bone Joint Surg Am 74:747–752.
6. Banwart JC, Asher MA, Hassanein RS (1995) Iliac crest bone graft harvest donor site morbidity. A statistical evaluation. Spine 20:1055–1060.

7. Borchers RE, Gibson LJ, Burchardt H, Hayes WC (1995) Effects of selected thermal variables on the mechanical properties of trabecular bone. Biomaterials 16: 545–551.
8. Boyce T, Edwards J, Scarborough N (1999) Allograft bone. The influence of processing on safety and performance. Orthop Clin North Am 30:571–581.
9. Brantigan JW, Cunningham BW, Warden K, McAfee PC, Steffee AD (1993) Compression strength of donor bone for posterior lumbar interbody fusion. Spine 18:1213–1221.
10. Buck BE, Malinin TI, Brown MD (1989) Bone transplantation and human immunodeficiency virus. An estimate of risk of acquired immunodeficiency syndrome (AIDS). Clin Orthop 240:129–136.
11. Burstein AH, Reilly DT, Martens M (1976) Aging of bone tissue: mechanical properties. J Bone Joint Surg Am 58:82–86.
12. Conrad EU, Gretch DR, Obermeyer KR, Moogk MS, Sayers M, Wilson JJ, Strong DM (1995) Transmission of the hepatitis-C virus by tissue transplantation. J Bone Joint Surg Am 77:214–224.
13. Currey JD, Foreman J, Laketic I, Mitchell J, Pegg DE, Reilly GC (1997) Effects of ionizing radiation on the mechanical properties of human bone. J Orthop Res 15:111–117.
14. Davy DT (1999) Biomechanical issues in bone transplantation. Orthop Clin North Am 30:553–563.
15. Dickenson RP, Hutton WC, Stott JR (1981) The mechanical properties of bone in osteoporosis. J Bone Joint Surg Br 63B:233–238.
16. Ehrler DM, Vaccaro AR (2000) The use of allograft bone in lumbar spine surgery. Clin Orthop 371:38–45.
17. Einhorn TA (2003) The structural properties of normal and osteoporotic bone. Instr Course Lect 52: 533–539.
18. Fernyhough JC, Schimandle JJ, Weigel MC, Edwards CC, Levine AM (1992) Chronic donor site pain complicating bone graft harvesting from the posterior iliac crest for spinal fusion. Spine 17:1474–1480.
19. Fideler BM, Vangsness CT Jr, Lu B, Orlando C, Moore T (1995) Gamma irradiation: effects on biomechanical properties of human bone-patellar tendon-bone allografts. Am J Sports Med 23:643–646.
20. Fideler BM, Vangsness CT Jr, Moore T, Li Z, Rasheed S (1994) Effects of gamma irradiation on the human immunodeficiency virus. A study in frozen human bone-patellar ligament-bone grafts obtained from infected cadavera. J Bone Joint Surg Am 76:1032–1035.
21. Glazer PA, Colliou O, Lotz JC, Bradford DS (1996) Biomechanical analysis of lumbosacral fixation. Spine 21:1211–1222.
22. Goldberg VM (2000) Selection of bone grafts for revision total hip arthroplasty. Clin Orthop 381:68–76.
23. Hamer AJ, Strachan JR, Black MM, Ibbotson CJ, Stockley I, Elson RA (1996) Biochemical properties of cortical allograft bone using a new method of bone strength measurement. A comparison of fresh, fresh-frozen and irradiated bone. J Bone Joint Surg Br 78: 363–368.
24. Hirn M, Laitinen M, Pirkkalainen S, Vuento R (2004) Cefuroxime, rifampicin and pulse lavage in decontamination of allograft bone. J Hosp Infect 56:198–201.
25. Jensen TB, Overgaard S, Lind M, Rahbek O, Bunger C, Soballe K (2002) Osteogenic protein 1 device increases bone formation and bone graft resorption around cementless implants. Acta Orthop Scand 73:31–39.
26. Journeaux SF, Johnson N, Bryce SL, Friedman SJ, Sommerville SM, Morgan DA (1999) Bacterial contamination rates during bone allograft retrieval J Arthroplasty 14:677–681.
27. Kainer MA, Linden JV, Whaley DN, Holmes HT, Jarvis WR, Jernigan DB, Archibald LK (2004) Clostridium infections associated with musculoskeletal-tissue allografts. N Engl J Med 350:2564–2571.
28. Kang JS, Kim NH (1995) The biomechanical properties of deep freezing and freeze drying bones and their biomechanical changes after in vivo allograft. Yonsei Med J 36:332–335.
29. Komiya K, Nasuno S, Uchiyama K, Takahira N, Kobayashi N, Minehara H, Watanabe S, Itoman M (2003) Status of bone allografting in japan—nationwide survey of bone grafting performed from 1995 through 1999. Cell Tissue Bank 4:217–220.
30. Kreibich DN, Scott IR, Wells JM, Saleh M (1994) Donor site morbidity at the iliac crest: comparison of percutaneous and open methods. J Bone Joint Surg Br 76:847–848.
31. Kummer FJ, Chen D, Spivak JM (1998) Optimal selection and preparation of fresh frozen corticocancellous allografts for cervical interbody spinal fusion. Spine 23:2295–2298.
32. Li CM, Ho YR, Liu YC (2001) Transmission of human immunodeficiency virus through bone transplantation: a case report. J Formos Med Assoc 100:350–351.
33. Lietman SA, Tomford WW, Gebhardt MC, Springfield DS, Mankin HJ (2000) Complications of irradiated allografts in orthopaedic tumor surgery. Clin Orthop 375:214–217.
34. Lill CA, Hesseln J, Schlegel U, Eckhardt C, Goldhahn J, Schneider E (2003) Biomechanical evaluation of healing in a non-critical defect in a large animal model of osteoporosis. J Orthop Res 21:836–842.
35. Lim EV, Lavadia WT, Roberts JM (1996) Superior gluteal artery injury during iliac bone grafting for spinal fusion. A case report and literature review. Spine 21:2376–2378.
36. Lobo Gajiwala A (2003) Tissue banking in India: gamma-irradiated allografts. Cell Tissue Bank 4:203–211.
37. Lohmann CH, Andreacchio D, Koster G, Carnes DL Jr, Cochran DL, Dean DD, Boyan BD, Schwartz Z (2001) Tissue response and osteoinduction of human bone grafts in vivo Arch Orthop Trauma Surg 121:583–590.
38. Malinin TI, Buck BE, Temple HT, Martinez OV, Fox WP (2003) Incidence of clostridial contamination in donors' musculoskeletal tissue. J Bone Joint Surg Br 85:1051–1054.
39. McCalden RW, McGeough JA, Barker MB, Court-Brown CM (1993) Age-related changes in the tensile properties of cortical bone. The relative importance of changes in porosity, mineralization, and microstructure. J Bone Joint Surg Am 75:1193–1205.
40. McGee MA, Findlay DM, Howie DW, Carbone A, Ward P, Stamenkov R, Page TT, Bruce WJ, Wildenauer CI, Toth C (2004) The use of OP-1 in femoral impaction grafting in a sheep model. J Orthop Res 22:1008–1015.

41. Nather A, Thambyah A, Goh JC (2004) Biomechanical strength of deep-frozen versus lyophilized large cortical allografts. Clin Biomech 19:526–533.
42. Navas J, Soto C (2003) The Colombian experience in tissue banking: the bone and tissue bank of the Cosmos and Damian Foundation, Bogota. Cell Tissue Bank 4:157–161.
43. Palmer SH, Gibbons CL, Athanasou NA (1999) The pathology of bone allograft. J Bone Joint Surg Br 81:333–335.
44. Pelker RR, Friedlaender GE (1987) Biomechanical aspects of bone autografts and allografts. Orthop Clin North Am 18:235–239.
45. Rasmussen TJ, Feder SM, Butler DL, Noyes FR (1994) The effects of 4 Mrad of gamma irradiation on the initial mechanical properties of bone-patellar tendon-bone grafts. Arthroscopy 10:188–197.
46. Reilly DT, Burstein AH (1974) The mechanical properties of cortical bone. J Bone Joint Surg Am 56:1001–1022.
47. Roder W, Muller H, Muller WE, Merz H (1992) HIV infection in human bone. J Bone Joint Surg Br 74:179–180.
48. Salzman NP, Psallidopoulos M, Prewett AB, O'Leary R (1993) Detection of HIV in bone allografts prepared from AIDS autopsy tissue. Clin Orthop Relat Res 292: 384–390.
49. Sanzen L, Carlsson A (1997) Transmission of human T-cell lymphotrophic virus type 1 by a deep-frozen bone allograft. Acta Orthop Scand 68:72–74.
50. Schwartz Z, Somers A, Mellonig JT, Carnes DL Jr, Dean DD, Cochran DL, Boyan BD (1998) Ability of commercial demineralized freeze-dried bone allograft to induce new bone formation is dependent on donor age but not gender. J Periodontol 69:470–478.
51. Siddiqui SA, Lipton JF, Vigorita VJ, Evangelista J, Bryk E (2004) Bone biopsy as a screening technique for bone bank allograft donation. Am J Orthop 33: 123–126.
52. Siff TE, Kamaric E, Noble PC, Esses SI (1999) Femoral ring versus fibular strut allografts in anterior lumbar interbody arthrodesis. A biomechanical analysis. Spine 24:659–665.
53. Simonds RJ, Holmberg SD, Hurwitz RL, Coleman TR, Bottenfield S, Conley LJ, Kohlenberg SH, Castro KG, Dahan BA, Schable CA, et al (1992) Transmission of human immunodeficiency virus type 1 from a seronegative organ and tissue donor. N Engl J Med 326:726–732.
54. Simonian PT, Conrad EU, Chapman JR, Harrington RM, Chansky HA (1994) Effect of sterilization and storage treatments on screw pullout strength in human allograft bone Clin Orthop 302:290–296.
55. Smith CB, Smith DA (1976) Relations between age, mineral density and mechanical properties of human femoral compacta. Acta Orthop Scand 47: 496–502.
56. Soballe K, Jensen TB, Mouzin O, Kidder L, Bechtold JE (2004) Differential effect of a bone morphogenetic protein-7 (OP-1) on primary and revision loaded, stable implants with allograft. J Biomed Mater Res 71A:569–576.
57. Stevenson S (1999) Biology of bone grafts. Orthop Clin North Am 30:543–552.
58. Sugihara S, van Ginkel AD, Jiya TU, van Royen BJ, van Diest PJ, Wuisman PI (1999) Histopathology of retrieved allografts of the femoral head. J Bone Joint Surg Br 81:336–341.
59. Summers BN, Eisenstein SM (1989) Donor site pain from the ilium. A complication of lumbar spine fusion. J Bone Joint Surg Br 71:677–680.
60. Thoren K, Aspenberg P (1995) Ethylene oxide sterilization impairs allograft incorporation in a conduction chamber. Clin Orthop 114:259–264.
61. Thoren K, Aspenberg P, Thorngren KG (1995) Lipid extracted bank bone. Bone conductive and mechanical properties. Clin Orthop 311:232–246.
62. Tomford WW, Mankin HJ (1999) Bone banking. Update on methods and materials. Orthop Clin North Am 30:565–570.
63. Tomford WW, Mankin HJ, Friedlaender GE, Doppelt SH, Gebhardt MC (1987) Methods of banking bone and cartilage for allograft transplantation. Orthop Clin North Am 18:241–247.
64. Tomford WW, Thongphasuk J, Mankin HJ, Ferraro MJ (1990) Frozen musculoskeletal allografts. A study of the clinical incidence and causes of infection associated with their use. J Bone Joint Surg Am 72:1137–143.
65. Vehmeyer S, Wolkenfelt J, Deijkers R, Petit P, Brand R, Bloem R (2002) Bacterial contamination in postmortem bone donors. Acta Orthop Scand 73:678–683.
66. Wittenberg RH, Moeller J, Shea M, White AA 3rd, Hayes WC (1990) Compressive strength of autologous and allogenous bone grafts for thoracolumbar and cervical spine fusion. Spine 15:1073–1078.
67. Woll JE, Kasprisin D (2001) Standards for Tissue Banking. McLean, Virginia: American Association of Tissue Banks.
68. Zhang Y, Homsi D, Gates K, Oakes K, Sutherland V, Wolfinbarger L Jr (1994) A comprehensive study of physical parameters, biomechanical properties, and statistical correlations of iliac crest bone wedges used in spinal fusion surgery. IV. Effect of gamma irradiation on mechanical and material properties. Spine 19:304–308.

# 4. Biodegradable Orthopedic Implants

*Hansoo Park, Johnna S. Temenoff, and Antonios G. Mikos*

## List of Abbreviations

ECM: extracellular matrix
GAG: glycosaminoglycan
HA: hyaluronic acid
MMP: matrix metalloproteinase
OPF: oligo(poly(ethylene glycol) fumarate)
PBS: phosphate-buffered saline
PCL: poly(ε-caprolactone)
PEG: poly(ethylene glycol)
PEG-DA: poly(ethlylene glycol)-diacrylate
PEG-DM: poly(ethlylene glycol)-dimethacrylate
PGA: poly(glycolic acid)
PLA: poly(lactic acid)
PLGA: poly(lactic-co-glycolic acid)
POE: poly(orthoester)
PPF: poly(propylene fumarate)
PPF-DA: poly(propylene fumarate)-diacrylate
rhBMP-2: recombinant human bone morphogenetic protein 2
TGF-β1: transforming growth factor β1

## 4.1 Introduction

Over the past 30 years, there have been significant advances in the development of biodegradable materials [79]. In particular, these materials have received attention for use as implants to aid regeneration of orthopedic defects [49, 91]. Every year more than 3.1 million orthopedic surgeries are performed in the United States alone [1]. However, although current treatments using nondegradable fixation materials have proven efficacious, tissue-engineering approaches with biodegradable implants are being considered as promising future alternatives [8, 49]. One possible advantage of these systems is that biodegradable implants can be engineered to provide temporary support for bone fractures, and because they can degrade at a rate matching new tissue formation, their use can eliminate the need for a second surgery [49]. In addition to providing support for the tissue surrounding a defect, the scaffold can serve as a substrate for seeded cells, facilitating new tissue formation at the site of injury [35, 100]. The incorporation of drugs or bioactive molecules may also accelerate new tissue formation, or can be used to treat specific conditions, such as osteomyelitis [4, 10].

In designing biodegradable orthopedic implants, several important factors should be considered. First, the material should degrade over an appropriate time, so that the scaffold functions as a temporary support, but allows space for newly generated tissue to replace the defect [49, 91]. Second, neither the initially implanted biomaterials nor the degraded materials and related products, such as monomers, initiators, and residual solvents, should elicit a serious inflammatory or immunogenic response in the body [28]. Finally, the material should possess sufficient mechanical strength to sustain loads applied to defects during the healing process. Additionally, the material should show a decrease in mechanical strength as defects are replaced with new tissue to encourage force transfer in load-bearing defects. In this way, mechanical signals are gradually transmitted to the resident cells, thus encouraging tissue remodeling via exposure to dynamic loading conditions [2, 106].

Over several decades, a number of biomaterials for orthopedic applications have been investigated and developed. In this chapter, applications, important properties, and different types of biodegradable materials will be discussed in order to provide an overview of the state of the art in orthopedic biomaterials.

## 4.2 Background

Before developing biomaterials for a particular orthopedic tissue-engineering application, it is important first to understand the basic properties of the different musculoskeletal tissues such as bone, cartilage, ligament, and tendon. This basic information allows developing materials and strategies that are specifically tailored for each type of tissue defect.

### 4.2.1 Bone

The main function of bone tissue is to support the body. Bone tissue is maintained by the balance in activity between bone-forming and bone-resorbing cells. The collagen fibers impart tensile strength, and the mineral salts, a form of calcium phosphate (hydroxyapatite), increase the toughness and hardness of the tissue [7]. Three types of cells coexist in bone: osteoblasts, osteoclasts, and osteocytes. Osteoblasts are bone-forming cells responsible for the formation of the hard extracellular matrix, whereas osteocytes are fully mature embedded bone cells that maintain the tissue structure. Osteoclasts selectively resorb bone in certain areas in response to a biochemical or biomechanical stimulus [21].

Human bones are described as compact (cortical) or spongy (cancellous), depending on their density. Compact bone consists of central canals and perforating canals surrounded by concentric rings of matrix. Spongy bone is much less dense, having irregular lattice structures where spaces are filled with bone marrow [101].

### 4.2.2 Cartilage

Cartilage is an avascular tissue composed of chondrocytes embedded in an extracellular matrix consisting of water and a solid matrix. The solid matrix consists of proteoglycans and collagens, as well as glycoproteins in lesser amounts. Three types of cartilage have been described, which differ in composition: hyaline cartilage, elastic cartilage, and fibrocartilage [36, 45].

Hyaline cartilage is a glassy and homogeneous cartilage composed primarily of type II collagen fibers and proteoglycans. This unique combination of collagen fibers and hydrophilic proteoglycans gives cartilage important viscoelastic properties that allow it to disperse forces while acting as a lubricator. Elastic cartilage is similar to hyaline cartilage; however, it also contains elastic fibers and an interconnecting sheet of elastic material. It is often found in the external ears and the walls of the acoustic meatus. Fibrocartilage possesses properties that are intermediate between those of dense connective tissue and hyaline cartilage and contains both type I and II collagen. Fibrocartilage is the main constituent of tissues such as the meniscus of the knee [36, 45].

### 4.2.3 Tendon

Tendons are dense tissues that connect muscle to bone. Tendon tissue consists of fibroblasts surrounded by type I collagen, a small amount of type III collagen, and small quantities of proteoglycans (dermatan sulfate and hyaluronic acid). Triple-helical collagen molecules are assembled into fibrils that are cross-linked through aldol or Schiff base adducts between aldehydes on one or more of the $\alpha$-chains of collagen molecules and aldehydes or amino groups on adjacent chains. This cross-linking imparts the high tensile strength needed for proper tendon function [5, 31].

### 4.2.4 Ligament

Ligaments are made up of closely packed fibers and are in many respects similar to tendons. However, the relative amounts of the various extracellular matrix (ECM) components are not the same as in tendons. Specifically, ligaments have less total collagen and more proteoglycans than tendons. Ligaments are less organized in structure but have higher DNA content than corresponding tendons [5, 31].

## 4.3 Applications of Biodegradable Orthopedic Implants

In designing scaffolds for orthopedic implants, the envisioned final application must be a primary concern from the beginning. Scaffolds may be used as internal fixation devices to support the defect site. Alternatively, scaffolds may be implanted to induce cell migration and proliferation to aid in tissue repair. Another potential strategy is the use of scaffolds to provide localized delivery of bioactive molecules, cells, or a combination to enhance defect healing.

### 4.3.1 Systems for Mechanical Support

In many cases, biodegradable orthopedic materials have been applied during the healing process in the form of fixation implants such as screws, staples, pins, rods, and suture anchors to support areas weakened by bone fracture, sports injury, or osteoporosis [14, 37, 98]. High mechanical strength and stiffness are extremely important in designing biodegradable devices for orthopedic procedures in which high loads are applied after the devices have been implanted. Long degradation times for the biomaterials are also often desired for these applications [17, 20]. A study comparing a biodegradable interference screw made of poly(L-lactide) with a titanium interference screw in the porcine anterior cruciate ligament demonstrated that the poly(L-lactide) screw could provide a promising alternative in terms of primary fixation strength [84]. A mixture of poly(propylene fumarate) (PPF) and poly(propylene fumarate)-diacrylate (PPF-DA) has been molded into a biodegradable fixation plates (Fig. 4.1A) and a bone allograft interbody fusion spacer (Fig. 4.1B) with acceptable mechanical properties for use in these applications [98].

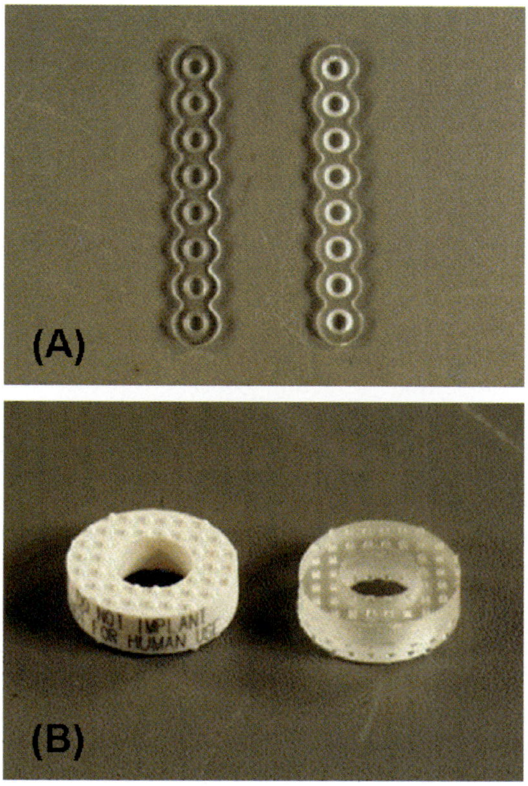


**Figure 4.1.** Photographs of a biodegradable fixation plate and an interbody fusion spacer fabricated by the use of transparent silicone molds. (A) 1.5-mm, eight-hole adaptation plate manufactured with 70:30 P(L/DL-LA) (left) and PPF/PPF-DA (double-bond ratio of 0.5) (right). (B) Plastic model (left) and PPF/PPF-DA (double-bond ratio of 0.5) replicate (right) of a 5-mm lordotic anterior cervical fusion (ACF) spacer. Reproduced with permission from Timmer et al. [98]. Copyright 2003, with permission from Elsevier.

### 4.3.2 Systems for Delivery of Cells or Bioactive Factors

#### 4.3.2.1 Bioactive Factors

In addition to providing physical support, scaffolds have been employed to introduce bioactive molecules at the defect site [39, 66]. In one strategy, scaffolds can be used to control the release of bioactive molecules, thus accelerating the healing process [41]. In other cases, the effectiveness of less stable drugs may be extended by encapsulating them inside a matrix [50]. Several delivery systems have been developed, including nano- or microparticles and hydrogel-based implants.

#### 4.3.2.1.1 Nano- or Microparticles

Nano- or microparticles are among the most common types of delivery vehicle for bioactive molecules. A variety of microparticles fabricated with polymers such as poly(ε-caprolactone) (PCL), poly(lactic acid) (PLA), poly(lactic-co-glycolic acid) (PLGA), or blends of PLGA with poly(ethylene glycol) (PEG) have been investigated as delivery matrices for orthopedic applications. These microparticles can be formed by several methods, such as a single/double emulsion technique or a solvent evaporation-extraction process. Because the mechanism by which bioactive molecules are released in these systems is mainly diffusion, the release rate and total amount released can be adjusted by altering fabrication parameters such as loading concentration, polymer molecular weight, copolymer ratio, and particle structure [24, 47, 89, 104].

Alternatively, release from nano- or microparticles made of naturally derived materials can be controlled through directed degradation rather than a diffusion mechanism, as in the polymeric systems described above. When transforming growth factor β1 (TGF-β1) was incorporated into gelatin microspheres, the release profiles depended on the presence of a gelatinase enzyme in the medium. In this case, it is likely that the polyionic complexation between the growth factor and the gelatin retards its release until the gelatin microparticle is degraded by the enzyme. Because enzymes such as matrix metalloproteinases (MMPs) are up-regulated in injured cartilage, this system may provide a unique mechanism to encourage drug release in areas undergoing tissue remodeling [41]. An additional advantage of gelatin microspheres is that when they are encapsulated in hydrogels, they can serve as porogens, thus providing additional space for tissue formation at the defect site (Fig. 4.2) [41, 77].

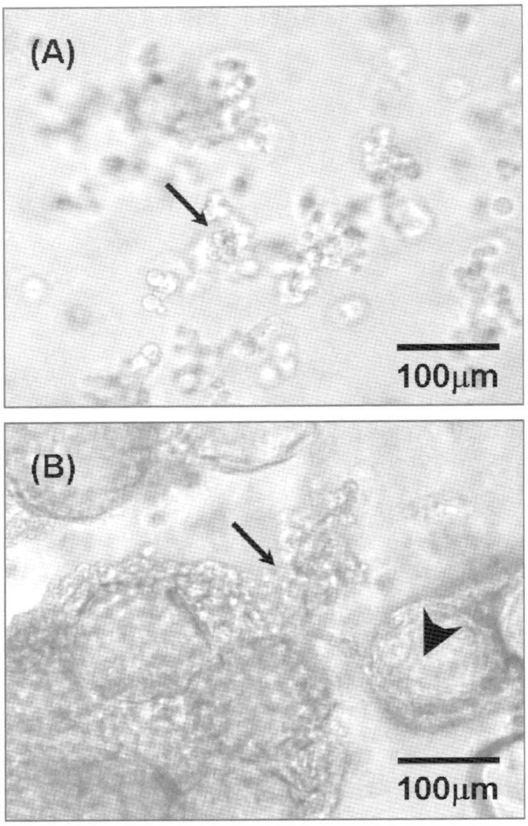


**Figure 4.2.** Light microscopy of oligo(poly(ethylene glycol) fumarate) (OPF) hydrogel composites containing chondrocytes at day 21. Arrows indicate encapsulated chondrocytes, and arrow heads indicate encapsulated microparticles. OPF hydrogel composites containing only chondrocytes are depicted in (A), while (B) shows OPF hydrogel composites containing chondrocytes and TGF-β1-loaded microparticles. Reproduced with permission from Park et al. [77]. Copyright 2005, with permission from Elsevier.

#### 4.3.2.1.2 Hydrogels

Hydrogels are three-dimensional polymers physically or chemically cross-linked and swollen by water. This enables them to entrap various drugs and later release them in a controlled manner. The release kinetics of drugs from hydrogels can be modulated by external stimuli such as changes in pH [52], temperature [50], or protein levels [80]. For the treatment of orthopedic defects, hydrogels have the advantage that they can be designed to function as biomimetic support materials, as well as drug-delivery matrices [85]. Moreover, depending on their composition, hydrogels may be injectable, allowing for their use in minimally invasive procedures. In one study, PEG-based macromers were photopolymerized to encapsulate DNA. By changing the monomer chemistry in this system, the DNA release profile was tailored to provide release over 6 to 100 days [82]. Another PEG-based oligomer, oligo(poly (ethylene glycol) fumarate), has also been developed as an injectable hydrogel carrier for growth factors useful for both bone and cartilage tissue engineering [40, 41, 54].

scaffolds because they are currently FDA-approved for use as suture materials and as drug-delivery systems. PGA can be highly crystalline (46%–50%), depending on its preparation method, and is hydrophilic in nature. Its high crystallinity makes it nonsoluble in many organic solvents except for those that are highly halogenated. PGA is mainly synthesized by methods employing ring-opening polymerization, and, like all polyesters, is degraded primarily by bulk hydrolysis of ester linkages at random sites. PGA crystallinity has a large impact on material degradation rate, because the more crystalline portions retard water entry and thus hydrolytic cleavage [8, 71].

PLA is another type of biodegradable and biocompatible poly(α-ester). It is also synthesized by ring opening polymerization and has two isomeric forms, D(−) and L(+). Like PGA, it is degraded by bulk hydrolysis of the ester linkage catalyzed by the presence of the degradation product, lactic acid [65]. PLA can also occur in crystalline forms, with the degree of crystallinity ranging as high as 37%. It is more hydrophobic than PGA and therefore has a slower degradation rate and a higher modulus [8, 72]. This high mechanical strength makes it a desirable material for orthopedic fixation devices [19]; however, the release of degrading crystal-like particles can be problematic.

Lactic acid and glycolic acid are often copolymerized at various ratios yielding poly(lactic-co-glycolic acid) (PLGA), with different properties from those of either of the homopolymers. The major difference is that the copolymer is amorphous within a wide range of copolymer ratios because of the disruption of the crystalline phases and therefore has a faster degradation rate and lower elastic modulus than PGA or PLA alone[8, 42, 76]. A study using two-dimensional and three-dimensional PLGA scaffolds impregnated with recombinant human bone morphogenetic protein 2 (rhBMP-2) and seeded with rabbit bone marrow stromal cells has reported in vitro osteogenic differentiation and ECM production over 2 months [44].

#### 4.5.2.2 Poly(ε-Caprolactone)

Poly(ε-caprolactone) (PCL) is a semicrystalline polymer with a melting temperature of 59 to 64C and a glass temperature of −60C. PCL is also synthesized by ring-opening polymerization of the cyclic monomer ε-caprolactone and is degraded by bulk hydrolysis. This material has a slower degradation rate than PLA and is easily copolymerized with other polymers [3, 70]. Recently, poly(ε-caprolactone) was used to fabricate three-dimensional nanofibrous scaffolds, allowing for in vitro chondrogenesis of seeded mesenchymal stem cells over 3 weeks [63].

#### 4.5.2.3 Poly(Orthoesters)

Poly(orthoesters) (POEs) are hydrophobic polymers that are degraded by surface erosion. Different degradation rates can be achieved by the addition of lactide groups, because carboxylic acids released by the degradation of the lactide segments facilitate the degradation of the orthoester [32]. An in vivo comparison between POE and PLGA scaffolds for bone tissue engineering found that POE scaffolds maintained their structural integrity after 6 and 12 weeks, whereas PLGA scaffolds partially collapsed after 6 weeks [6].

#### 4.5.2.4 Poly(Anhydrides)

Poly(anhydrides) are prepared by a melt condensation reaction of diacid molecules. They degrade by surface erosion and thus have been widely investigated as vehicles for biocompatible controlled release [90]. Poly(anhydrides), however, are not strong enough to be used as orthopedic materials, so photocross-linking or combination with other polymers such as polyimides has been used to improve the overall mechanical properties of implants [32].

#### 4.5.2.5 Poly(Ethylene Glycol)-Based Materials

Poly(ethylene glycol) (PEG) is a hydrophilic, highly biocompatible polymer with a variety of biomedical applications. Many different types of PEG-based materials have been developed as hydrogel scaffolds, including poly(ethylene glycol)-diacrylate (PEG-DA) and poly(ethylene glycol)-dimethacrylate (PEG-DM) [23, 67, 102]. Work with PEG-DM has demonstrated that it could encourage cartilage-like ECM production from encapsulated bovine chondrocytes over 4 weeks in vitro [12, 23, 67]. Although these derivatives often have limitations as scaffold materials because of their lack of degradability, PEG of low molecular weight can readily be excreted by humans and therefore can be copolymerized with other polymers such as PLA and PPF to be used as a biodegradable scaffold material [26, 83].

### 4.5.2.6 Poly(Amino Acids)

Poly(amino acids) have been considered as promising materials for biomedical applications because of their composition. However, the polymerization of pure poly(amino acids) is hard to control precisely. Furthermore, depending on the combination of amino acids, these materials can evoke an immune response in vivo [34, 76]. For these reasons, synthetic pseudo poly(amino acids), such as tyrosine-based polycarbonate, have been investigated recently. The polycarbonate not only exhibits good biocompatibility, but also supports the attachment of osteoblasts and osteoprogenitor cells. In addition, by varying the structure of the repeating unit, this material is easily modified to exhibit a range of mechanical properties, degradation rates, and bioactivity [81].

### 4.5.2.7 Fumarate-Based Polymers

Poly(propylene fumarate) (PPF) is a biodegradable poly(ester) whose degradation generates 1,2-propanediol and fumaric acid, the latter of which is a naturally-occurring material produced in the Krebs cycle [2, 25, 27]. A number of methods can be used to synthesize PPF, and each produces polymers with unique physical properties [25, 32]. The backbone of this polymer contains double bonds, which lead to the formation of a three-dimensional network either by photocross-linking with bis(2,4,6-trimethylbenzoyl) phenylphosphine oxide (BAPO) or by thermal cross-linking with benzoyl peroxide [25, 98]. PPF has been investigated for use in injectable orthopedic implants because it possesses, in its cross-linked form, mechanical properties similar to those of cancellous bone [98]. Its mechanical properties can be further improved by the alteration of cross-linking agents or by the incorporation of a nanophase or microphase [43]. In an in vivo study using rabbits, photocross-linked PPF scaffolds with different pore sizes and porosities exhibited good biocompatibility [27]. Additionally, P(PF-co-EG) has been evaluated for use as a thermoreversible hydrogel scaffold for the delivery of chondrocytes for articular cartilage replacement in tissue engineering [26].

Oligo(poly(ethylene glycol) fumarate) (OPF) is yet another novel biodegradable fumarate-based polymer. It is synthesized by the combination of PEG and fumaryl chloride in the presence of triethylamine [51]. Both in vitro and in vivo studies using this material demonstrated good biocompatibility, with a minimal inflammatory response observed after implantation for 12 weeks in cranial defects in rats and 14 weeks in osteochondral defects in rabbits [38, 86, 95]. High water absorption and mild in situ cross-linking conditions enable OPF to encapsulate living cells or bioactive growth factors for orthopedic tissue regeneration [41, 94]. Recently, OPF has been explored as a cell carrier for marrow stromal cells. After 4 weeks of culture in vitro, cells remained alive. Evidence of osteoblastic differentiation, including calcified ECM production throughout the hydrogel, was observed (Fig. 4.3) [93].

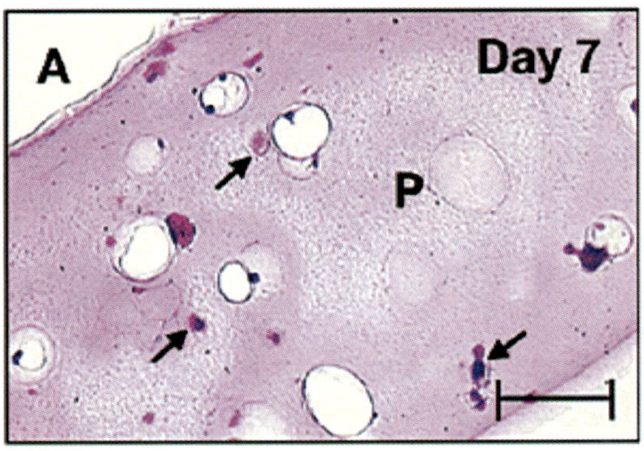


**Figure 4.3.** Histology of oligo(poly(ethylene glycol) fumarate) hydrogels containing rat marrow stromal cells after 7 (A), 21(B), and 28(C) days of in vitro culture with media supplemented with dexamethasone. Polymer is labeled P, mineralized matrix is labeled M, and arrows indicate the location of some of the cells found throughout the hydrogel. Reproduced with permission from Temenoff et al. [93]. Copyright 2004, American Chemical Society.

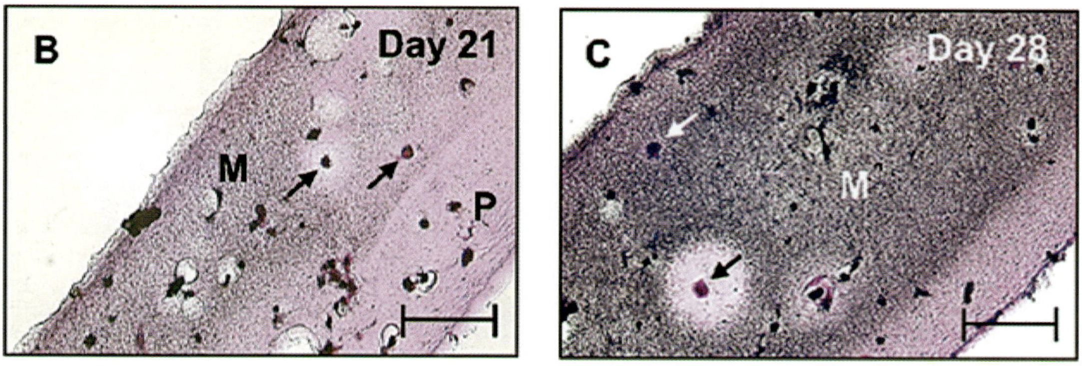


Figure 4.3. *Continued*

## 4.6 Summary

In this chapter, various applications, important properties, and different types of biodegradable materials that are candidates for use in orthopedic applications have been reviewed. For this purpose, both natural materials and synthetic polymers have been used to fabricate various types of orthopedic implants that include simple fixation devices, scaffolds for delivery of bioactive molecules, and carriers for delivery of living cell populations. In many cases, unique materials or strategies can be combined to produce a more optimal outcome that is compatible with the intended purpose. Even though they are still in the early stage of development, biodegradable scaffolds have already proven to aid in the repair of orthopedic defects. Thus, further research in this field holds great promise to effect complete regeneration of a variety of orthopedic tissues.

## Acknowledgment

We acknowledge funding from the National Institutes of Health (R01-AR42639, R01-AR48756, and R01-DE15164) for tissue-engineering applications using biodegradable polymers.

## References

1. American Academy of Orthopaedic Surgeons website, www.aaos.org.
2. Ahsan T, Sah RL (1999) Biomechanics of integrative cartilage repair. Osteoarthritis Cartilage 7:29–40.
3. Ali SA, Zhong SP, Doherty PJ, Williams, DF (1993) Mechanisms of polymer degradation in implantable devices. I. Poly(caprolactone). Biomaterials 14:648–656.
4. Ambrose CG, Clyburn TA, Louden K, Joseph J, Wright J, Gulati P, et al. (2004) Effective treatment of osteomyelitis with biodegradable microspheres in a rabbit model. Clin Orthop 421:293–299.
5. Amiel D, Frank C, Harwood F, Fronek J, Akeson W (1984) Tendons and ligaments: a morphological and biochemical comparison. J Orthop Res 1:257–265.
6. Andriano KP, Tabata Y, Ikada Y, Heller J (1999) In vitro and in vivo comparison of bulk and surface hydrolysis in absorbable polymer scaffolds for tissue engineering. J Biomed Mater Res 48:602–612.
7. Athanasiou K, Zhu C, Lanctot D, Agrawal C, Wang X (2000) Fundamentals of biomechanics in tissue engineering of bone. Tissue Eng 6:361–381.
8. Athanasiou KA, Agrawal CM, Barber FA, Burkhart SS (1998) Orthopaedic applications for PLA-PGA biodegradable polymers. Arthroscopy 14:726–737.
9. Awad HA, Erickson GR, Guilak F (2002) Biomaterials for cartilage tissue engineering. In: Lewandrowski K, Wise D, Trantolo D, Gresser J, Yaszemski M, Altobelli D, eds. Tissue Engineering and Biodegradable Equivalents: Scientific and Clinical Applications. Marcel Dekker, New York.
10. Barbensee JE, McIntire LV, Mikos AG (2000) Growth factor delivery for tissue engineering. Pharm Res 17:497–504.

11. Bostman O, Pihlajamaki H (2000) Clinical biocompatibility of biodegradable orthopaedic implants for internal fixation: a review. Biomaterials 21:2615–2621.
12. Bryant SJ, Anseth KS (2002) Hydrogel properties influence ECM production by chondrocytes photoencapsulated in poly(ethylene glycol) hydrogels. J Biomed Mater Res 59:63–72.
13. Butler DL, Goldstein SA, Guilak F (2000) Functional tissue engineering: the role of biomechanics. J Biomech Eng 122:570–575.
14. Caborn DN, Urban WPJ, Johnson DL, Nyland J, Pienkowski D (1997) Biomechanical comparison between BioScrew and titanium alloy interference screws for bone-patellar tendon-bone graft fixation in anterior cruciate ligament reconstruction. Arthroscopy 13: 229–232.
15. Chen VJ, Ma PX (2004) Nano-fibrous poly(-lactic acid) scaffolds with interconnected spherical macropores. Biomaterials 25:2065–2073.
16. Chenite A, Chaput C, Wang D, Combes C, Buschmann MD, Hoemann CD, et al (2000) Novel injectable neutral solutions of chitosan form biodegradable gels in situ. Biomaterials 21:2155–2161.
17. Claes LE, Ignatius AA, Rehm KE, Scholz C (1996) New bioresorbable pin for the reduction of small bony fragments: design, mechanical properties and in vitro degradation. Biomaterials 17:1621–1626.
18. Daamen WF, Nillesen STM, Hafmans T, Veerkamp JH, van Luyn MJA, van Kuppevelt TH (2005) Tissue response of defined collagen-elastin scaffolds in young and adult rats with special attention to calcification. Biomaterials 26:81–92.
19. Daniels AU, Chang MK, Andriano KP (1990) Mechanical properties of biodegradable polymers and composites proposed for internal fixation of bone. J Appl Biomater 1:57–78.
20. Disegi JA, Wyss H (1989) Implant materials for fracture fixation: a clinical perspective. Orthopedics 12:75–79.
21. Donahue HJ, Chen Q, Jacobs CR, Saunders MM, Yellowley CE (2001) Bone cells and mechanotransduction. In: Rosier R, Evans C, eds. Molecular Biology in Orthopaedics. American Academy of Orthopaedic Surgeons, Scottsdate, pp 179–190.
22. Drury JL, Dennis RG, Mooney DJ (2004) The tensile properties of alginate hydrogels. Biomaterials 25: 3187–3199.
23. Elisseeff J, McIntosh W, Anseth KS, Riley S, Ragan P, Langer R (2000) Photoencapsulation of chondrocytes in poly(ethylene oxide)-based semi-interpenetrating networks. J Biomed Mater Res 51:164–171.
24. Elisseeff J, McIntosh W, Fu K, Blunk BT, Langer R (2001) Controlled-release of IGF-I and TGF-beta1 in a photopolymerizing hydrogel for cartilage tissue engineering. J Orthop Res 19:1098–1104.
25. Fisher JP, Holland TA, Dean D, Engel PS, Mikos AG (2001) Synthesis and properties of photocross-linked poly(propylene fumarate) scaffolds. J Biomater Sci Polym Ed 12:673–687.
26. Fisher JP, Jo S, Mikos AG, Reddi AH (2004) Thermoreversible hydrogel scaffolds for articular cartilage engineering. J Biomed Mater Res 71A:268–274.
27. Fisher JP, Vehof JW, Dean D, van der Waerden JP, Holland TA, Mikos AG, et al. (2002) Soft and hard tissue response to photocrosslinked poly(propylene fumarate) scaffolds in a rabbit model. J Biomed Mater Res 59:547–556.
28. Fleming JE, Muschler GF, Boehm C, Lieberman IH, McLain RF (2004) Intraoperative harvest and concentration of human bone marrow osteoprogenitors for enhancement of spinal fusion. In: Goldberg V, Caplan A, eds. Orthopedic Tissue Engineering: Basic Science and Practice. Marcel Decker, New York, pp 51–65.
29. Goodstone NJ, Cartwright A, Ashton B (2004) Effects of high molecular weight hyaluronan on chondrocytes cultured within a resorbable gelatin sponge. Tissue Eng 10:621–631.
30. Gordon TD, Schloesser L, Humphries DE, Spector M (2004) Effects of the degradation rate of collagen matrices on articular chondrocyte proliferation and biosynthesis in vitro. Tissue Eng 10:1287–1295.
31. Goulet F, Germain L, Rancourt D, Caron C, Normand A, Auger FA (1997) Tendons and ligaments. In: Lanza R, Langer R, eds. Principles of Tissue Engineering. Academic Press, San Diego, pp 633–644.
32. Gunatillake PA, Adhikari R (2003) Biodegradable synthetic polymers for tissue engineering. Eur Cell Mater 5:1–16.
33. Halstenberg S, Panitch A, Rizzi S, Hall H, Hubbell JA (2002) Biologically engineered protein-graft-poly(ethylene glycol) hydrogels: a cell adhesive and plasmin-degradable biosynthetic material for tissue repair. Biomacromolecules 3:710–723.
34. Hanson SR, Harker LA (1994) Blood coagulation and blood–material interactions. In: Ratner B, Hoffman A, Schoen F, Lemons J, eds. Biomaterials Science: An Introduction to Materials in Medicine. Hanser, New York.
35. Heath CA (2000) Cells for tissue engineering. Trends Biotechnol 18:17–19.
36. Heinegard D, King K, Morgelin M, Rosenberg K, Wiberg C (2001) Matrix molecules with roles in cartilage assembly. In: Rosier R, Evans C, eds. Molecular Biology in Orthopaedics. American Academy of Orthopaedic Surgeons, Scottsdale, pp 315–323.
37. Higashi S, Yamamuro T, Nakamura T, Ikada Y, Hyon SH, Jamshidi K (1986) Polymer-hydroxyapatite composites for biodegradable bone fillers. Biomaterials 7:183–187.
38. Holland TA, Bodde EW, Baggett LS, Tabata Y, Mikos AG, Jansen JA (2005) Osteochondral repair in the rabbit model utilizing bilayered, degradable oligo(poly(ethylene glycol) fumarate) hydrogel scaffolds. J Biomed Mater Res A 75:156–167.
39. Holland TA, Mikos AG (2003) Advances in drug delivery for articular cartilage. J Control Release 86:1–14.
40. Holland TA, Tabata Y, Mikos AG (2003) In vitro release of transforming growth factor-beta 1 from gelatin microparticles encapsulated in biodegradable, injectable oligo(poly(ethylene glycol) fumarate) hydrogels. J Control Release 91:299–313.
41. Holland TA, Tessmar JKV, Tabata Y, Mikos AG (2003) Transforming growth factor-1 release from oligo (poly(ethylene glycol) fumarate) hydrogels in conditions that model the cartilage wound healing environment. J Control Release 94:101–114.
42. Hollinger JO (1983) Preliminary report on the osteogenic potential of a biodegradable copolymer of polyactide (PLA) and polyglycolide (PGA). J Biomed Mater Res 17:71–82.

43. Horch RA, Shahid N, Mistry AS, Timmer MD, Mikos AG, Barron AR (2004) Nanoreinforcement of poly (propylene fumarate)-based networks with surface modified alumoxane nanoparticles for bone tissue engineering. Biomacromolecules 5:1990–1998.
44. Huang W, Carlsen B, Wulur I, Rudkin G, Ishida K, Wu B, et al. (2004) BMP-2 exerts differential effects on differentiation of rabbit bone marrow stromal cells grown in two-dimensional and three-dimensional systems and is required for in vitro bone formation in a PLGA scaffold. Exp Cell Res 299: 325–334.
45. Hunziker EB (1992) The different types of chondrocytes and their function in vivo. In: Adolphe M, ed. Biological Regulation of the Chondrocytes. CRC Press, Boca Raton, pp 1–31.
46. Ibusuki S, Fujii Y, Iwamoto Y, Matsuda T (2003) Tissue-engineered cartilage using an injectable and in situ gelable thermoresponsive gelatin: fabrication and in vitro performance. Tissue Eng 9:371–84.
47. Iooss P, Le Ray AM, Grimandi G, Daculsi G, Merle C (2001) A new injectable bone substitute combining poly(epsilon-caprolactone) microparticles with biphasic calcium phosphate granules. Biomaterials 22:2785–2794.
48. Ishaug-Riley SL, Crane-Kruger GM, Yaszemski MJ, Mikos AG (1998) Three-dimensional culture of rat calvarial osteoblasts in porous biodegradable polymers. Biomaterials 19:1405–1412.
49. Jackson DW, Simon TM (1999) Tissue engineering principles in orthopaedic surgery. Clin Orthop 367S: 31–45.
50. Jeong B, Kim SW, Bae YH (2002) Thermosensitive sol-gel reversible hydrogels. Adv Drug Deliv Rev 54:37–51.
51. Jo S, Shin H, Shung AK, Fisher JP, Mikos AG (2001) Synthesis and characterization of oligo(poly(ethylene glycol) fumarate) macromer. Macromolecules 34: 2839–2845.
52. Kang SI, Bae YH (2004) pH-dependent elution profiles of selected proteins in HPLC having a stationary phase modified with pH-sensitive sulfonamide polymers. J Biomater Sci Polym Ed 15:879–894.
53. Karp JM, Sarraf F, Shoichet MS, Davies JE (2004) Fibrin-filled scaffolds for bone-tissue engineering: an in vivo study. J Biomed Mater Res 71A:162–171.
54. Kasper FK, Seidlits SK, Tang A, Crowther RS, Carney DH, Barry MA, et al (2005) In vitro release of plasmid DNA from oligo(poly(ethylene glycol) fumarate) hydrogels. J Control Release 104:521–539.
55. Kisiday J, Jin M, Kurz B, Hung H, Semino C, Zhang S, et al. (2002) Self-assembling peptide hydrogel fosters chondrocyte extracellular matrix production and cell division: implications for cartilage tissue repair. Proc Natl Acad Sci USA 99:9996–10001.
56. Kohn J, Langer R (1994) Bioresorbable and bioerodible materials. In: Ratner B, Hoffman A, Schoen F, Lemons J, eds. Biomaterials Science: An Introduction to Materials in Medicine. Hanser, New York, pp 65–73.
57. Kojima K, Ignotz RA, Kushibiki T, Tinsley KW, Tabata Y, Vacanti CA (2004) Tissue-engineered trachea from sheep marrow stromal cells with transforming growth factor [beta]2 released from biodegradable microspheres in a nude rat recipient. J Thorac Cardiovasc Surg 128:147–153.
58. Leddy HA, Awad HA, Guilak F (2004) Molecular diffusion in tissue-engineered cartilage constructs: effects of scaffold material, time, and culture conditions. J Biomed Mater Res 70B:397–406.
59. Lee CH, Singla A, Lee Y (2001) Biomedical applications of collagen. Int J Pharm 221:1–22.
60. Lee CR, Grodzinsky AJ, Spector M (2001) The effects of cross-linking of collagen-glycosaminoglycan scaffolds on compressive stiffness, chondrocyte-mediated contraction, proliferation and biosynthesis. Biomaterials 22:3145–3154.
61. Lee JE, Kim KE, Kwon IC, Ahn HJ, Lee SH, Cho H, et al. (2004) Effects of the controlled-released TGF-[beta]1 from chitosan microspheres on chondrocytes cultured in a collagen/chitosan/glycosaminoglycan scaffold. Biomaterials 25:4163–4173.
62. Lee KY, Mooney DJ (2001) Hydrogels for tissue engineering. Chem Rev 101:1869–1880.
63. Li WJ, Tuli R, Okafor C, Derfoul A, Danielson KG, Hall DJ, et al. (2005) A three-dimensional nanofibrous scaffold for cartilage tissue engineering using human mesenchymal stem cells. Biomaterials 26: 599–609.
64. Lima EG, Mauck RL, Han SH, Park S, Ng KW, Ateshian GA, et al. (2004) Functional tissue engineering of chondral and osteochondral constructs. Biorheology 41:577–590.
65. Lu L, Peter SJ, Lyman MD, Lai HL, Leite SM, Tamada JA, et al. (2000) In vitro degradation of porous poly(-lactic acid) foams. Biomaterials 21:1595–1605.
66. Luginbuehl V, Meinel L, Merkle HP, Gander B (2004) Localized delivery of growth factors for bone repair. Eur J Pharm Biopharm 58:197–208.
67. Lutolf MP, Hubbell JA (2003) Synthesis and physicochemical characterization of end-linked poly (ethylene glycol)-co-peptide hydrogels formed by Michael-type addition. Biomacromolecules 4:713–722.
68. Ma Z, Gao C, Gong Y, Shen J (2005) Cartilage tissue engineering PLLA scaffold with surface immobilized collagen and basic fibroblast growth factor. Biomaterials 26:1253–1259.
69. Mauck RL, Wang CC, Oswald ES, Ateshian GA, Hung CT (2003) The role of cell seeding density and nutrient supply for articular cartilage tissue engineering with deformational loading. Osteoarthritis Cartilage 11:879–890.
70. Middleton JC, Tipton AJ (2000) Synthetic biodegradable polymers as orthopedic devices. Biomaterials 21:2335–2346.
71. Mikos AG, Bao Y, Cima LG, Ingber DE, Vacanti JP, Langer R (1993) Preparation of poly(glycolic acid) bonded fiber structures for cell attachment and transplantation. J Biomed Mater Res 27:183–189.
72. Mikos AG, Thorsen AJ, Czerwonka LA, Bao Y, Langer R, Winslow DN, et al. (1994) Preparation and characterization of poly(L-lactic acid) foams. Polymer 35:1068–1077.
73. Moreira PL, An YH, Santos AR Jr, Genari SC (2004) In vitro analysis of anionic collagen scaffolds for bone repair. J Biomed Mater Res 71B:229–237.
74. O'Driscoll SW (1999) Articular cartilage regeneration using periosteum. Clin Orthop 367S:S186–203.
75. Ouyang HW, Goh JC, Thambyah A, Teoh SH, Lee EH (2003) Knitted poly-lactide-co-glycolide scaffold loaded with bone marrow stromal cells in repair and

regeneration of rabbit Achilles tendon. Tissue Eng 9:431–439.
76. Pachence JM, Kohn J (1997) Biodegradable polymers for tissue engineering. In: Lanza R, Langer R, eds. Principles of Tissue Engineering. Academic Press, San Diego, pp 273–293.
77. Park H, Temenoff JS, Holland TA, Tabata Y, Mikos AG (2005) Delivery of TGF-β and chondrocytes via injectable, biodegradable hydrogels for cartilage tissue engineering applications. Biomaterials 2005: 26:7095–7103.
78. Park SN, Park JC, Kim HO, Song MJ, Suh H (2002) Characterization of porous collagen/hyaluronic acid scaffold modified by 1-ethyl-3-(3-dimethylaminopr opyl)carbodiimide cross-linking. Biomaterials 23: 1205–1212.
79. Peppas NA, Langer R (1994) New challenges in biomaterials. Science 263:1715–1720.
80. Podual K, Doyle FJ III, Peppas NA (2000) Preparation and dynamic response of cationic copolymer hydrogels containing glucose oxidase. Polymer 41:3975–3983.
81. Pulapura S, Kohn J (1992) Tyrosine-derived polycarbonates: backbone-modified "pseudo"-poly (amino acids) designed for biomedical applications. Biopolymers 32:411–417.
82. Quick DJ, Anseth KS (2004) DNA delivery from photocrosslinked PEG hydrogels: encapsulation efficiency, release profiles, and DNA quality. J Control Release 96:341–351.
83. Rice MA, Anseth KS (2004) Encapsulating chondrocytes in copolymer gels: bimodal degradation kinetics influence cell phenotype and extracellular matrix development. J Biomed Mater Res 70A:560–568.
84. Rupp S, Krauss PW, Fritsch EW (1997) Fixation strength of a biodegradable interference screw and a press-fit technique in anterior cruciate ligament reconstruction with a BPTB graft. Arthroscopy 13: 61–65.
85. Shin H, Jo S, Mikos AG (2003) Biomimetic materials for tissue engineering. Biomaterials 24:4353–4364.
86. Shin H, Quinten-Ruhe P, Mikos AG, Jansen JA (2003) In vivo bone and soft tissue response to injectable, biodegradable oligo(poly(ethylene glycol) fumarate) hydrogels. Biomaterials 24:3201–3211.
87. Shin H, Zygourakis K, Farach-Carson MC, Yaszemski MJ, Mikos AG (2004) Modulation of differentiation and mineralization of marrow stromal cells cultured on biomimetic hydrogels modified with Arg-Gly-Asp containing peptides. J Biomed Mater Res 69A:535–543.
88. Suh JK, Matthew HW (2000) Application of chitosan-based polysaccharide biomaterials in cartilage tissue engineering: a review. Biomaterials 21:2589–2598.
89. Tabata Y, Ikada Y (1988) Macrophage phagocytosis of biodegradable microspheres composed of L-lactic acid/glycolic acid homo- and copolymers. J Biomed Mater Res 22:837–858.
90. Tamada J, Langer R (1992) The development of polyanhydrides for drug delivery applications. J Biomater Sci Polym Ed 3:315–353.
91. Temenoff JS, Mikos AG (2000) Injectable biodegradable materials for orthopaedic tissue engineering. Biomaterials 21:2405–2412.
92. Temenoff JS, Mikos AG (2000) Review: tissue engineering for regeneration of articular cartilage. Biomaterials 21:431–440.
93. Temenoff JS, Park H, Jabbari E, Conway DE, Sheffield TL, Ambrose CG, et al. (2004) Thermally cross-linked oligo(poly(ethylene glycol) fumarate) hydrogels support osteogenic differentiation of encapsulated marrow stromal cells in vitro. Biomacromolecules 5:5–10.
94. Temenoff JS, Park H, Jabbari E, Sheffield TL, LeBaron RG, Ambrose CG, et al. (2004) In vitro osteogenic differentiation of marrow stromal cells encapsulated in biodegradable hydrogels. J Biomed Mater Res 70A:235–244.
95. Temenoff JS, Shin H, Conway DE, Engel PS, Mikos AG (2003) In vitro cytotoxicity of redox radical initiators for cross-linking of oligo(poly(ethylene glycol) fumarate) macromers. Biomacromolecules 4:1605–1613.
96. Temenoff JS, Steinbis ES, Mikos AG (2004) Biodegradable scaffolds. In: Goldberg V, Caplan A, eds. Orthopedic Tissue Engineering: Basic Science and Practice. Marcel Decker, New York, pp 77–103.
97. Thornton AJ, Alsberg E, Albertelli M, Mooney DJ (2004) Shape-defining scaffolds for minimally invasive tissue engineering. Transplantation 77:1798–1803.
98. Timmer MD, Carter C, Ambrose CG, Mikos AG (2003) Fabrication of poly(propylene fumarate)-based orthopaedic implants by photo-crosslinking through transparent silicone molds. Biomaterials 24:4707–4714.
99. Tuli R, Nandi S, Li WJ, Tuli S, Huang X, Manner PA, et al. (2004) Human mesenchymal progenitor cell-based tissue engineering of a single-unit osteochondral construct. Tissue Eng 10:1169–1179.
100. Vacanti JP, Langer R, Upton J, Marler JJ (1998) Transplantation of cells in matrices for tissue regeneration. Adv Drug Deliv Rev 33:165–182.
101. Weiner S, Traub W (1992) Bone structure: from angstroms to microns. FASEB J 6:879–885.
102. Williams CG, Kim TK, Taboas A, Maliak AN, Manson P, Elisseeff JH (2003) In vitro chondrogenesis of bone marrow-derived mesenchymal stem cells in a photopolymerizing hydrogel. Tissue Eng 9:679–688.
103. Williamson AK, Chen AC, Sah RL (2001) Compressive properties and function-composition relationships of developing bovine articular cartilage. J Orthop Res 19:1113–1121.
104. Yamamoto M, Takahashi Y, Tabata Y (2003) Controlled release by biodegradable hydrogels enhances the ectopic bone formation of bone morphogenetic protein. Biomaterials 24:4375–4383.
105. Yamane S, Iwasaki N, Majima T, Funakoshi T, Masuko T, Harada K, et al. (2005) Feasibility of chitosan-based hyaluronic acid hybrid biomaterial for a novel scaffold in cartilage tissue engineering. Biomaterials 26:611–619.
106. Yaszemski MJ, Payne RG, Hayes WC, Langer R, Mikos AG (1996) In vitro degradation of a poly(propylene fumarate)-based composite material. Biomaterials 17:2127–2130.
107. Zhang SM, Cui FZ, Liao SS, Zhu Y, Han L (2003) Synthesis and biocompatibility of porous nano-hydroxyapatite/collagen/alginate composite. J Mater Sci Mater Med 14:641–645.

# 5. Titanium Fiber Mesh: A Nondegradable Scaffold Material

*Juliette van den Dolder and John A. Jansen*

## 5.1 Introduction

The grafting of bone in skeletal reconstruction has become a common task of the orthopedic surgeon. The need for reconstruction or replacement is often the result of trauma, congenital malformations, or cancer. Reconstructive surgery is based upon the principle of replacing defective tissue with viable, functioning alternatives. Various materials have been used to treat the defects, including autogenous bone and alloplastic materials. Grafting materials are necessary to bridge defects or to increase the bone volume. At present, autologous bone is the gold standard, but it has important disadvantages, including donor-site morbidity, limited availability, and unpredictable resorption characteristics. These factors have stimulated the search for other materials that can replace autogenous bone. Allografts and xenografts, although suitable in texture and content, have limitations that include the capacity of transmitting disease and generating an immunogenic response. A recently developed approach to the reconstruction or regeneration of lost or damaged body tissues is tissue engineering, which involves the fabrication of a so-called three-dimensional autologous tissue construct.

Although tissue engineering can be applied in many clinical situations, much attention is paid to the engineering of bone tissue. Two different strategies can be followed to achieve this goal, including both growth-factor–based and cell-based approaches. In the growth-factor–based approach, the scaffold material is loaded with specific bone-inductive growth factors prior to implantation. These exogenous growth factors are then released at the implant site, where they can act upon locally resident cells as well as recruiting other, more distant cells to form new bone tissue. A significant number of growth factors are commercially available for this purpose. In the second approach, the scaffold material is preseeded with osteogenic cells to promote bone formation. When these cell/scaffold constructs are cultured in vitro for an extended period of time, the seeded cells secrete matrix as well as other growth factors into the scaffold. At the implant site, these cell/scaffold constructs contribute to bone formation. In the cell-based approach, bone marrow cells often are used for creating a cell/scaffold construct.

The objective of this chapter is to summarize the results achieved with the nondegradable scaffold, titanium fiber mesh, for use as a bone-engineered construct. This material has been used in both strategies, the so-called cell-based and the growth-factor–based approaches.

## 5.2 Scaffolds

In the field of bone tissue engineering, various combinations of naturally derived and synthetic polymers, composites, ceramics, and growth factors as well as cellular systems are being studied. The development of bone fill materials for replacement should be based on an understanding of the natural structure to be substituted. The demands upon the material properties largely depend on the site of application and the function that must be restored. The ideal scaffold material is biocompatible and biodegradable. This means that it is nontoxic and does not elicit a foreign-body giant-cell reaction. Further, the scaffold material also must degrade to biocompatible products. A second characteristic of an ideal scaffold material is the absence of inflammatory reactions and disease transmission. This means that the scaffold must be nonimmunogenic and free of transmittable diseases. The third characteristic is the architectural quality of the scaffold material. The scaffold must be shaped easily. Further, it must be porous, and the pores must have interconnectivity to allow tissue ingrowth and stabilization. The fourth characteristic is osteoconduction, which means that the scaffold should possess surface characteristics that optimize bone ingrowth. Chemotaxis and delivery and control of osteoinductive proteins are the fifth characteristic of an ideal scaffold. Therefore, the scaffold needs to have the right surface charge and affinity for cells to attach and for osteoinductive proteins to adhere to the surface of the scaffold. The sixth characteristic is that a scaffold material also must promote rapid angiogenesis and vascularization of the device to achieve a solid vascularized bond to host bone [47].

Applied scaffold materials include degradable and nondegradable ceramics (e.g., hydroxyapatite and tricalcium phosphate), polymers (e.g., poly(lactic acid), poly(glycolic acid) or copolymers), bioglass, and nondegradable metals. Composite materials have also been created, such as poly(glycolic acid) with polyethylene glycol or hydroxyapatite with collagen. It has to be emphasized that none of the currently used materials have all of the properties postulated. Some of the materials produce an undesirable inflammatory response or a foreign-body reaction. These reactions are associated with a reduced osteoinductive response. On the other hand, other materials show a lack of structural support and poor mechanical characteristics.

# 5.3 Porous Metallic Scaffolds

## 5.3.1 Titanium

Titanium is well known for its excellent biocompatibility. Besides the bulk material of titanium, spongelike titanium has been used for tissue-engineering purposes. Spongelike titanium consists of titanium fibers with a defined diameter that are sintered together to create a mesh structure. By varying the diameter and density of the fibers, the porosity of the mesh can be varied. The advantages of using the mesh material for tissue-engineering purposes are its flexibility, strength, porosity, and interconnectivity. Strength allows the implant to bear the mechanical load. Flexibility eliminates focal stresses by distributing the stresses between implant and tissue over a larger area. Finally, porosity and interconnectivity allow tissue ingrowth and stabilization of the implant. The titanium fiber meshes used in the studies mentioned in this chapter have a volumetric porosity of 86%, a density of $600\,g/m^2$, and a fiber diameter of $50\,\mu m$, resulting in an average pore diameter of $250\,\mu m$.

The biocompatibility of titanium is demonstrated by two major observations: the very favorable response of tissues to titanium surfaces, and the absence of allergic reactions to titanium [28]. Jansen et al. compared the tissue responses of three mesh materials: titanium, stainless steel 316 L, and Fecralloy®. Histological analysis in this study revealed a material-related difference in tissue biocompatibility; it was observed that titanium mesh induced a better tissue response than did the other materials [16]. A tissue capsule was almost absent; only a thin, uniformly oriented tissue layer could be observed. Further histological studies showed that bone integrated titanium implants. Bone cells and mineralized bone matrix were deposited on titanium surfaces without interposition of other tissues, although coverage of the implant surface with organic molecules occurred first [28].

The second major observation indicating biocompatibility is the absence of allergic reactions to titanium. The long-term clinical experience with titanium and the use of titanium dioxide in many ointments and cosmetics has demonstrated that titanium does not trigger allergies. Titanium oxide forms spontaneously on titanium surfaces as long as oxygen is present and protects it from corrosion by forming a thin film, the so-called passive film. This passive film regenerates immediately after mechanical destruction and thus protects the surface instantaneously [28].

Another advantage of a metal such as titanium is strength, which makes it very useful for bone replacement. Further, the flexibility of a scaffold material, with reference to stress shielding, is an important factor. The work of Jansen et al. demonstrated a relationship between flexibility of the mesh structure and tissue response. Flexibility presumably eliminates focal stresses by distributing the stresses between implant and tissue over a larger area [16].

Finally, the tissue-engineered scaffold should have porosity and interconnectivity to allow tissue ingrowth and stabilization. These requirements limit the number of available candidates. In view of this, titanium fiber mesh is a possible candidate material. Porous and nonporous titanium implants with varying geometrical properties have been produced and their properties investigated. During the fabrication of the porous surfaces, porosity, pore size, and pore shape can be varied, which influences the amount of bone ingrowth into the porous surfaces. Comparison of porous metal with conventional solid metals used in the manufacture of orthopedic devices shows that porosity allows a more normal restoration of the bone than occurs with the use of nonporous implant materials. Proper bone growth requires initial stability. The frictional properties of porous titanium fiber mesh in contact with bone exceed those of the solid-metal materials that are available today. In the early postoperative period, these frictional and structural properties provide the construct with a high initial. In the long term, the porous metal serves as a scaffold for bone while allowing proper loading and maintenance of vascularity in surrounding and ingrown bone [7]. In summary, porous titanium fiber mesh offers several advantages over other materials owing to its uniformity and structural continuity, as well as to its strength, low stiffness, high porosity, and high coefficient of friction [7].

## 5.3.2 Other Nondegradable Metals

Tantalum is an elemental metal that is biocompatible and corrosion resistant. Like titanium, tantalum can be designed as a porous structural scaffold with the same advantages as those of titanium fiber mesh. The orthopedic applications of porous tantalum are diverse and include primary and revision joint-reconstruction implants, spinal interbody fusion devices, and trauma void-filling structural applications [8].

Cobalt-based alloys are generally described as nonmagnetic and resistant to wear, corrosion, and heat. Cobalt-based alloys are used for surgical applications, including orthopedic prostheses for the knee, shoulder, and hip, as well as for fracture fixation devices. Unfortunately, cobalt-based alloys are difficult to fabricate, which limits their use as nondegradable porous scaffolds. Furthermore, the properties of these cobalt-based alloys are less desirable than those of stainless steel and titanium [21].

Stainless steel, like tantalum and titanium, has several advantages including uniformity, structural continuity, strength, low stiffness, high porosity, and high coefficient of friction. The advantages of using stainless steel rather than titanium are its lower cost and higher fracture toughness [29]. However, Paquay et al. showed that porous stainless steel 316 L released significantly more corrosion products than did titanium mesh, which explained the better performance of titanium mesh, as judged by the good tissue reaction to the titanium meshes when they were placed subcutaneously [27]. Further observations showed that meshes fabricated of fibers of small diameter released significantly more corrosion products than did meshes with of larger fiber diameter. This phenomenon was attributed to the larger surface area of the meshes that contained fibers of small diameter. These results can be related to the results of an in vivo study [27] that found a correlation between the tissue behavior and the fiber diameter of the various 316 L stainless steel meshes. This experiment showed that although the bulk material was the same in the various mesh materials, the amount of corrosion products surrounding the implant markedly influenced the behavior of the tissue.

In summary, porous titanium fiber mesh offers several advantages over other materials by virtue of its uniformity and structural continuity, as well as by its strength, low stiffness, high porosity, corrosion resistance, and high coefficient of friction [7].

The rest of this chapter will focus on titanium fiber mesh as a scaffold material for bone tissue replacement. As a scaffold material, titanium fiber mesh can be used in both cell-based and growth-factor–based approaches.

## 5.4 Cell-Based Approach and Titanium Fiber Mesh

Mesenchymal stem cells (MSCs) can be used for the functional repair or regeneration of large bone defects. In this approach, a three-dimensional scaffold material is used to deliver the cells to the bone defect site. The final success of this tissue-engineered strategy is determined by the number of responsive MSCs loaded in the scaffold as well as by the material characteristics of the delivery vehicle.

MSCs or osteoprogenitor cells, which are precursors of osteogenic cells, are present in bone marrow, where they represent only a small fraction of the total number of bone marrow cells [12]. In view of this, methods have been developed to culture-expand and select the osteoprogenitor cell fraction from the total bone marrow. Cultured bone marrow-derived MSCs have proven more effective in bone formation than total cells from fresh marrow [18].

It is important to note that both the proliferation and differentiation of osteoprogenitor cells can be directed during culture by the use of several factors, such as dexamethasone and bone morphogenetic proteins (BMPs), that are known to direct the differentiation of MSCs into the osteoblast lineage in vitro [6]. The addition of agents stimulating proliferation (basic fibroblast growth factor, bFGF) and differentiation (recombinant human BMP-2, rhBMP-2) during culture enhances the in vivo osteogenic potential [11, 22]. The osteogenic potential of cell-loaded scaffolds can be increased further by modification of the conditions used during seeding, optimization of the number of loaded cells [13, 22, 42], and use of dynamic rather than static loading prior to culturing [10, 34].

### 5.4.1 Cell Seeding

The method used to seed the marrow cells into the nondegradable porous scaffold can define the success of the final three-dimensional bone graft. Several techniques have been used by various research groups to optimize the bone-generating properties of scaffold materials. These methods focus on improvement of the seeding or loading efficacy of the cells in the scaffold. The major seeding methods used to inoculate cells in a three-dimensional scaffold are: cell chamber [9], spinner flask [34, 46], static (droplet) [14, 22], cell suspension [13, 15] and perfusion systems [10]. Taken together, the combined results indicate that a dynamic system improves cell attachment and evens the distribution of cells throughout the scaffold. Static loading has cell-loading limitations. When cells are seeded in a droplet or cell suspension, cells are left on the surface of the porous constructs and penetrate only partly into the scaffold.

The cell density during cell seeding also plays an important role in cell attachment and distribution into the scaffold. Various studies with polymers have shown an improved seeding efficacy when a high initial cell density is used [14]. The same effect on seeding efficacy and cell differentiation was expected for porous titanium scaffolds. The effect on seeding efficacy was found only during the first 24 hours after inoculation. Thereafter, the same number of cells was present in all scaffolds. It appears that cells seeded in a low-cell-density suspension grew more rapidly than cells seeded in a high-cell-density suspension. Other research groups [14] reported increased expression of osteoblastic markers when a low-cell-density suspension was used for seeding. This contrasts with other results that showed increased expression of osteoblastic markers when a high-cell-density suspension was used [13].

Van den Dolder et al. [38] focused on the effect of seeding and loading techniques on the osteogenic differentiation of rat bone marrow cells grown in titanium fiber mesh in vitro. The meshes were seeded by various methods, i.e., the droplet, cell-suspension (high and low cell density), and rotating-plate methods. Osteogenic cells were cultured for several days into titanium fiber mesh. Statistical analysis of the results revealed that high cell density and low rotational speed always resulted in a significantly higher DNA content. Calcium measurements, used as a marker for matrix mineralization, and osteocalcin analysis, used to assess osteogenic differentiation, showed that using high cell densities during inoculation of the scaffolds produced more uniform results among experimental runs. Scanning electron microscopy (SEM) examination showed that for both droplet and cell-suspension samples, cells were present only at one side of the mesh. When rotation was used, no cell sheet was formed, and cells invaded the meshes and grew to surround the titanium fibers.

## 5.4.2 Nutritional Conditions

Another variable affecting implant outcome is the optimization of the nutrient conditions and oxygen supply so that the osteogenic capacity of the cultured cells is enhanced. Cell culture in three-dimensional scaffolds occurs under completely different conditions from those present in conventional planar two-dimensional conditions, in which all cells are continuously exposed to the culture medium. An inverse relationship between proliferation and differentiation in bone cell cultures resulting from a decline in the nutritional state during mineralized matrix deposition has been observed [26].

Dynamic culturing of cells (bioreactor, rotating-wall vessel, and spinner flask) after seeding of the scaffolds has been reported to have a positive effect on cell proliferation and differentiation. Furthermore, Goldstein et al. [10] demonstrated that use of flow-perfusion techniques enhanced the early differentiation and three-dimensional distribution of marrow stromal cells seeded on poly(DL-lactic-co-glycolic acid) scaffolds in comparison with scaffolds cultured in a spinner-flask bioreactor, in a rotating-vessel bioreactor, or under conventional static conditions. Static cultured constructs exhibited uneven cell distribution and low cellularity in the center of the constructs, with most cells growing near the periphery of the construct. In contrast, constructs cultured under dynamic conditions showed higher cellularity and a more uniform distribution of cells throughout the constructs. Interestingly, the production of extracellular matrix (ECM) was increased when a dynamic culture method replaced a static method.

Van den Dolder et al. [36] investigated the effect of a dynamic culture method on cell proliferation and differentiation from a seeded cell suspension of rat bone marrow stromal cells in a titanium fiber mesh. After seeding, the constructs were cultured under static conditions or in a flow-perfusion system for several days (Fig. 5.1). Cell proliferation and matrix mineralization increased in the flow-perfusion system. Examination by SEM revealed that the samples subjected to flow-perfusion culture were completely covered with layers of cells and mineralized matrix. In addition, this matrix extended deep into the scaffolds. In contrast, meshes cultured under static conditions had only a thin sheet of matrix present on the upper surface of the meshes. Evaluation of the light microscopy sections confirmed the SEM observations (Fig. 5.2).

Subsequent studies examined the influence of fluid flow and fluid shear forces on cell-loaded titanium fiber meshes in a flow-perfusion system [2, 33]. In the first study [2], the authors used different rates of flow for an extended period to permit osteoblast differentiation. Histological analysis showed that an increased flow rate produced a more uniform distribution of cells and matrix mineralization throughout the scaffold. Also, an increased flow rate produced an accelerated osteoblastic differentiation pathway. The osteoblast marker osteopontin appeared earlier, as did the late osteoblast differentiation event, calcium deposition.

In a second study [33], these researchers kept the fluid flow rate constant, but cultured the cell-loaded titanium fiber meshes in the flow-perfusion system using culture media of different viscosities. This strategy exposed cultured

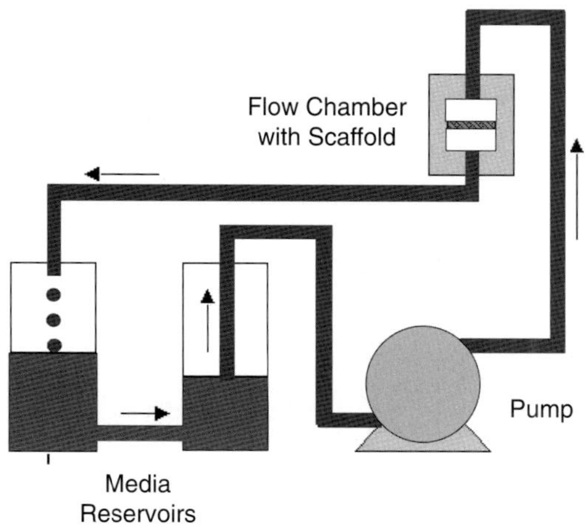


**Figure 5.1.** Schematic figure of the flow-perfusion system.

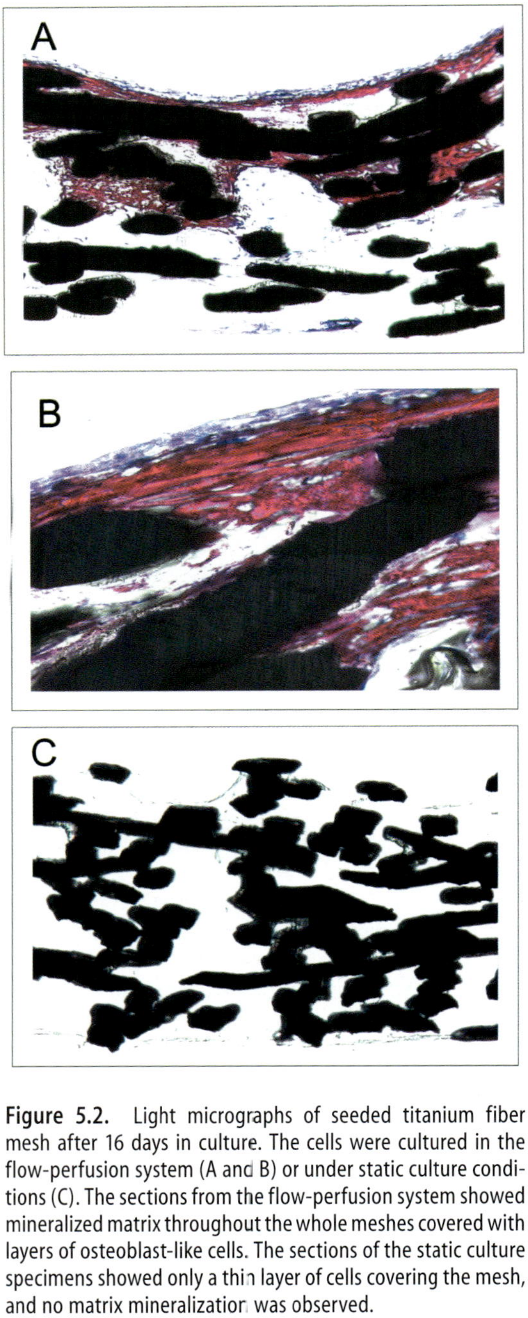


**Figure 5.2.** Light micrographs of seeded titanium fiber mesh after 16 days in culture. The cells were cultured in the flow-perfusion system (A and B) or under static culture conditions (C). The sections from the flow-perfusion system showed mineralized matrix throughout the whole meshes covered with layers of osteoblast-like cells. The sections of the static culture specimens showed only a thin layer of cells covering the mesh, and no matrix mineralization was observed.

cells to increasing levels of mechanical stimulation, in the form of fluid shear stress, whereas chemotransport conditions for nutrient delivery and waste removal remained constant. Increased shear forces produced an enhancement of mineralized matrix deposition and improved spatial distribution. Thus, fluid flow-induced shear forces clearly provide important biological stimulation of osteoblastic cells residing in three-dimensional metal scaffolds.

A level of shear stress in the range of 2 to 10 dynes/cm$^2$ appears sufficient to stimulate osteoblasts [1, 17, 19], i.e. by increased secretion of nitric oxide and prostaglandin $E_2$ after only a short period of exposure to fluid shear stress. It should be noted that exposure of cells to high levels of shear stresses may cause cell detachment or damage.

## 5.4.3 ECM Proteins

Cellular interactions with the ECM are thought to orchestrate tissue organization by regulating cell differentiation and function. The ECM produced by osteoblasts is complex and consists of several different classes of molecules that regulate the modeling and remodeling of bone. The ECM contains structural components, including type I collagen and fibronectin, as well as proteases that degrade the matrix. The ECM also serves as a reservoir for growth factors, including members of the transforming growth factor β (TGF-β) superfamily. These components of the ECM, produced by the osteoblasts, act alone or in synergy with other factors to affect cell differentiation and survival by means of autocrine feedback mechanisms that regulate the rate of bone formation.

Surface chemistry and precoating of implant materials are key components necessary to establish a proper biomaterial–bone interface. However, information concerning the behavior of cells on implants precoated with ECM proteins remains scarce. Several investigations using type I collagen-coated implants found that type I collagen enhances proliferation and accelerates differentiation and mineralization of osteoblastic cells [5, 20].

Another ECM protein that may provide information to osteoblasts during their differentiation is fibronectin. Fibronectin expression is highly localized to bone surfaces in vivo and occurs at the periphery of nodules in vitro [24]. Acting in this way, fibronectin can support the recruitment or migration of preosteoblasts. Furthermore, fibronectin also may promote the synthesis and organization of the ECM produced by osteoblasts that respond to signaling by growth factors present in the ECM of bone.

The ECM proteins interact with integrins, a heterodimeric cell-membrane receptor family. The $\alpha_4\beta_1$ and $\alpha_5\beta_1$ integrins appear to be the integrins found on osteoblasts that interact specifically with fibronectin, whereas $\alpha_1\beta_1$ and $\alpha_2\beta_1$ integrins recognize type I collagen [23, 25, 30]. The integrin-mediated contact of osteoblasts with fibronectin or type I collagen increases the expression of growth factors of the TGF-β/BMP family, which in turn stimulates osteoblast differentiation. The induction of osteoblastic differentiation requires interaction of the fibronectin coating or type I collagen coating with integrin receptors on the cells. Integrins mediate the transmission of information from the ECM by serving as a direct link between the ECM and the intracellular actin cytoskeleton. Signal transduction involves tyrosine phosphorylation, which is linked to the mitogen-activated protein kinase (MAPK) pathway and other pathways [31]. Focal adhesion kinase and MAPK are involved in the induction of alkaline phosphatase activity in osteoblasts.

Van den Dolder et al. [35] investigated the effect of fibronectin or a type I collagen coating on the differentiation of primary osteoblasts in titanium fiber mesh scaffolds. Rat bone marrow cells were cultured for several days in plain and coated titanium fiber meshes in the presence of antibodies against fibronectin and type I collagen integrins. The results showed no significant effects of the coatings on cellular proliferation, as indicated by DNA quantification. When antibodies against fibronectin and type I collagen integrins were used, a significant reduction in cell proliferation was observed for the uncoated titanium meshes, meshes coated with collagen, and meshes coated with collagen and fibronectin. The different coatings also did not affect the alkaline phosphatase activity, an early marker for differentiation, of the cells seeded on the coated and uncoated meshes. However, the presence of antibodies against fibronectin or type I collagen integrins significantly delayed the expression of alkaline phosphatase activity for uncoated titanium meshes, meshes coated with collagen, and meshes coated with collagen and fibronectin. No significant effect of fibronectin or type I collagen coating on matrix mineralization was observed. Furthermore, no difference in matrix mineralization was observed in the uncoated titanium meshes and meshes coated with fibronectin when antibodies against fibronectin or type I collagen integrins were present. Meshes coated with both type I collagen and fibronectin showed significantly higher calcium content, a marker for matrix mineralization, when cultured in the presence of antibodies against collagen or fibronectin integrins. A similar phenomenon also was observed for collagen-coated meshes cultured in the presence of antibodies against fibronectin integrins. No significant differences in osteocalcin content were observed among the treatment groups. However, all groups exposed to antibodies against fibronectin integrins showed a significant decrease in osteocalcin content. Therefore, fibronectin or type I collagen coating does not significantly stimulate the differentiation of rat bone marrow cells seeded in titanium fiber mesh. On the other hand, interactions between fibronectin and type I collagen integrins and the substratum seem to be important during the proliferation and early osteoblastic differentiation of rat marrow stromal osteoblasts on titanium surfaces.

## 5.5 In Vivo Bone Engineering: Cell-Based Approach

Previous studies with cell-loaded titanium fiber mesh showed that culture time is important for the final in vivo bone formation. Vehof et al. [41] found that cell-loaded titanium fiber mesh, without prolonged culture, produced a very limited amount of bone after subcutaneous implantation in rats. On the other hand, prolonged culture (8 days) of cell-loaded titanium fiber mesh resulted in abundant mineralization without a bonelike tissue organization [45]. To resolve the contradiction between short and long culture, we independently evaluated the effect of culture time on bone formation by rat bone marrow cells seeded in titanium fiber mesh. Osteogenic cells were cultured for 1, 4, and 8 days on titanium fiber mesh and then implanted subcutaneously in rats [39]. Analysis of DNA as an index of cell growth in the in vitro experiment revealed a lag phase from days 1 through 4, but a 42% increase in DNA content occurred between days 4 and 8. SEM and calcium measurements showed an increase in calcium from days 1 through 4, with only a small, although significant, increase between days 4 and 8 (Fig. 5.3). Histological analysis

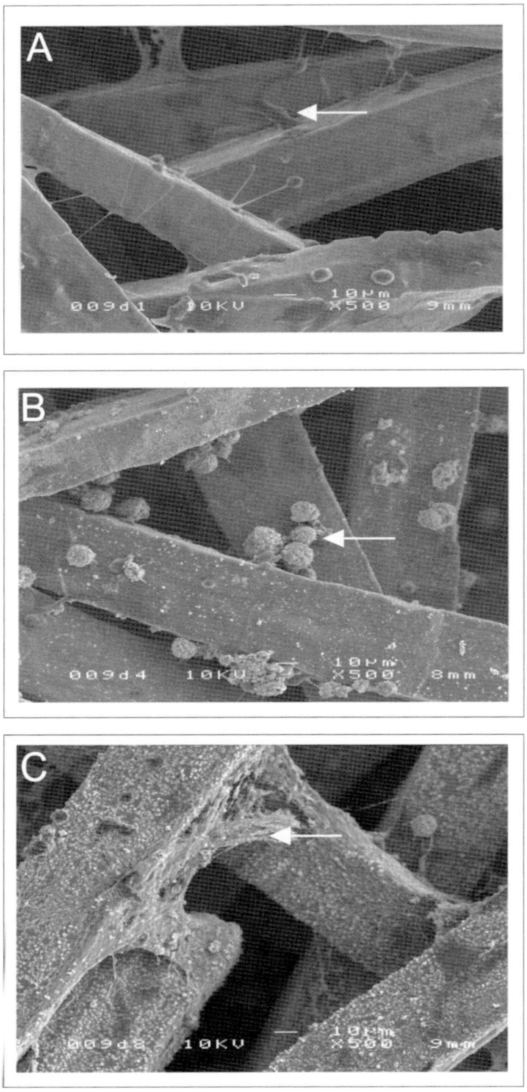


**Figure 5.3.** Scanning electron micrographs show that (A) after 1 day of culture, fibers were covered with layers of well-spread osteogenic cells (arrow). (B) After 4 days of culture, the deposition of a calcified matrix, characterized by the occurrence of globular accretions (arrow), could already be recognized. (C) After 8 days of culture, calcification appeared to increase, and large and small globular accretions as well as collagen bundles (arrow) covered the fibers almost completely.

demonstrated that bone had formed in all day-1 and day-4 implants, and that the bonelike tissue was present uniformly through the meshes. The bony tissue was morphologically characterized by the occurrence of osteocytes embedded in a mineralized matrix with a layer of osteoid and osteoblasts at the surface. In the day-8 implants calcium phosphate had deposited only in the titanium fiber mesh. Calcium measurements of the implants revealed that calcification in day-1 implants was significantly higher than in day-4 and day-8 implants. No significant difference in calcium content was observed between day-4 and day-8 implants. On the basis of these results, we concluded that bone formation was enhanced by a short culture time of osteogenic cells after seeding in titanium fiber mesh and that dynamic cell seeding is probably more effective than static cell seeding.

Although the cell-loaded meshes demonstrated osteoinductive properties in a subcutaneous model, it was important to evaluate the bone regenerative properties of cell-loaded titanium fiber meshes in a bony environment [37]. Therefore, meshes with cells were subcultured for 1 day under standard conditions. Cell-loaded implants and controls then were placed in an 8-mm cranial defect. After 3 days of implantation, mineralized-like matrix deposition and blood cells were observed inside the mesh porosity of both groups. In addition to blood cells, blood vessels were visible in two out of six of the cell-loaded specimens. After 15 days of implantation, only one out of six control implants showed bone formation inside the implant porosity, but bone was present uniformly throughout all cell-loaded meshes. Blood vessels and bone marrow were also observed. Only two cell-loaded implants showed union at the cranial defect perimeter. After 30 days of implantation, all cell-loaded implants showed bone formation inside the mesh, but in the control group only four of six implants had produced new bone (Fig. 5.4). Bone marrow and bone union at the bone defect borders were found only in five out of eight of the cell-loaded implants. The histomorphometrical evaluation found that no bone tissue was present in either implant group after 3 days of implantation, and that after 15 and 30 days, significantly more bone was present in cell-loaded implants than in unseeded control implants. On the basis of these results, we con-

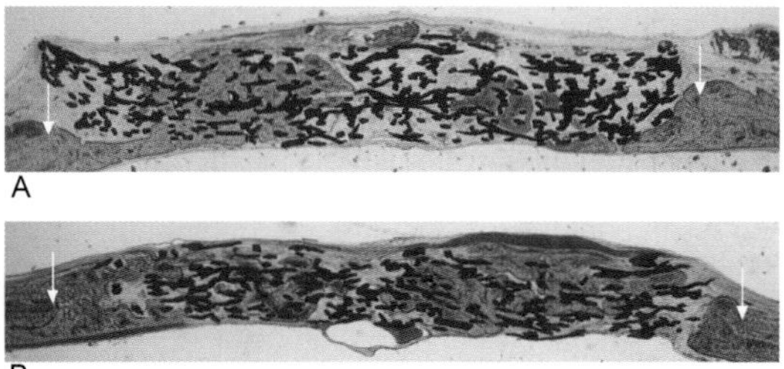


**Figure 5.4.** Light micrographs of implants loaded or not loaded, with rat bone marrow cells after implantation for 30 days. Arrows indicate the edges of the bone defects. A. Unloaded implants. B. Cell-loaded implants. Some of the titanium implants showed bone formation, morphologically characterized by the occurrence of osteocytes embedded in a mineralized matrix (1.6× magnification). Bonelike tissue was distributed uniformly in all titanium-rat bone marrow implants. Union of skull bone with newly formed bone in the implant was also observed at one side of the implant (1.6× magnification).

cluded that inoculating titanium fiber mesh with bone marrow cells improves the bone-healing capacity of this material.

Earlier in vitro studies had shown that dynamic culturing, especially the flow-perfusion system, enhanced the osteogenic differentiation and growth of cells inside the meshes. In vivo studies had not yet investigated the effect of this osteogenic improvement on the final osteoinductive properties of the titanium mesh constructs. Therefore, cell-seeded meshes were precultured for 1, 4, and 8 days under static conditions or with the flow-perfusion system [32]. After culture, cell-loaded implants were placed in an 8-mm cranial defect and retrieved after 7 and 30 days of implantation for both histological and histomorphometrical examinations. After 7 days of implantation, bone formation was absent in all groups. Further, the fiber mesh porosity was filled with fibrous tissue containing capillaries. After 30 days of implantation, most implants showed bone formation, except for one implant precultured for 4 days under flow-perfusion conditions and one implant precultured for 8 days under static conditions. Both blood vessels and bone marrow were observed. The results of the histomorphometrical measurement showed that preculturing cells for 1 day in the flow-perfusion system produced a significantly higher percentage of bone present in the implant than preculturing them for 4 days. For the other groups, no significant differences were observed. Furthermore, no significant differences were observed between implants cultured under various conditions, including static and flow perfusion. However, it seems that preculturing cells for 1 day under flow perfusion enhanced bone formation more than preculture under static conditions. The results of this study are consistent with those of a previous study demonstrating that bone formation in an orthotopic site was more effectively induced by a short preculture of osteogenic cells after seeding in titanium fiber mesh. However, these results provide only weak evidence that flow perfusion in the present form has the potential to increase bone formation in an orthotopic site. Comprehensive testing and verification in a modified experimental setting are needed before flow perfusion can be assumed to increase bone formation.

## 5.6 Growth-Factor-Based Approach: Titanium Fiber Mesh

Numerous in vivo experiments have been performed to evaluate the effect of growth factor-coated titanium fiber mesh on bone formation. The osteoinductive properties of porous titanium fiber mesh with a calcium phosphate coating loaded with rhBMP-2 were subcutaneously placed in Wistar rats and implanted for 3 to 40 days. Histological analysis demonstrated the induction of ectopic cartilage and bone formation by 5 and 7 days, respectively. At 9 days, cartilage was seen together with trabecular bone. At 20 days, bone formation had increased and was characterized by the presence of trabecular bone and bone marrow-like tissue. At 40 days, more lamellar bone and hematopoietic bone marrow-like tissue were present. Thus, calcium phosphate-coated titanium fiber mesh containing rhBMP-2 can induce ectopic endochondral-like bone formation in a rat model over short implantation periods [43, 44].

In another study, rhTGF-$\beta_1$-loaded titanium fiber meshes were implanted in a New Zealand white rabbit noncritical-size cranial-defect

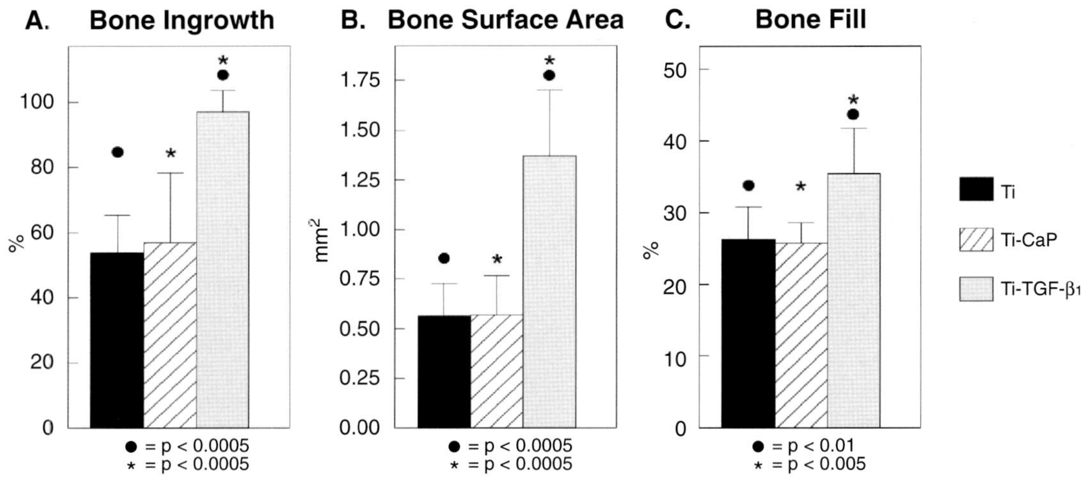


**Figure 5.5.** (A) Bone ingrowth in various titanium implants, (B) bone surface area, and (C) bone fill. The results of the paired *t*-test comparing titanium (Ti) with titanium-calcium phosphate (Ti-CaP) and of the *t*-tests comparing titanium with Ti-TGF-$\beta_1$ and Ti-CaP with Ti-TGF-$\beta_1$ are indicated. Significant differences between Ti and Ti-TGF-$\beta_1$ (•) and between Ti-CaP and Ti-TGF-$\beta_1$ (*) are indicated. No significant difference was found between Ti and Ti-CaP implants for any parameter ($p > 0.05$).

model. Calcium phosphate-coated and -noncoated porous titanium implants, half of them loaded with rhTGF-$\beta_1$, were bilaterally implanted and left to ingrow. Histological analysis demonstrated that in the TGF-$\beta_1$-loaded implants, bone had formed throughout the implant up to the center, whereas in the absence of growth factor, only partial ingrowth of bone was observed. The bone had a trabecular appearance and was present along with bone marrow-like tissue. All histological findings were confirmed by image analysis: 97% ingrowth was seen in the rhTGF-$\beta_1$-loaded implants, whereas only 57% and 54% ingrowth was observed in the nonloaded calcium phosphate-coated and -noncoated implants, respectively. Bone surface area and bone fill were significantly higher in the rhTGF-$\beta_1$-loaded implants (1.37 $mm^2$ and 36%, respectively) than in the nonloaded implants (0.57 $mm^2$ and 26%) (Fig. 5.5). There were no statistically significant differences in any parameter between the calcium phosphate-coated and -noncoated implants. Quadruple fluorochrome labeling showed that in the titanium and titanium-calcium phosphate implants, bone guidance had occurred from the former defect edge, whereas in the titanium-TGF-$\beta_1$ implants, bone formation had been initiated in the center of the pore and proceeded in a centrifugal manner.

A study of in vitro release of TGF-$\beta_1$ from the titanium fiber meshes showed a burst of release during the first 2 hours, when more than 70% release had occurred. Following the burst, a slower phase liberated 80% of the theoretical dose by 1 week. It thus seems that a dose-response relationship exists for TGF-$\beta_1$ release with respect to bone induction. Higher doses do not necessarily generate more bone formation; rather, there is an optimum dose [3, 4]. Taken together, these results show that the combination of titanium-mesh with TGF-$\beta_1$ can induce orthotopic bone formation [40].

## 5.7 Conclusions

Autologous bone or bone derivatives and substitutes for bone reconstruction have significant limitations in terms of availability, morbidity, efficacy, immunologic reaction, and disease transmission. As a result, novel tissue-engineering models have been designed to overcome these problems. The factors necessary for tissue engineering include cells, the scaffold for cell proliferation and differentiation, and growth factors. For example, one practical way to provide an environment suitable for induction of tissue regeneration at a defect involves placing a scaffold as an artificial

ECM. The implant temporarily supports initial cell attachment and subsequent proliferation and differentiation. Resident cells surrounding the scaffold, or cells preseeded in the scaffold, will proliferate and differentiate on the foundation of a compatible scaffold. In some cases, a scaffold and bonelike cells are not enough, and supplementation with growth factors is required.

In this chapter, we have focused on the metal titanium fiber mesh used as a scaffold material for bone reconstruction either by loading it with cells or by loading it with growth factors. The results of the various studies clearly demonstrated the excellent characteristics of titanium fiber mesh: biocompatibility, strength, low stiffness, high porosity, and high coefficient of friction. Further, cell-loaded meshes initiated bone formation at orthotopic and ectopic sites, and these cell-loaded meshes were shown to be further optimized by dynamic seeding, culturing, and addition of an ECM coating. The growth-factor-loaded meshes showed increased bone formation in comparison with unloaded meshes when implanted either subcutaneously or in a cranial defect. In summary, titanium fiber mesh is a useful scaffold material that warrants further investigation as a clinical tool for bone reconstructive surgery.

## References


1. Bakker AD, Soejima K, Klein-Nulend J, Burger EH (2001) The production of nitric oxide and prostaglandin E(2) by primary bone cells is shear stress dependent. J Biomech 34:671–677.
2. Bancroft GN, Sikavitsas VI, van den Dolder J, Sheffield TL, Ambrose CG, Jansen JA, Mikos AG (2002) Fluid flow increases mineralized matrix deposition in three-dimensional perfusion culture of marrow stromal osteoblasts in a dose-dependent manner. PNAS 99:12600–12605.
3. Beck LS, Amento EP, Xu Y, Deguzman L, Lee WP, Nguyen T, Gillett NA (1993) TGF-β1 induces bone closure of skull defects: temporal dynamics of bone formation in defects exposed to rhTGF-β1. J Bone Miner Res 8:753–761.
4. Beck LS, Deguzman L, Lee WP, Xu Y, McFatridge LA, Gillett NA, Amento EP (1991) TGF-β1 induces bone closure of skull defects. J Bone Miner Res 6:1257–1265.
5. Becker D, Geissler U, Hempel U, Bierbaum S, Scharnweber D, Worch H, Wenzel KW (2002) Proliferation and differentiation of rat calvarial osteoblasts on type I collagen-coated titanium alloy. J Biomed Mater Res 59:516–527.
6. Bruder SP, Jaiswal N, Haynesworth SE (1997) Growth kinetics, self-renewal, and the osteogenic potential of purified human mesenchymal stem cells during extensive subcultivation and following cryopreservation. J Cell Biochem 64:278–294.
7. Chang YS, Oka M, Kobayashi M, Gu HO, Li ZL, Nakamura T, Ikada Y (1996) Significance of interstitial bone ingrowth under load-bearing conditions: a comparison between solid and porous implant materials. Biomaterials 17:1141–1148.
8. Cohen R (2002) A porous tantalum trabecular metal: basic science. Am J Orthop 31:216–217.
9. Gao JM, Niklason L, Langer R (1998) Surface hydrolysis of poly(glycolic acid) meshes increases the seeding density of vascular smooth muscle cells. J Biomed Mater Res 42:417–424.
10. Goldstein AS, Juarez TM, Helmke CD, Gustin MC, Mikos AG (2001) Effect of convection on osteoblastic cell growth and function in biodegradable polymer foam scaffolds. Biomaterials 22:1279–1288.
11. Hanada K, Dennis JE, Caplan I (1997) Stimulatory effects of basic fibroblast growth factor and bone morphogenetic protein-2 on osteogenic differentiation of rat bone marrow derived mesenchymal stem cells. J Bone Miner Res 12:1606–1614.
12. Haynesworth SE, Goshima J, Goldberg VM, Caplan AI (1992) Characterization of cells with osteogenic potential from human marrow. Bone 13:81–88.
13. Holy CE, Shoichet MS, Davies JE (2000) Engineering three-dimensional bone tissue in vitro using biodegradable scaffolds: investigating initial cell-seeding density and culture period. J Biomed Mater Res 51:376–382.
14. Ishaug SL, Crane GM, Miller MJ, Yasko AW, Yaszemski MJ, Mikos AG (1997) Bone formation by three-dimensional stromal osteoblast culture in biodegradable polymer scaffolds. J Biomed Mater Res 36:17–28.
15. Jaiswal N, Haynesworth SE, Caplan AI, Bruder SP (1997) Osteogenic differentiation of purified, culture-expanded human mesenchymal stem cells in vitro. J Cell Biochem 64:295–312.
16. Jansen JA, von Recum AF, van der Waerden JPCM, de Groot K (1992) Soft tissue response to different types of sintered metal fibre-web materials. Biomaterials 13(13):959–968.
17. Johnson DL, McAllister TN, Frangos JA (1996) Fluid flow stimulates rapid and continuous release of nitric oxide in osteoblasts. Am J Physiol 271:E205–208.
18. Kadiyala S, Young RG, Thiede MA, Bruder SP (1997) Culture expanded canine mesenchymal stem cells possess osteochondrogenic potential in vivo and in vitro. Cell Transplant 6:125–134.
19. Klein-Nulend J, Burger EH, Semeins CM, Reisz LG, Pilbeam CC (1997) Pulsating fluid flow stimulates prostaglandin release and inducible prostaglandin G/H synthase mRNA expression in primary mouse bone cells. J Bone Miner Res 12:45–51.
20. Lynch MP, Stein JL, Stein GS, Lian JB (1995) The influence of type I collagen on the development and maintenance of the osteoblast phenotype in primary and passaged rat calvarial osteoblasts: modification of expression of genes supporting cell growth, adhesion, and extracellular matrix mineralization. Exp Cell Res 216:35–45.

21. Marti A (2000) Cobalt-base alloys used in bone surgery. Injury 31(suppl 4):S-D18–21.
22. Mendes SC, Van den Brink I, De Bruijn JD, van Blitterswijk CA (1998) In vivo bone formation by human bone marrow cells: effect of osteogenic culture supplements and cell densities. J Mater Sci Mater Med 9:855–858.
23. Mizuno M, Kuboki Y (2001) Osteoblast-related gene expression of bone marrow cells during the osteoblastic differentiation induced by type I collagen. J Biochem (Tokyo) 129:133–138.
24. Moursi AM, Damsky CH, Lull J, Zimmerman D, Doty SB, Aota S, Globus RK (1996) Fibronectin regulates calvarial osteoblast differentiation. J Cell Sci 109(Pt 6):1369–1380.
25. Moursi AM, Globus RK, Damsky CH (1997) Interactions between integrin receptors and fibronectin are required for calvarial osteoblast differentiation in vitro. J Cell Sci 110:2187–2196.
26. Owen TA, Aronow M, Shalhoub V, Barone LM, Wilming L, Tassinari MS, Kennedy MB, Pockwinse S, Lian JB, Stein GS (1990) Progressive development of the rat osteoblast phenotype in vitro: reciprocal relationships in expression of genes associated with osteoblast proliferation and differentiation during formation of the bone extracellular matrix. J Cell Physiol 143:420–430.
27. Paquay YCGJ, de Blieck-Hogervorst JMA, Jansen JA (1996) Corrosion behaviour of metal fibre web structures. J Mater Sci Mater Med 7:585–589.
28. Pohler OEM (2000) Unalloyed titanium for implants in bone surgery. Injury 31:S-D7–13.
29. Rancourt D, Shirazi-Adl A, Drouin G, Paiement AG (1990) Friction properties of the interface between porous-surfaced metals and tibial cancellous bone. J Biomed Mater Res 24:1503–1519.
30. Saito T, Albelda SM, Brighton CT (1994) Identification of integrin receptors on cultured human bone cells. J Orthop Res 12:384–394.
31. Siebers MC, Ter Brugge PJ, Walboomers XF, Jansen JA (2005) Integrins as linker proteins between osteoblasts and bone replacing materials. A critical review. Biomaterials 26:137–146.
32. Sikavitsas V, van den Dolder J, Bancroft GN, Jansen JA, Mikos AG (2003) Influence of the in vitro culture period on the in vivo performance of cell/titanium bone tissue engineered constructs using a rat cranial critical size defect. J Biomed Mater Res 67A:944–951.
33. Sikavitsas VI, Bancroft GN, Holtorf HL, Jansen JA, Mikos AG (2003) Mineralized matrix deposition by marrow stromal osteoblasts in 3D perfusion culture increases with increasing fluid shear forces. PNAS 100:14683–14688.
34. Sikavitsas VI, Bancroft GN, Mikos AG (2002) Formation of three-dimensional cell/polymer constructs for bone tissue engineering in a spinner flask and a rotating wall vessel bioreactor. J Biomed Mater Res 62:136–148.
35. Van den Dolder J, Bancroft GN, Sikavitsas V, Spauwen PH, Mikos AG, Jansen JA (2003) The effect of fibronectin and collagen I coated titanium fiber mesh on proliferation and differentiation of osteogenic cells. Tissue Eng 9:505–516.
36. Van den Dolder J, Bancroft GN, Sikavitsas VI, Spauwen PH, Jansen JA, Mikos AG (2003) Flow perfusion culture of marrow stromal osteoblasts in titanium fiber mesh. J Biomed Mater Res 64:235–241.
37. Van den Dolder J, Farber E, Spauwen PHM, Jansen JA (2003) Bone tissue regeneration using titanium fiber mesh combined with rat bone marrow cells for the treatment of bone defects. Biomaterials 24:1745–1750.
38. Van den Dolder J, Spauwen PHM, Jansen JA (2003) Evaluation of various seeding techniques for culturing osteogenic cells on titanium fiber mesh. Tissue Eng 9:315–326.
39. Van den Dolder J, Vehof JWM, Spauwen PHM, Jansen JA (2002) Bone formation by rat bone marrow cells cultured on titanium fiber mesh: effect of in vitro culture time. J Biomed Mater Res 62:350–358.
40. Vehof JWM, Fisher JP, Dean D, van der Waerden JP, Spauwen PH, Mikos AG, Jansen JA (2002) Bone formation in transforming growth factor β1-loaded titanium fiber mesh implants. Clin Oral Implants Res 13:94–102.
41. Vehof JWM, Spauwen PHM, Jansen JA (2000) Bone formation in calcium-phosphate-coated titanium mesh. Biomaterials 21:2003–2009.
42. Vehof JW, de Ruijter AE, Spauwen PH, Jansen JA (2001) Influence of rhBMP-2 on rat bone marrow stromal cells cultured on titanium fiber mesh. Tissue Eng 7:373–383.
43. Vehof JWM, Mahwood J, Takita H, van't Hof MA, Kuboki Y, Spauwen PH, Jansen JA (2001) Ectopic bone formation in bone morphogenetic protein loaded calcium phosphate coated titanium fiber mesh. Plast Reconstr Surg 108:434–443.
44. Vehof JWM, Takita H, Kuboki Y, Spauwen PH, Jansen JA (2002) Histological characterization of the early stages of bone morphogenetic protein-induced osteogenesis. J Biomed Mater Res 61:440–446.
45. Vehof JWM, van den Dolder J, de Ruijter JE, Spauwen PH, Jansen JA (2003) Bone formation in Ca-P coated and non-coated titanium fiber mesh. J Biomed Mater Res 64:417–426.
46. Xiao YL, Riesle J, van Blitterswijk CA (1999) Static and dynamic fibroblast seeding and cultivation in porous PEO/PBT scaffolds. J Mater Sci Mater Med 10:773–777.
47. Yang S, Leong KF, Du Z, Chua CK (2001) The design of scaffolds for use in tissue engineering. Part 1. Traditional factors. Tissue Eng 7:679–690.

# 6. Engineering Polymeric Scaffolds for Bone Grafts

*Martha W. Betz, Diana M. Yoon, and John P. Fisher*

## 6.1 Introduction

Orthopedic injuries resulting from trauma or improper development often require surgical intervention to restore natural tissue function. Currently, over one million operations are performed annually for the surgical reconstruction of bone [50]. The well-known limitations associated with autografts, allografts, and bone cements have led to the investigation of synthetic polymers as support matrices for bone tissue engineering. Polymers are long-chain molecules that are formed by linking repetitive monomer units. They have been extensively studied for tissue-engineering applications. Constructs designed from these polymers can act as a support matrix to deliver cell populations or induce surrounding tissue ingrowth. The properties of scaffolds directly determine their success in tissue engineering and must be designed specifically for each application. A successful scaffold provides initial support, growth factors, and transitions through degradation to allow tissue regeneration and return of function. This chapter will discuss the fabrication and properties of polymeric tissue-engineering scaffolds, including curing methods, polymer assembly, scaffold fabrication, surface properties, macrostructure, mechanical properties, biodegradation, and biocompatibility, as well as current synthetic polymers under investigation.

## 6.2 Scaffold Formation

### 6.2.1 Curing Methods

The method of curing requires knowledge of polymer chain formation into bulk material and the chemical nature of the polymer, specifically polymer length and functionality [5]. Two major curing methods often used are polymer entanglement and cross-linking.

Polymer entanglement is based on the principle that many polymers associate with one another in solution. This is common with long, linear polymers as well as branched polymers. The polymer is dissolved in an appropriate solvent and placed in a mold. The solvent is removed by evaporation, leaving the polymer in the shape of the mold. This is accomplished with the aid of pressure, temperature, or both [53]. The process is relatively simple, but the result may lack mechanical stability.

Cross-linking of individual polymers through chemical bonds to form a bulk material is another curing method. Individual polymer chains can form hydrogen or ionic bonds with one another through noncovalent interactions [5]. For the formation of covalent bonds, the polymer must contain a reactive site for cross-linking, such as a carbon–carbon double bond. Covalent cross-linking is generally induced by a free radical that is initiated by heat, light, chemical accelerant, or time [33].

For example, photopolymerization is a commonly used technique based on photopolymer polymerization initiated by electromagnetic radiation [98]. The photopolymers used are typically low-molecular-weight monomers that react to form long-chain polymers when activated by a specific wavelength. In addition, since scaffold formation occurs in response to a signal, the polymer may be used as an injectable material and can form in situ when exposed to the signal. However, the chemical reactions necessary for cross linking are often associated with unreacted components and reaction by-products that may harm the surrounding tissue.

## 6.2.2 Polymer Assembly

Polymer assembly may occur before or during implantation of the scaffold into the body [97]. Prefabrication is the term applied to assembly before implantation. The scaffold is formed outside of the body, and any cytotoxic or non-biocompatible by-products can be removed prior to transplantation. Prefabrication also allows for cell encapsulation and in vitro culture before implantation. However, the structure of the prefabricated construct generally may not fit the host site precisely. An imperfect match may lead to host immune reactions such as fibrosis and thus to construct failure. In situ fabrication techniques address this concern and involve curing the construct at the tissue defect site [33]. Liquid components are injected into the desired site, and their deformability allows for improved integration into the host tissue. Furthermore, because this method uses liquid components, it is less invasive than the surgical procedures sometimes necessary for prefabricated constructs. However, in situ fabrication does not allow for the removal of harmful by-products, and the surrounding tissue may therefore be exposed to toxic components. As a result, the variety of chemical components that can be used to form the construct in situ is restricted.

## 6.2.3 Conventional Scaffold Fabrication Methods

Fabrication is the process of forming a cured or curing polymer into a scaffold. Scaffold fabrication can be performed by conventional or rapid prototyping (solid free-form) practices (Table 6.1). A number of conventional techniques are used to create porous scaffolds, fiber bonding, solvent-casting particulate leaching, phase separation, melt molding, freeze drying, and gas foaming.

### 6.2.3.1 Fiber Bonding

Fibers are commonly processed from semi-crystalline polymers, including poly(glycolic acid) (PGA). These fibers can be used to create

**Table 6.1.** Fabrication methods, associated characteristics, and synthetic polymers used in bone tissue-engineering scaffolds

| Fabrication Method | Scaffold Attributes | Polymers | References |
|---|---|---|---|
| Fiber bonding | High porosity | PGA, PCL | [31, 56, 64] |
| Solvent-casting particulate leaching | Controlled porosity<br>Controlled pore size | PLA, PLGA, PPF | [12, 67, 73, 87, 95] |
| Phase separation | Porous<br>Biomolecule incorporation | PLLA, PLGA, PLA | [40, 69, 70, 101] |
| Melt molding | Controlled porosity<br>Controlled pore size<br>Biomolecule incorporation | PLGA | [90, 97] |
| Freeze drying | Controlled pore size | PLGA | [39, 78, 94] |
| Gas foaming | Controlled porosity<br>Controlled pore structure | PLLA, PLGA, PLA | [61, 71, 82] |
| Sheet lamination | Porous | PLA, PLGA | [66] |
| Three-dimensional printing | Controlled mechanical strength | PCL, PEO, PLGA, PLA | [30, 47, 96] |
| Laser stereolithography | Biomolecule incorporation | PPF, PEGDA | [20, 60] |
| Fused deposition | Controlled pore size | PCL | [16, 43, 77, 100] |

a fiber mesh or a three-dimensional patterned structure with variable pore size by weaving or knitting. These mesh constructs have a large surface area with high porosity, which induces greater cell attachment and better nutrient diffusion and waste removal [97]. However, the increased porosity of these scaffolds causes them to be mechanically unstable. To alleviate this difficulty, a fiber-bonding technique has been evolved to dissolve poly(lactic acid) (PLA) in a solvent and to cast it over a PGA mesh that is aligned in the desired shape [64, 97]. Heating the construct above the melting temperature of PGA evaporates the solvent. The PGA mesh becomes connected at fiber cross-points when the construct is cooled and PLA is redissolved. Fiber bonding has also been used to fabricate scaffolds from poly(ε-caprolactone) (PCL) [31]. Although this technique allows for greater structural stability, it has some disadvantages. The porosity varies and cannot be finely controlled. In addition, the solvent used to dissolve the polymer may be harmful to an incorporated cell population and the surrounding tissue.

### 6.2.3.2 Solvent-Casting Particulate Leaching

Solvent-casting particulate leaching is a technique by which dispersed particles such as sodium chloride, tartrate, citrate, or saccharose are mixed in solution with a polymer and cast in a mold [67, 97]. Casting or freeze drying is performed to evaporate the solvent. The dispersed particles are leached out of the scaffold, leaving void spaces that form a porous and highly interconnected structure. This process allows the independent control of porosity and pore size [97]. This technique has been used to form constructs with PLA, poly(D,L-lactic acid-co-glycolic acid) (PLGA), and poly(propylene fumarate) (PPF) [12, 67, 73, 87, 95].

### 6.2.3.3 Phase Separation

Phase separation is used to isolate components of a heterogeneous mixture. In this process the polymer is first dissolved in a solvent such as molten phenol, naphthalene, or dioxane [40, 97]. The polymer solvent solution is cooled, causing liquid–liquid or solid–liquid phase separation, with the polymer in a separate phase from the solvent. The solvent is then evaporated, leading to the formation of a porous polymer membrane [59, 70, 97]. One considerable advantage of this technique is that biomolecules can be incorporated into the scaffold without exposure to harsh chemical or thermal conditions. In addition, changes in the polymer composition, polymer concentration, and solvent-to-nonsolvent ratio can be utilized to augment the scaffold structure. However, the effect of these modifications may be difficult to predict. Phase separation has been used to create scaffolds of poly(L)lactic acid (PLLA), PLGA, and PLA [40, 69, 70, 101].

### 6.2.3.4 Melt Molding

Melt molding combines a polymer powder and microspheres to form a scaffold [97]. This technique has been used with a fine PLGA powder and gelatin microspheres heated in a Teflon mold [90, 97]. Heating the polymer above the glass transition temperature allows the polymer powder to melt. The molded polymer is then removed and placed in water, where the entrapped microspheres are removed; this results in a three-dimensional porous structure. There are several advantages to this technique. Pore size is directly related to the microsphere diameter, and changing the polymer-to-gelatin ratio modifies the porosity. Furthermore, biomolecules can be incorporated into the scaffold, since this process occurs under moderate conditions without the use of organic solvents. Also, by changing the shape of the mold, a defined construct shape can be developed. However, this technique may involve very high temperatures if semicrystalline polymers are heated beyond their glass transition temperature [97].

### 6.2.3.5 Freeze Drying

Freeze drying uses temperature change to create a porous structure [39]. In this technique, synthetic polymers, such as PLGA, are dissolved in cold solvents, glacial acetic acid, or benzene, and water is added to create an emulsion [39, 78, 94]. The emulsion is quick-frozen and the resulting ice crystals are sublimed by the freeze-drying technique. This leaves a highly connected porous matrix. With this technique, pore size can be controlled by altering the freezing rate; in general, faster freezing leads to smaller pores [78]. However, it is difficult to control pore structure with freeze drying alone. Better results can be achieved by combining with other techniques, such as the above-described particulate-leaching method [17].

#### 6.2.3.6 Gas Foaming

In gas foaming, scaffold pores are formed by gases that are under pressure or undergoing a chemical reaction [71]. It is the bubbles in the polymer that cause pore formation in the construct. Variations in gas volume and in the rates of gas nucleation and diffusion modify the porosity and pore structure of the scaffold. By this method the scaffold is formed in a moderate environment without the use of organic solvents. The results of gas foaming, like those of freeze drying, can be improved by combining it with particulate leaching [82]. Gas foaming has been used with PLLA, PLGA, and PLA to create scaffolds for bone tissue [61, 71, 82].

### 6.2.4 Rapid Prototyping (Solid Free-Form Fabrication)

All of the techniques described above are limited in how they regulate scaffold parameters such as pore size, pore shape, pore interconnectivity, and pore wall thickness. This lack of fine control has led to the development of new techniques to produce scaffolds directly from a computer-aided design model. Rapid prototyping, also known as solid free-form fabrication, has been used to guide surgical procedures based on computerized topography of the patient in question [98]. These techniques generally produce three-dimensional scaffolds in a layer-by-layer fashion and can be designed to form very specific shapes. These techniques can be carried out at room temperature, thereby allowing for cell encapsulation and biomolecule incorporation without significantly affecting viability. However, these methods involve processes that alter some polymers limiting their use in fabrication. Rapid prototyping techniques include sheet lamination, three-dimensional printing, laser stereolithography, and fused deposition modeling [20, 42, 57, 98].

#### 6.2.4.1 Sheet Lamination

Sheet lamination is a technique that creates scaffolds with a layer-by-layer approach. A three-dimensional cross-section of the scaffold is built out of a roll of sheets that have been lined with an adhesive [98]. The layers are cut by a carbon dioxide laser and bonded by heat and/or pressure. The technique does not allow formation of small inner holes within the scaffold, a disadvantage with respect to nutrient and waste transport [98].

#### 6.2.4.2 Three-Dimensional Printing

Three-dimensional printing forms sequential powder layers of the scaffold by ink-jet printing a binder [42, 57]. In this technique, a computer model is used to create a slicing algorithm that defines the structure of each scaffold layer. Powder is thinly layered over a bed, and binder material is printed on top where the scaffold is to be formed. A piston is lowered to allow the next layer of powder to be spread and bonded. In this technique, the packing density of the powder particles can be used to control the adhesive bonding of the material and thus the resulting mechanical strength [42]. This technique has been used to create scaffolds from polyethylene oxides (PEOs), PLA, PCL, and PLGA [30, 42, 47, 57, 96].

#### 6.2.4.3 Laser Stereolithography

Laser stereolithography is another computer-aided design method that allows for three-dimensional scaffold formation. This method is similar to the three-dimensional printing described above, but utilizes a liquid polymer to fabricate a scaffold [57]. The computer model creates two-dimensional slices of the scaffold model which modulate a platform submerged in a liquid photopolymer. The liquid is then exposed to a focused laser light, which cures the polymer, forming a solid at specific points. A significant advantage of this technique is the ability to produce complex internal architecture. Furthermore, liquid solutions containing different biomolecules can be used for each scaffold layer [57]. When this technique was used with cross-linking of diethyl fumarate (DEF) and PPF, it resulted in pore sizes that ranged from 150 to 800 μm, with porosity as high as 90% [20]. Laser stereolithography has also been used to create scaffolds with PEG diacrylate [60].

#### 6.2.4.4 Fused Deposition Modeling

In fused deposition modeling, the polymer is deposited in thin layers on a base that solidifies as it attaches to the previous layer [42]. Initially the technique was used only with nonresorbable materials, but it has recently been applied to PCL and PCL/hyaluronic acid scaffolds [16, 42, 77, 100]. As with the other computer techniques, this process is highly reproducible. Fused deposition modeling also supports incorporation of pores into the scaffold to modulate mechanical strength and molecular diffusion.

## 6.2.5 Synthetic Polymers for Scaffolds

The molecular structure and properties of synthetic polymers allow specific cell and tissue processes to become part of engineered bone. This is an advantage over natural polymers, whose variable molecular structure makes for less precise modification. Synthetic polymers are most often present in a semicrystalline or an amorphous state. A semicrystalline polymer contains dense chain regions randomly distributed throughout the material. These regions act as physical cross-links and contribute to the mechanical strength of the polymer network. Amorphous polymers are similar to glass when they are below their glass transition temperature and act like rubber when heated above that temperature. The structure of amorphous polymers can be altered by chemical bonding, copolymerization, physical mixing, or blending [63]. In their unmodified form, synthetic polymers lack biomolecules that can aid cell attachment in some natural polymers. However, synthetic polymer surfaces can be made to include biomolecules that stimulate cell attachment and proliferation [92]. Common synthetic polymers include polyesters, polyanhydrides, polyphosphazenes, polycarbonates. and poly(ethylene glycol).

### 6.2.5.1 Polyesters

#### 6.2.5.1.1 Poly(D,L-Lactic Acid-Co-Glycolic Acid)

Poly(D,L-lactic acid-co-glycolic acid) (PLGA), is a copolymer of poly(lactic acid) (PLA) and poly(glycolic acid) (PGA), with properties distinct from the two homopolymers [94]. PGA and PLA are semicrystalline, whereas PLGA is an amorphous solid. PGA degrades slowly, whereas PLGA can degrade rapidly [1, 89, 90, 97]. The co-monomer ratio can be varied to produce different mechanical, physical, and degradation properties [82]. Degradation times vary from 6 to 12 months when the monomer ratio is 85:15 but only 1 to 2 months when the ratio is 50:50. The polymer can therefore be readily engineered to an appropriate degradation rate [63]. Owing to its ester linkages, which affect mechanical properties as PLGA degrades, the polymer can also undergo bulk degradation (see Section 6.3.4 below) [63]. The degradation products include glycolic and lactic acids, both of which can be removed via the body's metabolic pathways [78, 82, 94].

The possibility of modulating PLGA properties significantly, as well as the fact that it can support a variety of cell types, has led to great interest in this polymer for tissue engineering. Osteoblasts attach to PLGA [26, 44], and extracellular matrix (ECM) components, such as osteopontin and osteonectin, along with collagen, fibronectin, vitronectin, and laminin, are abundantly associated with PLGA scaffolds [26]. These molecules are important for the extracellular environment that osteoblasts require.

#### 6.2.5.1.2 Poly(ε-Caprolactone)

Poly(ε-caprolactone) (PCL) is an aliphatic polyester with a repeating molecular structure of five nonpolar methylene groups and a single polar ester group [97]. A semicrystalline polymer, PCL has a melting point of approximately 60°C and is formed by the ring-opening polymerization of ε-capolactone [97]. PCL is known to be highly water soluble and is hydrolyzed under physiologic conditions [34]. Degradation to caproic acid occurs by either a bulk or a surface mechanism. Caproic acid alters the scaffold degradation rate, therefore the by-product concentration should be kept low [97]. PCL is known to degrade very slowly, with a degradation time of approximately 2 years [97]. To shorten the degradation rate for certain applications, PCL has been copolymerized with collagen, PGA, PLA and PEG [9, 22, 74]. In addition, PCL may support load-bearing applications and can maintain mechanical strength for an extended period of time [1].

PCL has been used as a scaffold to support osteoblast growth. A porous PCL scaffold facilitates osteoblast production of alkaline phosphatase, a marker of bone mineralization, and favors attachment and proliferation of osteoblasts [19]. PCL has also been combined with hyaluronic acid to improve the compressive strength associated with the polymer, thus enhancing its application in bone tissue engineering [19].

#### 6.2.5.1.3 Poly(Propylene Fumarate)

Poly(propylene fumarate) (PPF) is an aliphatic linear polyester composed of repeating units of two ester groups and one central unsaturated carbon–carbon double bond [29]. The polymer degrades by hydrolysis of an ester bond that leads to formation of fumaric acid and propylene glycol [29]. These by-products cause mild and short inflammation, and therefore the polymer is likely to be biocompatible [29]. The double bonds of PPF allow it to be covalently cross-linked. Cross-linking in response to a trigger allows scaffold fabrication in situ, thus making PPF an injectable biomaterial [28]. In addition, the cured form of PPF has significant compressive and tensile strengths and may therefore constitute scaffold material for bone tissue engineering [28].

PPF scaffolds with varying porosities and pore sizes have been investigated to analyze tissue response in cranial defects. In all cases the scaffolds only induced a mild tissue response and allowed for vascularization of the area [29]. In addition, PPF scaffolds coated with TGF-β1 induced significant bone formation in cranial defects [92].

#### 6.2.5.1.4 Polyorthoester

Polyorthoesters (POEs) are a family of biodegradable polymers [8]. They are formed through a reaction of ketene acetals with hydroxy-containing molecules, such as diols [24]. POEs are hydrophobic substances and undergo surface degradation [24, 35]. However, the properties of POEs can be modified by copolymerization. For example, degradation of the polymer can be adjusted to an appropriate rate by incorporating short acid groups such as glycolic or lactic acid [4, 35]. In addition, the orthoester linkages present within POEs have been found to be more susceptible to hydrolytic cleavage in acids than bases, which demonstrates another method of degradation control [34, 63].

POE polymers are desirable for bone tissue engineering because they undergo surface degradation yet maintain mechanical stability. They can therefore be used in load-bearing applications even while tissue formation is incomplete. Scaffolds constructed of POEs have been implanted into calvarial defects and have been shown to promote new bone formation [4].

### 6.2.5.2 Other Synthetic Polymers

#### 6.2.5.2.1 Polyanhydrides

Polyanhydrides have a polymer backbone containing an anhydride bond [48]. They contain bonds that easily react with water, causing degradation via surface erosion [49]. Polyanhydrides are synthesized by a dehydration reaction of diacids, and degrade into these nontoxic diacid monomers, which are removed from the body within weeks to months [48]. The polyanhydride degradation rate can be modified by changing the monomer concentrations: increasing hydrophobicity decreases degradation rate. For example, polyanhydrides synthesized with carboxyphenoxypropane degrade over a period of 3 to 4 years. However, when synthesized with 79% sebacic acid, the construct degrades over 2 weeks [49]. Furthermore, polyanhydride synthesis can be activated by a trigger such as photocross-linking, and therefore curing can occur in situ [13, 14].

Polyanhydrides were first studied in an attempt to regulate the release of bioactive molecules [32, 48]. Polyanhydrides have limited mechanical stability and therefore are inappropriate for load-bearing applications. However, when imides were incorporated to form cross-linkable networks, the mechanical stability of the construct was increased [1, 91]. This is thought to be due to the rigidity of the aromatic imide group [91]. Specifically, scaffolds containing succinic acid have shown compressive strengths of 50 to 60 MPa and were degraded by hydrolysis of the anhydride bonds plus the imide bonds [32, 91]. The mechanical properties of the polymer have been increased by photocross-linking [6, 13, 68].

#### 6.2.5.2.2 Polyphosphazene

Polyphosphazene contains a backbone composed of alternating nitrogen and phosphorus atoms with two side groups attached to each atom [2]. Polyphosphazene is hydrophobic and degrades by surface degradation into phosphate and ammonium salt by-products. Variation in polyphosphazene constructs can be achieved by adding various hydrolytically labile substituents to the phosphorus atoms [75]. The degradation rate of phosphazenes cannot be altered significantly. These polymers generally degrade slowly in vivo [75].

Polyphosphazenes have been of interest because they can be readily modified. Their slow degradation rate makes them attractive for long-term controlled-release devices [75]. Polyphosphazenes have also been used in orthopedic applications because of their high strength and surface degradation properties [52]. Osteoblast cells seeded onto three-dimensional polyphosphazene scaffolds have been shown to support proliferation and skeletal tissue formation [51].

#### 6.2.5.2.3 Polycarbonate

Tyrosine-derived polycarbonate (P(DTR carbonate)) is an amorphous polycarbonate that is modifiable due to the presence of alkyl ester pendant groups located within its linear chain [88]. P(DTR carbonate) contains three bonds that can be hydrolytically degraded: amide, carbonate, and ester [88]. The carbonate bonds have been found to degrade faster than the ester bonds, and the amide bond is stable to hydrolysis at body temperature [27, 88]. The ester bond degrades into carboxylic acid and alcohol, whereas the carbonate bond by-products include two alcohols and carbon dioxide [88]. P(DTR carbonate) is thought to be a biocompatible material because it is based on the natural amino acid tyrosine and degrades mostly into nonacidic by-products [18].

P(DTR carbonate) can be modified to degrade over months or years [18]. It has been investigated as a bone scaffold and shown to elicit a response of bone ingrowth at the bone–polymer interface [18]. In addition, research has demonstrated the ability of osteoblast cells to attach onto the surface of P(DTR carbonate) and maintain their phenotype [54]. Other investigations with poly(deasminotyrosyl-tyrosine ethyl ester carbonate) (poly(DTE carbonate)) demonstrated that bone ingrowth occurred in cranial defects and that the patterns of bone formation mimicked the structure of the scaffold [85]. This suggests that polycarbonate scaffolds can be designed to reflect bone tissue morphology and thus can induce growth appropriate to the specific site. Further studies with poly(DTE carbonate) have show that it elicits more direct bone apposition than other polycarbonates. This may be due to the ethyl ester pendant group in the polymer [45]. The hydrolysis of these groups produces calcium chelation sites on the polymer surface that appear to be related to polymer–bone bonding [45].

#### 6.2.5.2.4 Poly(Ethylene Glycol)

Poly(ethylene glycol) (PEG) is a linear-chained polymer with an oxygen-carbon-carbon repeating unit. By varying the number of units, the length and molecular weight of the polymer can be changed [15, 33]. PEG homopolymer is nondegradable. However, it can be copolymerized with degradable polymers to allow degradation [11]. PEG is highly water soluble due to the oxygen molecule present in the polymer backbone. Copolymerization of PEG with other materials causes the subsequent material to become more hydrophilic. This has led to investigation of its potential function as a hydrogel. However, linear PEG chains are susceptible to rapid diffusion and also have low mechanical stability [81]. Networks of PEG can be formed by attaching functional groups to the ends of PEG chains and initiating their cross-linking [72, 79, 86].

PEG has low mechanical stability and is therefore not often used in bone tissue engineering for load-bearing applications. However, because it can be cross-linked into a network with other synthetic materials and thereby affect degradation, it is attractive as a copolymer to obtain controlled erosion and degradation rates. PEG has been copolymerized with poly(lactic acid), combined with a hydroxyapatite ceramic, and used to deliver bone morphogenetic protein, resulting in complete repair of bone defects [46]. Similarly, PEG has been combined with PLA and p-dioxanone and used to deliver bone morphogenetic protein (BMP), exhibiting osteoconductive capacity [62]. PEG hydrogels have also been modified with cell adhesion peptides and used in tissue engineering. These gels delivered growth factors, resulting in efficient and highly localized bone regeneration [58]. In addition, PEG has been copolymerized with PLGA to form a foam that delivers periosteal cells in vivo and supports osteochondral repair [80].

# 6.3 Scaffold Design Properties

Scaffolds can be made to mimic the tissue that is being regenerated. Aspects of the scaffold that can be altered include the surface, the macrostructure, mechanical properties, biodegradation, and biocompatibility.

## 6.3.1 Surface Properties

The majority of cell types used in bone tissue engineering are anchorage dependent. The engineered scaffold should therefore facilitate cell attachment. The scaffold surface is the initial and primary site of interaction with the surrounding tissue. Scaffolds that cells can attach to abundantly and easily with large, accessible surface areas are favored. In addition, the scaffold surface should support cell proliferation. Strong cell adhesion promotes cell proliferation, and a rounded surface promotes differentiation [89]. Hydrophilic polymers have highly wettable surfaces. This allows cells to be encapsulated through capillary action [23]. However, the most significant surface property of polymers is that they provide an environment for scaffold–host interaction. Many natural polymers can facilitate attachment because they contain functional groups that vary in polarity, electrostatic charge, hydrophobicity, and the ability to interact by van der Waals forces. In addition, by utilizing covalent and noncovalent assembly, association constants can be varied and the structure of natural polymers can be precisely controlled [21]. In synthetic polymers, an attempt is made to mimic the characteristics of natural polymers. By altering polymer and side-chain architecture, functional groups can be made part of the surface or included within the scaffold. For example, modification of a polymer with short peptide sequences or long protein chains promotes interaction with the surrounding tissue [83]. Specifically, ligands that are common in the extracellular matrix, such as fibronectin, vitronectin, and laminin, have been used as surface molecules [83]. This surface modification technique is being widely investigated [21, 38, 83].

## 6.3.2 Macrostructure

A porous scaffold permits cells to become part of the porous void space. A porous scaffold is also important for diffusion of nutrients and waste removal. In general, it is advantageous for the scaffold to have a high surface-area-to-volume ratio. This promotes the formation of pores with a diameter that is small but still larger than the diameter of most cells. High-porosity scaffolds have poor mechanical integrity, and engineering for appropriate diffusion and mechanical strength is an important challenge in their construction. Fiber meshes, foam scaffolds, and hydrogels are examples of materials that provide added mechanical strength to porous scaffolds.

Fiber meshes are formed into three-dimensional structures by knitting or weaving individual polymer fibers, thus providing a large surface area that promotes cell attachment [97]. The structure of fiber-mesh scaffolds resembles that of the ECM, which allows for nutrient diffusion and waste removal. The use of fiber bonding helps strengthen the mechanical integrity of fiber-mesh scaffolds [64].

Foam scaffolds are generally prefabricated before implantation. Similar to fiber meshes, their structure allows for adequate nutrient diffusion and waste removal. Foam scaffolds tend to be more stable than fiber meshes but still lack sufficient mechanical integrity. Porosity and pore structure can be modified by using different processing techniques, such as solvent casting particulate leaching, melt molding, freeze drying, and gas foaming.

Hydrogels are formed from hydrophilic polymers by physical polymer entanglements or cross-linking [33, 36]. The hydrophilic polymers can absorb water up to a thousand times their own dry weight [36]. The aqueous environment created in hydrogels simulates the in vivo environment and therefore provides an ideal setting for cell encapsulation. In addition, the aqueous environment supports quick diffusion of nutrients, proteins, and waste, thus promoting cell growth and proliferation. Some hydrogels, including PEG-based hydrogels, are easily injectable and capable of being molded, allowing minimally invasive implantation [25]. The disadvantages of hydrogels are that they lack strong mechanical stability and are difficult to sterilize [25, 36].

## 6.3.3 Mechanical Properties

The ability of a scaffold to provide needed mechanical support is a critical component of the construct. However, the high mechanical strength of bone as compared with other tissues makes the design of a structure with this feature challenging. Compact bone is mechanically the equivalent of a semibrittle, viscoelastic, and orientation-dependent material [7]. The longitudinal strength of compact bone varies between 78.8 and 151 MPa for tension and 131 and 224 MPa for compression [99]. The elastic modulus for compact bone is 17.0 to 20.0 GPa in the longitudinal direction, with a shear modulus of 3.30 GPa and a structural density of 1.80 $g/cm^3$ [99]. In contrast to compact bone, cancellous bone is spongy and highly porous, with a structural density of 0.20 $g/cm^3$. In general, cancellous bone is oriented along the directions of the principal stresses imposed by the external loading environment [7]. The strength of cancellous bone is based upon its apparent density; it ranges from 2.00 to 5.00 MPa and from 90.0 to 400 MPa for strength and modulus, respectively [76].

For proper tissue regeneration without significant deformation, a scaffold should provide a mechanical modulus of 10 to 1500 MPa for hard tissues and 0.4 to 350 MPa for soft tissues [37]. Mechanical requirements are therefore very important for orthopedic hard tissues and dictate the method of fabrication of the polymer. For example, fabrication with particulate leaching and gas foaming leads to a maximum compressive modulus of 0.4 MPa and therefore is not appropriate for scaffolds to be used for hard-tissue regeneration [37]. The lack of mechanical stability associated with many of the conventional fabrication techniques emphasizes the utility of rapid prototyping techniques for engineered scaffolds. These more precise methods of fabrication result in scaffolds with significant mechanical stability.

Finally, scaffolds should provide interim support while the tissue regenerates. The scaffold material should therefore not degrade before the regenerated tissue provides sufficient load-bearing support and stress dissipation. Two common scaffold designs support bone ingrowth with proper mechanical support. In one the physical scaffold provides mechanical support for the polymer/cell/tissue construct from initial seeding to remodeling by the host [41]. The scaffold matrix must therefore provide enough support to withstand in vivo stresses and loading. The other strategy imposes transitional support. Here the scaffold provides mechanical support while the cells proliferate and differentiate in vitro [41]. Once implanted, the scaffold is designed to degrade at the same rate at which the cells produce the ECM for support.

## 6.3.4 Biodegradation

The majority of scaffolds are designed to degrade by the time the tissue is completely formed. Synthetic polymers degrade primarily by chemical hydrolysis of unstable polymer backbones [63]. The polymer can also be designed to degrade enzymatically, relying on body enzymes or catalysts embedded within the scaffold. Degradation can alter the mechanical properties of the construct; this in turn influences the effectiveness of the implant. Additionally, the degradation products can modify the implant environment, depending on their biocompatibility. Degradation products are a function of the structure, components, and fabrication techniques of the material and the rate of degradation. Degradation also depends on the location and geometry of the implant, as well as the presence of catalysts, impurities and other additives [63].

Hydrolysis of the polymer backbone occurs in two stages [63]. First, water penetrates the polymer, converting the long chains into shorter water-soluble degradation products by attacking the chemical bonds in the amorphous phase. Next, the fragments are enzymatically degraded, causing a rapid decrease in polymer mass. These two phases are part of two overall mechanisms of degradation.

Overall scaffold degradation has been well described. Polymeric scaffolds undergo bulk or surface degradation, or both. In bulk degradation, erosion at the surface is slower than in the interior [63]. Initially, the surface begins to degrade when the construct is in contact with water. Then, as water penetrates the inside of the material, the bulk of the scaffold begins to degrade. Bulk degradation is associated with a decrease in mass, while the volume of the construct stays the same. This results in a decrease in density and mechanical strength.

One concern with bulk degradation is a phenomenon known as the autocatalytic effect [55]. This often occurs with synthetic polymers whose degradation products are acidic. When degradation occurs, the interior degradation products cannot diffuse through the polymer network. This causes a local increase in acidity, which in turn causes a more rapid degradation resulting from hydrolysis of labile linkages.

Surface degradation is similar to soap dissolution. The material degrades at the surface at a constant rate. This causes the construct to thin out, yet bulk integrity and structure are maintained. Surface degradation is common with polyanhydrides and polyorthoesters, which, though hydrophobic, are highly susceptible to hydrolysis and degrade at the surface. As the material degrades, the size of the construct decreases as mass is lost, but the density remains unchanged. This feature allows the polymer to maintain mechanical integrity, a property critical for bone tissue engineering.

The preferred method of degradation is a function of the engineering requirements, the host tissue, and the need for mechanical integrity. The speed at which a scaffold degrades can be arrived at by varying the properties of the polymer. For example, a material with more hydrophilic monomers and acidic end groups and a more hydrolytically reactive backbone, less crystallinity, and smaller size would tend to have a higher degradation rate [63]. The site of the implant can also affect degradation. In a poorly vascularized area with low diffusion, degradation products will tend to remain longer, causing a rise in acidity. This is a situation similar to that of the scaffold interior during bulk degradation, when degradation is increased. All of these factors must be taken into account when aiming for a specific method and rate of degradation.

## 6.3.5 Biocompatibility

All implanted materials elicit a host reaction, but the intensity of the response varies. Tissue responses include inflammation, immune reactions, and variability in wound healing [84]. If a material produces minimal inflammatory and immune responses and functions without harm to the host, it may be considered biocompatible. Material intended for implantation should be such as to minimize the intensity and duration of the response.

The tissue response to an implanted scaffold involves three stages [84]. Stage 1 occurs during the first 1 to 2 weeks after implantation and is characterized by acute and chronic inflammatory responses. Acute inflammation generally lasts minutes to days and depends on the extent of the injury [3]. Chronic inflammation results from the long-term presence of inflammatory stimuli and is confined to the implantation site. In general, the stage 1 response is independent of the degradation rate of the polymer [84]. Stage 2 begins as the numbers of monocytes and macrophages increase. In stage 2, fibrous encapsulation of the foreign material is initiated. In contrast to stage 1, the length of stage 2 is a function of the rate of biodegradation of the scaffold [84]. Fibrous encapsulation continues in stage 3. The length of this stage depends on the degradation rate of the polymer. Slowly degrading polymers have a stage 3 response that lasts weeks to months, whereas with rapidly degrading polymers, stage 3 can be as short as 1 to 2 weeks [84].

The immune response is of major concern in bone tissue engineering, because degradation products cause failure in many orthopedic implants [93]. Degradation products less than 20 μm in diameter can be phagocytosed by macrophages [93]. Degradation particles act on bone cells indirectly through the secretory products of macrophages that are drawn to the area from the immune response [93]. Microparticles of PLLA and PLGA have been shown to suppress osteoblast differentiation early in culture [93]. Degradation particles also can interact directly with osteoblasts and affect their proliferation [65]. In addition, dense fibrous capsules composed of macrophages and foreign-body giant cells have formed in response to PLLA bone plates and screws [10]. The properties of biomaterials clearly affect the magnitude and duration of the host response. Characteristics that can alter the immune response include the size, shape, and chemical and physical properties of the material [84]. Therefore in designing a biomaterial, one must consider not only the initial properties of the scaffold, but also its degradation products and their effect on the host.

## 6.4 Summary

Scaffold design is an intricate process that must be custom tailored for different applications. Scaffolds must induce cell growth, support cell adhesion and proliferation, and provide the mechanical stability needed in different sites. Synthetic polymers can be readily modified so that their properties are appropriate for bone tissue engineering. Their properties are modulated by the fabrication method: both conventional and rapid-prototyping techniques have produced viable bone tissue scaffolds. Fundamental design parameters depend on the needs of the regenerated tissue and include polymer assembly, curing methods, surface properties, macrostructure, mechanical properties, biodegradation, and biocompatibility, both of the scaffold per se and its degradation products.

## References


1. Agrawal CM, Ray RB (2001) Biodegradable polymeric scaffolds for musculoskeletal tissue engineering. J Biomed Mater Res 55:141–150.
2. Allcock HR, Cameron CG (1994) Synthesis and characterization of photo-cross-linkable small-molecule and high-polymeric phosphazenes bearing cinnamate groups. Macromolecules 27:3125–3130.
3. Anderson JM (2001) Biological response to materials. Annu Rev Mater Res 31:81–110.
4. Andriano KP, Tabata Y, Ikada Y, Heller J (1999) In vitro and in vivo comparison of bulk and surface hydrolysis in absorbable polymer scaffolds for tissue engineering. J Biomed Mater Res 48:602–612.
5. Angelova N, Hunkeler D (1999) Rationalizing the design of polymeric biomaterials. Trends Biotechnol 17:409–421.
6. Anseth KS, Svaldi DC, Laurencin CT, Langer R (1997) Photopolymerization of novel degradable networks for orthopedic applications. Photopolymerization 673:189–202.
7. Athanasiou KA, Zhu C, Lanctot DR, Agrawal CM, Wang X (2000) Fundamentals of biomechanics in tissue engineering of bone. Tissue Eng 6:361–381.
8. Barr J, Woodburn KW, Ng SY, Shen HR, Heller J (2002) Post surgical pain management with poly(ortho esters). Adv Drug Deliv Rev 54:1041–1048.
9. Barralet JE, Wallace LL, Strain AJ (2003) Tissue engineering of human biliary epithelial cells on polyglycolic acid/polycaprolactone scaffolds maintains long-term phenotypic stability. Tissue Eng 9:1037–1045.
10. Bergsma JE, de Bruijn WC, Rozema FR, Bos RR, Boering G (1995) Late degradation tissue response to poly(L-lactide) bone plates and screws. Biomaterials 16:25–31.
11. Bourke SL, Kohn J (2003) Polymers derived from the amino acid L-tyrosine: polycarbonates, polyarylates and copolymers with poly(ethylene glycol). Adv Drug Deliv Rev 55:447–466.
12. Burdick JA, Frankel D, Dernell WS, Anseth KS (2003) An initial investigation of photocurable three-dimensional lactic acid based scaffolds in a critical-sized cranial defect. Biomaterials 24:1613–1620.
13. Burkoth AK, Anseth KS (2000) A review of photocrosslinked polyanhydrides: in situ forming degradable networks. Biomaterials 21:2395–2404.
14. Burkoth AK, Burdick J, Anseth KS (2000) Surface and bulk modifications to photocrosslinked polyanhydrides to control degradation behavior. J Biomed Mater Res 51:352–359.
15. Cai J, Bo S, Cheng R, Jiang L, Yang Y (2004) Analysis of interfacial phenomena of aqueous solutions of polyethylene oxide and polyethylene glycol flowing in hydrophilic and hydrophobic capillary viscometers. J Colloid Interface Sci 276:174–181.
16. Cao T, Ho KH, Teoh SH (2003) Scaffold design and in vitro study of osteochondral coculture in a three-dimensional porous polycaprolactone scaffold fabricated by fused deposition modeling. Tissue Eng 9 Suppl 1:S103–112.
17. Chen GP, Ushida T, Tateishi T (2001) Development of biodegradable porous scaffolds for tissue engineering. Materials Science and Engineering C-Biomimetic and Supramolecular Systems 17:63–69.
18. Choueka J, Charvet JL, Koval KJ, Alexander H, James KS, Hooper KA, Kohn J (1996) Canine bone response to tyrosine-derived polycarbonates and poly(L-lactic acid). J Biomed Mater Res 31:35–41.
19. Ciapetti G, Ambrosio L, Savarino L, Granchi D, Cenni E, Baldini N, Pagani S, Guizzardi S, Causa F, Giunti A (2003) Osteoblast growth and function in porous poly epsilon-caprolactone matrices for bone repair: a preliminary study. Biomaterials 24:3815–3824.
20. Cooke MN, Fisher JP, Dean D, Rimnac C, Mikos AG (2003) Use of stereolithography to manufacture critical-sized 3D biodegradable scaffolds for bone ingrowth. Journal of Biomedical Materials Research Part B-Applied Biomaterials 64B:65–69.
21. Cunliffe D, Pennadam S, Alexander C (2004) Synthetic and biological polymers-merging the interface. Eur Polym J 40:5–25.
22. Dai NT, Williamson MR, Khammo N, Adams EF, Coombes AG (2004) Composite cell support membranes based on collagen and polycaprolactone for tissue engineering of skin. Biomaterials 25: 4263–4271.
23. Dar A, Shachar M, Leor J, Cohen S (2002) Optimization of cardiac cell seeding and distribution in 3D porous alginate scaffolds. Biotechnol Bioeng 80: 305–312.
24. Davis KA, Anseth KS (2002) Controlled release from crosslinked degradable networks. Crit Rev Ther Drug Carrier Syst 19:385–423.
25. Drury JL, Mooney DJ (2003) Hydrogels for tissue engineering: scaffold design variables and applications. Biomaterials 24:4337–4351.
26. El-Amin SF, Lu HH, Khan Y, Burems J, Mitchell J, Tuan RS, Laurencin CT (2003) Extracellular matrix production by human osteoblasts cultured on biodegradable polymers applicable for tissue engineering. Biomaterials 24:1213–1221.

27. Ertel SI, Kohn J (1994) Evaluation of a series of tyrosine-derived polycarbonates as degradable biomaterials. J Biomed Mater Res 28:919–930.
28. Fisher JP, Timmer MD, Holland TA, Dean D, Engel PS, Mikos AG (2003) Photoinitiated cross-linking of the biodegradable polyester poly(propylene fumarate). Part I. Determination of network structure. Biomacromolecules 4:1327–1334.
29. Fisher JP, Vehof JW, Dean D, van der Waerden JP, Holland TA, Mikos AG, Jansen JA (2002) Soft and hard tissue response to photocrosslinked poly(propylene fumarate) scaffolds in a rabbit model. J Biomed Mater Res 59:547–556.
30. Giordano RA, Wu BM, Borland SW, Cima LG, Sachs EM, Cima MJ (1996) Mechanical properties of dense polylactic acid structures fabricated by three dimensional printing. J Biomater Sci Polym Ed 8:63–75.
31. Gomes ME, Sikavitsas VI, Behravesh E, Reis RL, Mikos AG (2003) Effect of flow perfusion on the osteogenic differentiation of bone marrow stromal cells cultured on starch-based three-dimensional scaffolds. J Biomed Mater Res A 67:87–95.
32. Gunatillake PA, Adhikari R (2003) Biodegradable synthetic polymers for tissue engineering. Eur Cell Mater 5:1–16; discussion 16.
33. Gutowska A, Jeong B, Jasionowski M (2001) Injectable gels for tissue engineering. Anat Rec 263: 342–349.
34. Hayashi T (1994) Biodegradable polymers for biomedical uses. Prog Polym Sci 19:663–702.
35. Heller J, Barr J, Ng SY, Abdellauoi KS, Gurny R (2002) Poly(ortho esters): synthesis, characterization, properties and uses. Adv Drug Deliv Rev 54:1015–1039.
36. Hoffman AS (2002) Hydrogels for biomedical applications. Adv Drug Deliv Rev 54:3–12.
37. Hollister SJ (2005) Porous scaffold design for tissue engineering. Nat Mater 4:518–524.
38. Holy CE, Fialkov JA, Davies JE, Shoichet MS (2003) Use of a biomimetic strategy to engineer bone. J Biomed Mater Res A 65:447–453.
39. Hsu YY, Gresser JD, Trantolo DJ, Lyons CM, Gangadharam PR, Wise DL (1997) Effect of polymer foam morphology and density on kinetics of in vitro controlled release of isoniazid from compressed foam matrices. J Biomed Mater Res 35:107–116.
40. Hua FJ, Kim GE, Lee JD, Son YK, Lee DS (2002) Macroporous poly(L-lactide) scaffold 1. Preparation of a macroporous scaffold by liquid-liquid phase separation of a PLLA-dioxane-water system. J Biomed Mater Res 63:161–167.
41. Hutmacher DW (2000) Scaffolds in tissue engineering bone and cartilage. Biomaterials 21:2529–2543.
42. Hutmacher DW (2001) Scaffold design and fabrication technologies for engineering tissues—state of the art and future perspectives. J Biomater Sci Polym Ed 12:107–124.
43. Hutmacher DW, Schantz T, Zein I, Ng KW, Teoh SH, Tan KC (2001) Mechanical properties and cell cultural response of polycaprolactone scaffolds designed and fabricated via fused deposition modeling. J Biomed Mater Res 55:203–216.
44. Ishaug-Riley SL, Okun LE, Prado G, Applegate MA, Ratcliffe A (1999) Human articular chondrocyte adhesion and proliferation on synthetic biodegradable polymer films. Biomaterials 20:2245–2256.
45. James K, Levene H, Parsons JR, Kohn J (1999) Small changes in polymer chemistry have a large effect on the bone–implant interface: evaluation of a series of degradable tyrosine-derived polycarbonates in bone defects. Biomaterials 20:2203–2212.
46. Kaito T, Myoui A, Takaoka K, Saito N, Nishikawa M, Tamai N, Ohgushi H, Yoshikawa H (2005) Potentiation of the activity of bone morphogenetic protein-2 in bone regeneration by a PLA-PEG/hydroxyapatite composite. Biomaterials 26:73–79.
47. Kim SS, Utsunomiya H, Koski JA, Wu BM, Cima MJ, Sohn J, Mukai K, Griffith LG, Vacanti JP (1998) Survival and function of hepatocytes on a novel three-dimensional synthetic biodegradable polymer scaffold with an intrinsic network of channels. Ann Surg 228:8–13.
48. Kumar N, Langer RS, Domb AJ (2002) Polyanhydrides: an overview. Adv Drug Deliv Rev 54:889–910.
49. Langer R (2000) Biomaterials in drug delivery and tissue engineering: one laboratory's experience. Acc Chem Res 33:94–101.
50. Langer R, Vacanti JP (1993) Tissue engineering. Science 260:920–926.
51. Laurencin CT, El-Amin SF, Ibim SE, Willoughby DA, Attawia M, Allcock HR, Ambrosio AA (1996) A highly porous 3-dimensional polyphosphazene polymer matrix for skeletal tissue regeneration. J Biomed Mater Res 30:133–138.
52. Laurencin CT, Norman ME, Elgendy HM, el-Amin SF, Allcock HR, Pucher SR, Ambrosio AA (1993) Use of polyphosphazenes for skeletal tissue regeneration. J Biomed Mater Res 27:963–973.
53. Lee SH, Kim BS, Kim SH, Kang SW, Kim YH (2004) Thermally produced biodegradable scaffolds for cartilage tissue engineering. Macromol Biosci 4:802–810.
54. Lee SJ, Choi JS, Park KS, Khang G, Lee YM, Lee HB (2004) Response of MG63 osteoblast-like cells onto polycarbonate membrane surfaces with different micropore sizes. Biomaterials 25:4699–4707.
55. Lu L, Garcia CA, Mikos AG (1999) In vitro degradation of thin poly(DL-lactic-co-glycolic acid) films. J Biomed Mater Res 46:236–244.
56. Lu L, Mikos AG (1996) The importance of new processing techniques in tissue engineering. MRS Bull 21:28–32.
57. Lu Y, Chen SC (2004) Micro and nano-fabrication of biodegradable polymers for drug delivery. Adv Drug Deliv Rev 56:1621–1633.
58. Lutolf MP, Weber FE, Schmoekel HG, Schense JC, Kohler T, Muller R, Hubbell JA (2003) Repair of bone defects using synthetic mimetics of collagenous extracellular matrices. Nat Biotechnol 21:513–518.
59. Ma PX, Zhang R (1999) Synthetic nano-scale fibrous extracellular matrix. J Biomed Mater Res 46:60–72.
60. Mapili G, Lu Y, Chen S, Roy K (2005) Laser-layered microfabrication of spatially patterned functionalized tissue-engineering scaffolds. J Biomed Mater Res B Appl Biomater.
61. Mathieu LM, Mueller TL, Bourban PE, Pioletti DP, Muller R, Manson JA (2005) Architecture and properties of anisotropic polymer composite scaffolds for bone tissue engineering. Biomaterials.

62. Matsushita N, Terai H, Okada T, Nozaki K, Inoue H, Miyamoto S, Takaoka K (2004) A new bone-inducing biodegradable porous beta-tricalcium phosphate. J Biomed Mater Res A 70:450–458.
63. Middleton JC, Tipton AJ (2000) Synthetic biodegradable polymers as orthopedic devices. Biomaterials 21:2335–2346.
64. Mikos AG, Bao Y, Cima LG, Ingber DE, Vacanti JP, Langer R (1993) Preparation of poly(glycolic acid) bonded fiber structures for cell attachment and transplantation. J Biomed Mater Res 27:183–189.
65. Mikos AG, McIntire LV, Anderson JM, Babensee JE (1998) Host response to tissue engineered devices. Adv Drug Deliv Rev 33:111–139.
66. Mikos AG, Sarakinos G, Leite SM, Vacanti JP, Langer R (1993) Laminated three-dimensional biodegradable foams for use in tissue engineering. Biomaterials 14:323–330.
67. Mikos AG, Thorsen AJ, Czerwonka LA, Bao Y, Langer R, Winslow DN, Vacanti JP (1994) Preparation and characterization of poly(L-lactic acid) foams. Polymer 35:1068–1077.
68. Muggli DS, Burkoth AK, Keyser SA, Lee HR, Anseth KS (1998) Reaction behavior of biodegradable, photo-cross-linkable polyanhydrides. Macromolecules 31:4120–4125.
69. Nam YS, Park TG (1999) Biodegradable polymeric microcellular foams by modified thermally induced phase separation method. Biomaterials 20: 1783–1790.
70. Nam YS, Park TG (1999) Porous biodegradable polymeric scaffolds prepared by thermally induced phase separation. J Biomed Mater Res 47:8–17.
71. Nam YS, Yoon JJ, Park TG (2000) A novel fabrication method of macroporous biodegradable polymer scaffolds using gas foaming salt as a porogen additive. J Biomed Mater Res 53:1–7.
72. Novikova LN, Novikov LN, Kellerth JO (2003) Biopolymers and biodegradable smart implants for tissue regeneration after spinal cord injury. Curr Opin Neurol 16:711–715.
73. Oh SH, Kang SG, Kim ES, Cho SH, Lee JH (2003) Fabrication and characterization of hydrophilic poly(lactic-co-glycolic acid)/poly(vinyl alcohol) blend cell scaffolds by melt-molding particulate-leaching method. Biomaterials 24:4011–4021.
74. Park YJ, Lee JY, Chang YS, Jeong JM, Chung JK, Lee MC, Park KB, Lee SJ (2002) Radioisotope carrying polyethylene oxide-polycaprolactone copolymer micelles for targetable bone imaging. Biomaterials 23:873–879.
75. Qiu LY, Zhu KJ (2000) Novel biodegradable polyphosphazenes containing glycine ethyl ester and benzyl ester of amino acethydroxamic acid as cosubstituents: syntheses, characterization, and degradation properties. J Appl Polym Sci 77:2987–2995.
76. Rohl L, Larsen E, Linde F, Odgaard A, Jorgensen J (1991) Tensile and compressive properties of cancellous bone. J Biomech 24:1143–1149.
77. Rohner D, Hutmacher DW, Cheng TK, Oberholzer M, Hammer B (2003) In vivo efficacy of bone-marrow-coated polycaprolactone scaffolds for the reconstruction of orbital defects in the pig. J Biomed Mater Res B Appl Biomater 66:574–580.
78. Sachlos E, Czernuszka JT (2003) Making tissue engineering scaffolds work. Review: the application of solid freeform fabrication technology to the production of tissue engineering scaffolds. Eur Cell Mater 5:29–39; discussion 39–40.
79. Sawhney AS, Pathak CP, Hubbell JA (1993) Bioerodible hydrogels based on photopolymerized poly(ethylene glycol)-co-poly(alpha-hydroxy acid) diacrylate macromers. Macromolecules 26:581–587.
80. Schaefer D, Martin I, Shastri P, Padera RF, Langer R, Freed LE, Vunjak-Novakovic G (2000) In vitro generation of osteochondral composites. Biomaterials 21:2599–2606.
81. Seal BL, Otero TC, Panitch A (2001) Polymeric biomaterials for tissue and organ regeneration. Materials Science and Engineering R-Reports 34: 147–230.
82. Sheridan MH, Shea LD, Peters MC, Mooney DJ (2000) Bioabsorbable polymer scaffolds for tissue engineering capable of sustained growth factor delivery. J Control Release 64:91–102.
83. Shin H, Jo S, Mikos AG (2003) Biomimetic materials for tissue engineering. Biomaterials 24:4353–4364.
84. Shive MS, Anderson JM (1997) Biodegradation and biocompatibility of PLA and PLGA microspheres. Adv Drug Deliv Rev 28:5–24.
85. Simon JL, Roy TD, Parsons JR, Rekow ED, Thompson VP, Kemnitzer J, Ricci JL (2003) Engineered cellular response to scaffold architecture in a rabbit trephine defect. J Biomed Mater Res A 66:275–282.
86. Sims CD, Butler PE, Casanova R, Lee BT, Randolph MA, Lee WP, Vacanti CA, Yaremchuk MJ (1996) Injectable cartilage using polyethylene oxide polymer substrates. Plast Reconstr Surg 98:843–850.
87. Taboas JM, Maddox RD, Krebsbach PH, Hollister SJ (2003) Indirect solid free form fabrication of local and global porous, biomimetic and composite 3D polymer-ceramic scaffolds. Biomaterials 24: 181–194.
88. Tangpasuthadol V, Pendharkar SM, Kohn J (2000) Hydrolytic degradation of tyrosine-derived polycarbonates, a class of new biomaterials. Part I: Study of model compounds. Biomaterials 21:2371–2378.
89. Thomson RC, Wake MC, Yaszemski MJ, Mikos AG (1995) Biodegradable polymer scaffolds to regenerate organs. Biopolymers II 122:245–274.
90. Thomson RC, Yaszemski MJ, Powers JM, Mikos AG (1995) Fabrication of biodegradable polymer scaffolds to engineer trabecular bone. J Biomater Sci Polym Ed 7:23–38.
91. Uhrich KE, Gupta A, Thomas TT, Laurencin CT, Langer R (1995) Synthesis and characterization of degradable poly(anhydride-co-imides). Macromolecules 28:2184–2193.
92. Vehof JW, Fisher JP, Dean D, van der Waerden JP, Spauwen PH, Mikos AG, Jansen JA (2002) Bone formation in transforming growth factor beta-1-coated porous poly(propylene fumarate) scaffolds. J Biomed Mater Res 60:241–251.
93. Wake MC, Gerecht PD, Lu L, Mikos AG (1998) Effects of biodegradable polymer particles on rat marrow-derived stromal osteoblasts in vitro. Biomaterials 19:1255–1268.
94. Whang K, Tsai DC, Nam EK, Aitken M, Sprague SM, Patel PK, Healy KE (1998) Ectopic bone formation

via rhBMP-2 delivery from porous bioabsorbable polymer scaffolds. J Biomed Mater Res 42:491–499.
95. Wolfe MS, Dean D, Chen JE, Fisher JP, Han S, Rimnac CM, Mikos AG (2002) In vitro degradation and fracture toughness of multilayered porous poly(propylene fumarate)/beta-tricalcium phosphate scaffolds. J Biomed Mater Res 61:159–164.
96. Wu BM, Borland SW, Giordano RA, Cima LG, Sachs EM, Cima MJ (1996) Solid free-form fabrication of drug delivery devices. J Control Release 40: 77–87.
97. Yang S, Leong KF, Du Z, Chua CK (2001) The design of scaffolds for use in tissue engineering. Part I. Traditional factors. Tissue Eng 7:679–689.
98. Yang S, Leong KF, Du Z, Chua CK (2002) The design of scaffolds for use in tissue engineering. Part II. Rapid prototyping techniques. Tissue Eng 8:1–11.
99. Yaszemski MJ, Payne RG, Hayes WC, Langer R, Mikos AG (1996) Evolution of bone transplantation: molecular, cellular and tissue strategies to engineer human bone. Biomaterials 17:175–185.
100. Zein I, Hutmacher DW, Tan KC, Teoh SH (2002) Fused deposition modeling of novel scaffold architectures for tissue engineering applications. Biomaterials 23:1169–1185.
101. Zhang K, Wang Y, Hillmyer MA, Francis LF (2004) Processing and properties of porous poly(L-lactide)/bioactive glass composites. Biomaterials; 25:2489–2500.

# 7. Injectable Scaffolds for Bone and Cartilage Regeneration

*Claudio Migliaresi, Antonella Motta, and Anthony T. DiBenedetto*

## 7.1 Introduction

Every year hundreds of thousands people worldwide receive hip prostheses, implants for bone repair, and surgical repair of degraded cartilage. "Over 15 million people worldwide suffer from knee-joint failure each year due to the breakdown of surrounding cartilage in the joint and the inability of this cartilage to repair itself through the natural regenerative processes of healing in the body" [27]. Additionally, at least 10 percent of the population suffers from periodontal disease, and one-third of these individuals will require a tooth implant during their lifetime. The standard procedure for repair of orthopedic injuries by tissue grafting is to harvest tissue from the iliac crest or femur of a patient and surgically placing it at the injury site [59]. A similar approach has been developed for the strengthening and rebuilding of a jaw bone by removing periosteal cells from the patient's jaw, cultivating them in autologous blood serum mixed with a matrix substance, and placing the mixture in the degraded jaw bone [42]. Although these procedures are ideal from the point of view of new bone growth in terms of osteoconductivity or osteoinductivity, both the harvesting of the tissue and the placement of the graft require invasive surgery that may result in significant complications. Furthermore, preformed scaffolds used as a host for cells in vitro normally have simple geometric shapes that do not readily conform to irregularly shaped defects when implanted at a site in the body [66].

Bone graft substitutes have been classified into five categories: allograft-based formulations using allograft bone, alone or in combination with other materials; factor-based formulations using natural and recombinant growth factors, alone or in combination with other materials; cell-based formulations using cells to generate new tissue, alone or seeded on a support matrix; ceramic-based formulations using calcium phosphate and other ceramics, alone or in combination; and polymer-based formulations using both biodegradable and nondegradable polymers, alone or in combination with other materials (see Chapter 5). Approximately 60% of the bone graft substitutes currently available contain ceramics, either alone or in combination with another material [59].

In many clinical situations involving replacement of hard or soft tissue, the aims of minimizing the need for invasive surgery, avoiding the medical complications associated with harvested tissue, and overcoming the limitations of preformed scaffolds have assumed primary importance. The use of noninvasive, injectable scaffolds responds to these concerns. When properly designed, an injectable scaffold can provide a structure that encapsulates a homogeneous distribution of cells and bioactive molecules that stimulate the regeneration of bone and cartilage in a biomimetic fashion. Silica-based ceramics, calcium- and phosphate-based solids, natural and synthetic polymers, and composites containing one or more of these materials have been fabricated as aqueous solutions, pastes, and gels that can be injected directly into an injured site and then solidified chemically, thermoreversibly, or by other means. In all cases, bone morphogenetic protein (BMP), basic fibroblast growth factor (β-FGF/FGF-2), cells to generate tissue formation, and a variety of other additives can be included in the scaffold mixture to enhance bioactivity. Reviews of materials for tissue replacement have been published [11, 14, 32, 37, 58, 66, 68, 69, 85, 88, 96].

The term "injectable scaffold" requires clarification, since not all such materials proposed in the literature are injectable in a noninvasive sense. For example, many of the injectable ceramic/polymer composites are mixtures of a particulate calcium phosphate compound dispersed in a moldable polymer matrix. The design of a one- or two-syringe/needle device for dispensing such a product is determined by the nature of the application, the size, shape, and concentration of the particles, and the viscoelastic response and chemical reactivity of the composite. The diameter of the needle required, for example, will certainly be limited by these factors. In addition, the manner in which the scaffold accomplishes the specific interaction between the biomaterial and the local and systemic tissues requires specification. The only officially accepted definition of a scaffold is (ASTM: F2150) [2A]"the support, delivery vehicle, or matrix for facilitating the migration, binding, or transport of cells or bioactive molecules used to replace, repair, or regenerate tissues."

## 7.2 Necessary Properties of an Injectable Scaffold

In order to develop an acceptable injectable scaffold for orthopedic applications in which the regeneration of bone and cartilage is stimulated by active cells within the scaffold, a number of fundamental biological, mechanical and morphological conditions must be satisfied, including biocompatibility, biological character, sterilizability, and viscous and viscoelastic properties.

### 7.2.1 Biocompatibility

As with all biomaterials used in the human body, the components of an injectable scaffold must be biocompatible. *This means that the material must not elicit an unresolved inflammatory response or demonstrate extreme immunogenicity or cytotoxicity* [96]. A biodegradable material is preferred in most cases. The degradation products of the scaffold must also be biocompatible so that they can be eliminated from the body in an appropriate period of time. A generally accepted definition of biocompatibility for tissue-engineered products is as follows: "The biocompatibility of a scaffold or matrix for a tissue-engineering product refers to the ability to perform as a substrate that will support the appropriate cellular activity, including the facilitation of molecular and mechanical signalling systems, in order to optimize tissue regeneration, without eliciting any undesirable effects in those cells, or inducing any undesirable local or systemic responses in the eventual host. They actively participate in the signalling process, usually with the requirement of safe degradation as part of the process." [100, 101].

### 7.2.2 Biological Character

Although the scaffold matrix itself need not be bioactive, it should provide a positive environment for cell activity, including enhanced cell adhesion, migration, and function, as well as vascularization (where appropriate) and free space for bone or other tissue growth. A primary objective should be that the scaffold creates a biomimetic system that resembles as closely as possible that of the host.

### 7.2.3 Sterilizability

As with all implanted materials, an injectable scaffold material must be easily sterilized to prevent infection when implanted. The method of sterilization must not negatively affect chemical composition, biocompatibility, or bioactivity [96].

### 7.2.4 Viscous and Viscoelastic Properties

Initially, an injectable scaffold material must be a fluid, stable liquid/solid dispersion or gel that can be injected through a needle of the size required by the specific application. Once the material is injected, its state must change to an elastic or viscoelastic solid in order for it to remain intact at the defect site and eventually be capable of supporting a load. There are several ways of accomplishing this physical change. One technique is to utilize a thermoreversible system that is a liquid solution (sol) at injection temperature and a viscoelastic solid or gel in situ at body temperature. Another possibility is to employ a thixotropic fluid, paste, or reversible gel that is sufficiently shear-thinned during injection to flow through the required needle yet maintains sufficient elasticity in situ to be retained at the defect site. Another technique is to chemically or ionically crosslink the injectable fluid during the placement procedure. In all cases, the reaction time must be short enough to set the material before it flows from the placement site. The temperature change occurring during the change in state must be small enough to avoid or minimize damage to the surrounding tissue. If the scaffold matrix is a charged, water-soluble polymer, it may undergo a sol–gel transition in response to a pH change. If appropriate, the pH-temperature phase behavior may be utilized as a mechanism for injection and hardening of the scaffold [37].

## 7.3 Ceramic-Based Injectable Scaffolds

Calcium phosphate ceramics, such as hydroxyapatite, tricalcium phosphate (TCP), biphasic calcium phosphate (BCP), and bioactive glasses (BG), in combination with a variety of biodegradable polymeric matrices, have been extensively studied and used during the past decades as alternatives to autogenous bone for repair, substitution, or augmentation [20, 39]. In the injectable versions of scaffolds, these biomaterials are dispersed in water or polymeric solutions that serve as modifiers of rheological properties or as binding agents. Alginates, chitosan, cellulose derivatives, silicone oils, biodegradable polyester copolymers, and a variety of other biocompatible, biodegradable polymers have been studied for this purpose [37, 46, 71, 90]. Since their description in the 1970s and 1980s, numerous formulations of resorbable calcium phosphate cements have been investigated and commercialized [13, 26, 39].

Resorbable calcium phosphate-based composite scaffolds generate bone trabeculae, provided the rate of resorption of the calcium phosphate is sufficiently slow for osteoblasts to be able to regenerate new bone [8]. Research carried out in the late 1980s and early 1990s [22, 23, 24, 36] has stimulated considerable interest in the use of particulate calcium phosphate-based ceramic mixtures for bone reconstructive surgery [22, 30, 31, 40, 55, 56, 63, 79, 88, 102]. When their compositions are designed to match as closely as possible the mineral composition of natural bone, their biological responses closely mimic those of the inorganic phase of natural bone.

With regard to the necessary properties of ceramic biomaterial scaffolds in general, two of the unique factors are porosity and bioresorbability [8]. The bioceramic material provides an osteoconductive scaffold for the growth of new bone. For many applications, the particulate phase consists of micron-sized porous particles with a broad distribution of pore size. Pores of less than about 5 μm are important for bioresorbability, whereas pores in the range of 400 to 600 μm facilitate the infiltration and differentiation of osteogenic cells necessary for bone reconstruction [8]. The morphology of the porous structure also influences that of the regenerated bone, inasmuch as the penetrated fibrovascular tissue moves in the direction of the pore channels. An interconnecting pore structure is thought to be superior to that of isolated pores, because it better provides for spatial continuity of the new bone [8, 79]. Since macropores are required for bone reconstruction, the range of particle sizes required of a scaffold is at least on the order of 100 to 500 μm. A scaffold material containing an appropriate amount of particulate matter of these dimensions is not easily injectable and therefore often has to be surgically implanted. To provide for both injectability and a more natural path for fibrovascular tissue, particles with diameters on the order of 10 to 100 μm have been used that allow the invading tissue to grow over and around the surfaces of the particles in response to the mechanical stresses that promulgate growth and remodeling processes [30, 56]. In these situations, the total particulate surface area partially replaces the role of macropores. Bioresorption is the process by which the ceramic material dissolves into its ionic components in the physiological fluid. The remodeling process occurs simultaneously, generating new bone–ceramic interfaces that control the rate of formation of new bone. If the resorption activity of osteoclasts and the associated osteoblastic activity are balanced by the proper choice of a ceramic, the remodeling process appears to be at its maximum effectiveness. Therefore, after complete resorption of the ceramic, the remodeled bone is stronger than new bone/ceramic mix formed with nonresorbing biomaterial [8, 20, 40, 88, 102]. The remodeling process of the host bone and the resorption of the ceramic are affected by the phagocytosing cells of the host and are a function of the microporosity and chemical makeup of the ceramic [8, 20]. These processes are modulated by the osteoclast-induced degradation of a calcium phosphate ceramic [40, 88, 102]. The attachment of osteoclasts to the ceramic surface is mediated largely by extracellular matrix (ECM) proteins. In addition, the capability of the osteoclasts to resorb calcium phosphate ceramic appears to be related to the solubility of the ceramic [88]. After attachment to the ceramic surface, the osteoclasts create a sealed extracellular microenvironment into which calcium ions are released. As resorption proceeds, $[Ca^{2+}]_i$ increases in the fluid inside the podosome membrane. Beyond a certain $[Ca^{2+}]_i$ value, resorption ceases and the osteoclast migrates away. A highly soluble ceramic such as β-TCP leads to ineffective remodelling of the defect site. Therefore mixtures of low-solubility hydroxyapatite (HA) with high solubility β-tricalcium phosphate (β-TCP) have been used to tailor the osteoclastic process to the requirements of a particular scaffold application [8, 102].

The "bioactive concept" for biphasic calcium phosphate ceramics (BCP) is based on an optimum balance of the more stable phase of HA and the more soluble TCP [20, 22, 36]. The biodegradation of implanted particles or blocks of BCP results in an increase in the HA/β-TCP ratio, a decrease in the average size of the BCP crystals, and an increase in macroporosity of the surface, as well as in the core of the ceramic. The formation of microcrystals with Ca/P ratios similar to those of bone apatite crystals occurs simultaneously. The abundance of these crystals is directly related to the HA/β-TCP ratio in the BCP. This indicates that it is possible to regulate the kinetics of dissolution and precipitation and therefore the bioactivity. "The coalescing interfacial zone of biological apatite and residual crystals provides a scaffold for bone cell adhesion and further bone ingrowth" [20, 63]. The "bioactive concept" is thus based on the assertion that the dissolution/transformation processes are applicable to bulk, coating and injectable ceramic-based biomaterials, and that the events at the calcium phosphate biomaterial/bone interfaces represent a dynamic process that ultimately contributes to the unique strength of these interfaces [20].

The above-mentioned work has led to a variety of commercially available BCP products for bone graft or bone substitutes in orthopedics and dentistry [21]. In vivo studies of a number of injectable bone substitutes (IBSs) have been carried out [30]. This work utilized a 50:50 weight ratio mineral phase of BCP granules, (40 to 80 μm or 80 to 200 μm in diameter) dispersed in a 3% aqueous solution of hydroxypropylmethylcellulose (HPMC). The injectability and properties of an IBS (with 40- to 80-μm particles) were compared with those of a calcium phosphate cement composed of ~10 μm calcium phosphate granules. Calcium phosphate cement was found to be more readily injectable than the IBS when needles with inside diameters of 0.84 or 0.61 mm were used; when needles of 0.51 or 0.41 mm were used, however, calcium phosphate cement was not injectable, whereas the IBS material remained injectable.

In vivo studies on the two IBSs were compared with those performed on calcium phosphate cement [30]. Ten New Zealand white rabbits were injected with bone substitutes implanted into critical-sized defects at the distal end of the femur. Three weeks after implantation, scanning electron microscopy (SEM) indicated that newly formed bone developed throughout the defect volume and in the intergranular spaces that surrounded the BCP granules in the IBS. With the calcium phosphate cement, on the other hand, newly formed bone only developed on the surface of the microporous cement. Bone formation was also significantly greater in the faster-degrading IBS. Microtomography was used to show, in a three-dimensional reconstruction, that newly formed bone developed homogeneously in the intergranular spaces in the IBSs, resulting in an interconnected new bone network (Fig. 7.1). In the calcium phosphate cement, on the other hand, new bone developed only on the cement surface and in peripheral fissures (Fig. 7.2). An in vivo study using the same biomaterials, but with BCP particle sizes of 200 to 500 μm and 80 to 200 μm, yielded similar results, but with greater emphasis on the analytical capabilities of three-dimensional microtomographic imaging and two-dimensional SEM and their correlation with biomechanical properties. "The study showed the ability of non-destructive techniques to investigate biological and mechanical aspects of bone replacement with injectable biomaterials." [31].

If the setting rate of the ceramic or the rate of bone regeneration is too slow, the scaffold may not have sufficient mechanical strength. The control of HA/β-TCP in biphasic calcium phosphates has helped overcome problems of this sort, whereas other investigators have chosen to use compositions of ceramic pastes that set rapidly with high initial compressive strength. The most studied of these are the dicalcium phosphate dihydrate (DCPD) cements and the calcium-deficient hydroxyapatite (CDHA) cements [1, 18, 33, 43, 95, 96].

An injectable, resorbable apatitic calcium phosphate substitute was found to undergo full resorption within 1 to 2 months following surgical implantation in the femur of young adult male beagle dogs [55]. Trabecular bone was formed within the first 4 weeks after implantation, and lamellar or cortical bone was well established after 12 weeks, with no noticeable demarcation between old and new bone. By 26 weeks, restoration of the native bone was virtually complete. No discernible differences were observed between the regenerated bone and those developed with the use of autografting and calcium phosphate bone substitute techniques. These materials are distributed by Etex Corporation in the United States with the tradename of α-BSM® and in Europe with the name of Biobon® [55].

An injectable calcium phosphate cement that hardens in vivo to form a carbonated apatite (dahllite), was developed [19] and commercial-

**Figure 7.1.** Reconstructed microtomographic images of injectable bone substitute (IBS) before (A) and 3 weeks after (B and C) implantation in femoral defects. New bone is shown in gray, biphasic calcium phosphate (BCP) ceramic in white, and soft tissues in black (bars: 100 μm). (A) General aspect of the composite IBS before implantation. The presence of the polymer confers its rheologic properties on the IBS and manages interconnected intergranular spaces. (B) Image showing bone ingrowth in a femoral defect filled with IBS containing 80 to 200 μm BCP particles. New bone formed an extensive interconnected network in the intergranular spaces. (C) Image showing bone ingrowth in a femoral defect filled with IBS containing 40 to 80 μm BCP particles. A new bone network joined the BCP particles to one another, incorporating the remaining BCP particles in a large amount of newly formed bone. Reprinted with permission from Gauthier et al. [30].

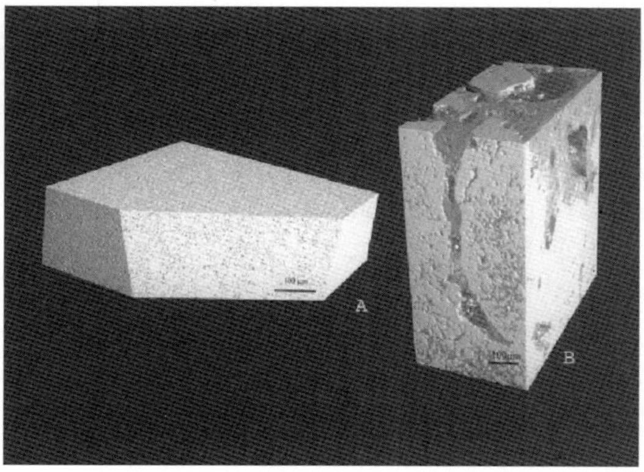


**Figure 7.2.** Reconstructed microtomographic images of the calcium phosphate content (CPC) 3 weeks after implantation in femoral defects. New bone is shown in gray, CPC in white, and soft tissues in black (bars: 100 mm). (A) General aspect of the central area of the CPC after implantation, showing the absence of bone substitution with only microporosity. (B) Bone ingrowth developed only on the surface of the cement and in peripheral fissures. Reprinted with permission from Gauthier et al. [30].

ized as a Skeletal Repair System (SRS®, Norian Corporation, Cupertino, CA). The product is used mostly in orthopedic and trauma applications. The material is said to allow normal fracture healing by resorption via normal cellular remodeling and to maintain strength during the remodeling process. Norian SRS is delivered to the surgery room as a mixture of calcium phosphate and carbonate. After being mixed with a sodium phosphate solution, it forms an injectable paste that hardens in about 10 minutes, with a compressive strength of about 10 MPa that then increases to about 55 MPa in 12 hours. Tensile and shear strengths, however, are much lower, attaining a maximum of the order of 2 to 3 MPa [19, 58]. Norian has also been used in craniofacial surgery [67].

A variety of water-soluble gelling agents have been employed to produce a fully injectable calcium phosphate cement. A necessary condition for their use is that they do not substantially affect the setting properties of the cement. Aqueous solutions of glycerine, derivatives of cellulose, and salts of alginic acid were examined as modifiers for an apatitic calcium phosphate cement [56]. The addition of glycerine or cellulose derivatives resulted in sticky pastes that did not set properly, nor was there satisfactory conversion of the ingredients to hydroxyapatite. The addition of salts of alginic acid imparted satisfactory flow properties to the cement and did not affect the setting reaction drastically, even though the phase conversion was retarded somewhat. A 2% w/w ratio of the gelling agent permitted smooth flow through an 18-gauge needle and resulted in a setting time of 20 minutes and a mean compressive strength of 11 to 12 MPa, a value comparable to that of trabecular bone. X-ray diffraction and Fourier transform infrared analyses indicated that the cement had been converted to a hydroxyapatite structure that was similar to vertebrate bone [56].

The utilization of aqueous solutions of natural polymers to modify the rheological properties of ceramic-based scaffolds has also led to the introduction of bioactive ingredients in ceramic bone scaffolds [9, 12, 56, 64, 77, 80, 84, 103].

Anorganic bone mineral (ABM) particles coated with the cell-binding domain of type I collagen (P-15 peptide) and suspended in injectable hyaluronate hydrogels were tested in vitro by using human osteosarcoma cells to evaluate cell–material interaction and osteoblastic activity. The addition of coated ABM particles promoted cell adhesion, enhanced osteoblastic activity, and increased matrix mineralization [77].

An in vitro study of the morphology of nucleus pulposus cells seeded onto gelatin, demineralized bone matrix (DBM), and polylactide scaffolds showed that cells attached to gelatin microcarriers and DBM fragments assumed an elongated, fibroblast-like morphology, retained metabolic activity, and expressed genes for major ECM components. Both the gelatin and DBM are said to have potential for use in injectable composites for intervertebral disc tissue engineering [12].

To provide preventive antibiotic therapy in vivo, tetracycline, at concentrations up to 7% wt, was added to an injectable calcium phosphate cement that contained silicone (2% by weight). The combination of tetracycline and silicone caused setting time to increase and mechanical properties to diminish [84].

All injectable materials to be used in bone repair must be sterilizable. Injectable calcium phosphate cements are often sterilized by gamma radiation, which does not affect chemical reactivity [103]. The natural and synthetic polymer gels and solutions used for modification of the viscoelastic properties of the IBSs, however, undergo changes because polymers tend to be subject to scission and/or cross linking under hydrolytic conditions. Polymers used in these applications must therefore be tested after sterilization with gamma radiation, steam, or ultrafiltration [103].

## 7.4 Hydrogel-Based Injectable Scaffolds

Hydrogels have proven to be effective as therapeutic delivery devices of cells and growth factors for soft-tissue engineering applications. From a biological viewpoint, aqueous gels make ideal porous scaffolds when load-bearing support is either unnecessary or otherwise available. These scaffolds possess the cohesive properties of soft solids and permit diffuse liquid transport. As viscoelastic solids, they have low static and dynamic moduli because of high water content and high permeability for oxygen, nutrients, and other water-soluble metabolites [78, 92]. If properly designed, natural and synthetic hydrogel scaffolds can function biomimetically, exhibit biocompatibility, and cause minimal inflammatory responses, thrombosis, and tissue damage [34, 35, 78].

An ideally designed hydrogel scaffold will behave like an extracellular matrix with an aqueous "matrix" to encapsulate the osmotically active components and provide a mechanical fiberlike network that supports extant mechanical stresses. Hydrophilic and hydrophilic/hydrophobic polymers absorb large quantities of water. By modifying the hydrophilic/hydrophobic ratio of the polymer, one can control the concentration of the aqueous phase over a wide range. The mechanical fiberlike network that determines the viscoelastic behavior of the hydrogel is generated by either a permanent, covalently cross-linked structure (an irreversible hydrogel) or a nonpermanent, hydrogen-bonded skeletal network (a reversible hydrogel) [92]. Appropriate viscoelastic characteristics can be developed at polymer concentrations as low as a few weight percent. These highly swollen structures retain high permeability for oxygen and other water-soluble nutrients and metabolites. Their use as injectable scaffolds for bone and tissue repair is particularly interesting because, when injected into an irregularly shaped defect, they can readily wet all surfaces of the injured site and create a low-density aqueous cavity that contains all the components necessary for bone and tissue regeneration. If the spreading of the fluid also promotes adhesion to the surfaces of the defect cavity, the hydrogel is likely to protect the defect surfaces from unwanted soft tissue that contains undesirable cellular elements, maintaining at the same time an osteoconductive and osteogenic-like environment within the scaffold. Under these conditions, new tissue can form at the old bone–tissue interface and on the skeletal network of the scaffold. Regenerated trabecular bone formed under these conditions seems to mesh cleanly with the original bone structure, with no visible transition between the old and the new bone, and a final structure that is close to that of the original bone (Fig. 7.3) [28].

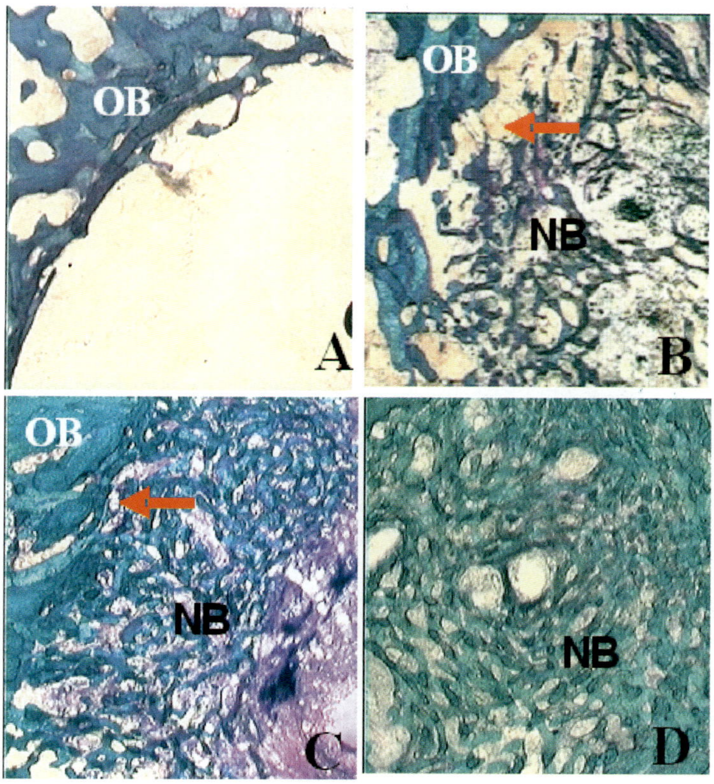


**Figure 7.3.** Histological sections of the untreated and treated defects at 4 weeks. (A) Untreated, (B) treated with commercial gel, (C) treated with silk fibroin hydrogel, (D) treated with silk fibroin hydrogel with full recovery. Red arrows indicate the interface between old bone (OB) and new bone (NB). (A) The formation of NB in the untreated cavity was restricted to the edge of the defect. (B) Newly formed bone was generated radially inward from the defect surface, leaving a distinct interphase between OB and NB. (C) Newly formed bone was generated radially inward from the defect surface, with no noticeable interface between OB and NB. (D) Defect cavity completely filled with NB with no noticeable interface between OB and NB. Unpublished figures from Fini, Motta, Torricelli, Giavaresi, Aldini, Giardino, and Migliaresi, based on work described in Fini et al. [28].

Synthetic and natural hydrogels have been used for a wide variety of medical applications [4, 6, 11, 37, 54, 62, 91]. Among the synthetic polymers considered for use as injectable carriers for tissue-engineering applications are a variety of hydrophilic/hydrophobic diblock and triblock copolymer combinations of poly(lactic acid) (PLA), poly(glycolic acid) (PGA), poly(lactic-co-glycolic) acid, (PLGA) and poly(ethylene glycol) (PEG) [46, 47, 48, 49, 50, 51, 61, 83] and copolymers of poly(ethyleneoxide) (PEO) and (poly)propyleneoxide (PPO), under the commercial names of Pluronics and Polyoxamer [15, 71, 94].

The use of cross-linkable polypropylene fumarate as an injectable polymer to fill defects in cancellous bone has been well documented in the literature [82, 96]. An oligomeric copolymer synthesized from fumaryl chloride and polyethylene glycol, oligo(poly(ethylene glycol)-fumarate) (OPF), has been studied as an injectable carrier for cartilage tissue regeneration [41, 81, 90, 97]. The repeating glycol units on the OPF impart water solubility to the material, and the repeating fumarate double bonds facilitate cross linking. The material is injectable, biodegradable, and cross-linkable in situ. Gelatin microparticles loaded with transforming growth factor $\beta_1$ (TGF-$\beta_1$) were mixed with OPF and a cross-linking agent poly(ethylene glycol)diacrylate into a buffered saline phosphate solution that contained thermal radical initiators and $9 \times 10^6$ chondrocyte cells/ml [81]. The suspensions were then injected into molds to form individual hydrogel constructs. The aim of the work was to determine the effect of the microparticles and TGF-$\beta_1$ on the in vitro proliferation and glycosaminoglycan (GAG) production of chondrocytes encapsulated in the hydrogels. A synergistic effect of having the two components together in the hydrogel was observed; the composite structures exhibited a 7.9-fold increase in cell numbers over a 28-day period, with significant production in the first 7 days, whereas control specimens containing unloaded microparticles did not show significant increases in the first 14 days, and their production was slower overall than that of the loaded microparticles. Control specimens without microparticles did not exhibit significant increases in cellularity over 28 days. The authors speculate that the gelatin microparticles may promote cell proliferation by providing sites for chondrocyte attachment [81].

In an earlier study, three OPF implants using scaffolds of PEG(MW 930)/OPF(MW4470) and one using a scaffold of PEG(MW 6090)/OPF(MW14430) were implanted into holes (~6.3 mm) drilled into the parietal cranial bones of mature female New Zealand White rabbits [90]. The animals were sacrificed at 4 and 12 weeks, and the bones were sectioned and analyzed histologically. In all implants, a limited fibrous capsule formation was observed after 4 weeks. Low numbers of inflammatory cells and macrophages were seen at implant–tissue interfaces; this observation confirms that the chemically cross-linked hydrogels were evoking a mild tissue response. At 12 weeks, only implants with a scaffold of PEG(MW 6090)/OPF(MW14430) exhibited a high number of inflammatory cells. The presence of inflammatory cells led to fragmentation of the hydrogel and extensive surface erosion. The authors claim that the degradation rate can be controlled by tailoring the macromolecular structure of the hydrogel, making it a desirable material for a biodegradable scaffold for tissue engineering [90].

A thermoreversible copolymer, poly(N-isopropylacrylamide-co-acrylic acid) [P(NiPAAm-co-AcA)], is a potential hydrogel carrier for agents that promote soft-tissue renewal, specifically chondrocytes [45]. The copolymer was synthesized in solution by a free radical polymerization. The dynamic viscoelastic properties of a 10 wt% solution of copolymer (in 0.1 M phosphate-buffered saline) indicated a lower critical solution temperature around 35°C and a relatively sharp sol–gel transition at around 35.5°C; the gel continued to harden further over the next 3°C. An important characteristic of this gel is that it exhibits stability upon dilution, so that when gelled in situ it will not revert to a liquid state. Fresh articular cartilage cells from 12-month-old New Zealand white rabbits were suspended at a final concentration of 5.5 $\times 10^5$ cells/ml in a 5 wt% P(NiPAAm-co-AcA) solution, placed in culture plates, and gelled at 37°C. Cells recovered from the hydrogel cultures over a 4-week period expressed the original chondrocyte phenotype and displayed typical chondrocytic morphology, in contrast to those recovered from monolayer cultures, which appeared to be more "fibroblast-like" [45]. The material is injectable through small-diameter needles and is not acutely toxic to living cells. It therefore shows promise for cartilage repair and other tissue-engineering applications.

A new class of thermogelling poly (organophosphazines) containing oligomeric side chains of hydrophilic methoxypoly (ethylene glycol) (MPEG) and hydrophobic tripeptide or tetrapeptide side groups shows promise for drug delivery and tissue engineering [89]. They have sol–gel transitions in the range of 35° to 43°C and form hydrogels with relatively high strength. The thermosensitivity of the polymers depends primarily on the structure of the hydrophobic parts of the oligopeptide side groups, which may form strong physical junctions in the polymer solution. The hydrophobic/hydrophilic ratio plays an important role in establishing suitable properties for scaffolding applications.

Among the natural polymers most frequently proposed for injectable tissue-engineering applications are alginates, collagen, chitosan, hyaluronates, fibrin, and fibroin [4, 6, 7, 16, 25, 28, 37, 57, 60, 91]. Alginates derived from brown seaweed are anionic linear polysaccharides composed of 1,4-linked β-D-mannuronate (M) and 1,4-linked α-L-guluronate (G) residues. A cross-linked alginate hydrogel based on polysaccharides from seaweeds has been found useful as a carrier for controlled release of therapeutic peptides and proteins when the cross-linking reaction is controlled in situ [37]. A recent study has shown that alginate dialdehyde (ADA), an oxidized form of alginate in combination with gelatin, can self-cross-link in a controlled manner in the presence of small concentrations of borax [4]. A material suitable for both drug-delivery and tissue-engineering applications was injected with a double-syringe fibrin glue applicator. One syringe was filled with gelatin solution that contained the therapeutic agents, and the other was filled with the oxidized alginate in the presence of borax. By varying the concentrations of the reactants, the gelation time within the hypodermic needle of the applicator was varied from a few seconds to less than a minute. Cells encapsulated in the gel retained their protein-producing viability, and the solidified gel was completely degraded after 5 weeks [4]. Another study confirmed that oxidized alginates rapidly degrade at physiological pH [10]. Cytotoxicity screening using mouse fibroblasts confirmed the nontoxic character of the gels. It has also been demonstrated that alginate gels support proliferation of chondrocytes both in vitro and in vivo. New cartilage tissue was formed when freshly isolated calf chondrocytes mixed with alginate solution were injected into mice subcutaneously [3, 37].

Chitosan is a polycationic biopolymer obtained by deacetylation of chitin, the main component of the shell of crustaceans [75]. When ionically cross-linked, chitosan forms a reversible hydrogel structure that is well suited to a wide variety of pharmaceutical applications [6, 16, 17, 93]. Chitosan can be used as an injectable carrier for tissue-engineering applications when covalently cross-linked in situ or when an active particulate filler is used as a reinforcing agent. Covalent cross linking of chitosan has had limited success, because most catalysts used for covalent cross linking are not biocompatible [6]. Use of an active particulate filler has met with some degree of success [37, 38]. Chitosan–calcium phosphate composites form injectable and moldable pastes at pH values below 6.5 and undergo a phase transition at physiological pH. The phase transition entraps calcium phosphate within the hydrogel matrix. Field emission micrography has shown the resulting scaffold to have a highly porous structure, with polymer strands that bind the micrometer-sized aggregates of the ceramic phase [37].

A photocross-linkable chitosan has been used as a noninjectable carrier to induce neovascularization in vivo and to regulate the release of growth factors [44]. By introducing lactose (lactobionic acid) moieties into the chitosan molecules by means of a condensation reaction with the amino groups, it proved possible to make an injectable chitosan/lactic acid (CH-LA) scaffold. This involved dissolving the polymer and mixing it with β-FGF/FGF-2 and nonanticoagulant ($IO_4$–) heparin. When the resulting hydrogel was injected into the right and left sides of the backs of mice, the β-FGF/FGF-2 molecules, encapsulated by the chitosan/($IO_4$–) heparin hydrogel, were gradually released as the gel was biodegraded, and there was a "substantial effect to induce vascularization and fibrous tissue formation" [29].

A thermosensitive chitosan hydrogel has been prepared by grafting more than about 40 wt% of PEG to the chitosan chains via covalent bonding [7]. The resulting copolymeric hydrogel was injectable at low temperature and became a semisolid at body temperature. When the semisolid hydrogel was cross-linked with genipin in situ, protein was released for up to 40 days. The hydrogel can be prepared in solutions at a physiological pH, allowing incorporation of a wide range of bioactive molecules used in tissue-engineering applications [7].

Hyaluronic acid (HA) is an important constituent of the ECM. It is a polyanionic glycosaminoglycan (GAG) that is required for cell proliferation and differentiation and to regulate cell adhesion [91]. Although HA is widely used medically [5, 60, 65, 76], its use as an injectable scaffold has been limited. Sodium hyaluronate has been evaluated as a matrix to deliver β-FGF/FGF-2 in the course of bone repair [37]. A single injection of the gel into a fresh rabbit fibula fracture caused an increase of bone and callus formation and restored the mechanical strength at the site [37].

Recent work on the development of an injectable, cross-linkable hydrogel of HA with 3-thiopropanoylhydrazide-poly(ethyleneglycoldiacrylate) (HA-DTPH/PEGDA) at a 2:1 molar ratio indicated potential utility for tissue-engineering uses [91]. A PEGDA solution seeded with T31 human tracheal scar fibroblasts was added to an HA-DTPH solution containing newborn calf serum, L-glutamine, and an antibiotic-antimycotic to form a cross-linkable hydrogel with 97.5% to 98.8% water content. The hydrogel was used to determine in vitro cell viability and proliferation and in vivo fibrous tissue generation. This was done by bilaterally transplanting the hydrogel into surgically prepared subcutaneous pockets on the backs of 4- to 6-week-old nude mice. T31 fibroblast viability was demonstrated by an almost 10-fold increase in the number of cells after 28 days of culture in vitro. The in vivo measurements were made at 2, 4, and 8 weeks after implantation. There was no evidence of necrosis or damage to the tissues, and the cells retained their initial phenotype and were actively secreting new ECM.

Regenerated silk fibroin has been used for the fabrication of films, nets, regenerated fibers, foams, creams, and hydrogels [53, 72, 73, 74, 86, 87, 98, 99]. Injectable fibroin hydrogels and their composites have also been evaluated for their effectiveness in bone regeneration [25, 28]. Human cell lines and primary cells isolated from biopsies were seeded on fibroin-based materials. Proliferation of fibroblasts, osteoblasts, epithelial cells, keratinocytes, and hepatocytes was observed [98, 99]. Silk threads from *Bombyx mori* have a fibrous core of fibroin, a biocompatible structural protein that favors cell adhesion and activation [2, 70, 98]. When dissolved in water, regenerated fibroin molecules act as hydrophilic-hydrophobic-hydrophilic polymers that form an emulsion of irregularly sized micelles by chain folding and hydrophobic interactions [52]. The intervention of hydrophilic blocks within the hydrophobic sequences prevents β-sheet crystallization and, as the concentration of micelles increases, results in the formation of a skeletal arrangement of micelles "cross-linked" by hydrogen bonding and hydrophobic interactions [52]. Regenerated fibroin is prepared by degumming *B. mori* cocoons in aqueous $Na_2CO_3$ solutions, dissolving the nearly pure fibroin in lithium bromide solution, and then dialyzing against distilled water to obtain aqueous solutions of the order of 2% to 5% weight/unit volume. An injectable hydrogel is formed directly from solution at a pH below the isoelectric point [25, 73].

MG63 human osteoblast cell-line bioactivity was examined on pure fibroin-based injectable hydrogels prepared by different methods. In vitro biocompatibility was evaluated by measuring lactic dehydrogenase release, cell proliferation (WST1, water-soluble tetrazolium salt), differentiation (ALP, alkaline phosphatase and OC, Osteocalcin) and synthetic activity (collagen I, TGF-$\beta_1$, transforming growth factor β1 and interleukin-6) [28]. In vitro tests also confirmed that the fibroin hydrogels were not cytotoxic [73]. In a series of in vivo tests, it was found that a 2.5 wt% fibroin hydrogel had the capacity to regenerate bone in critical-size holes drilled into the femoral condyle of rabbits, without the addition of cells, growth factors, or other components known for their bioactivity [28]. The in vivo studies were performed at the Rizzoli Orthopaedic Institute of Bologna [28] by implanting 2.5% fibroin hydrogels, brought to their isoelectric points by the addition of citric acid, into bilateral confined cancellous defects (10-mm depth and 6-mm diameter) in the femoral condyle of 10 adult New Zealand white disease-free rabbits. Four control specimens were used. The cancellous defects in two rabbits were left untreated; in two other rabbits a commercial scaffold of poly(D,L-glycolide) copolymer (ratio 50:50 mol%) dispersed in an aqueous solution of PEG and 15% dextran

(Sintbone Slurry Gel®) was inserted. Seven rabbits (5 + 2 controls) were sacrificed after 1 month and after 3 months. Histological, histomorphometric, and high-resolution x-ray investigations were carried out on sections 200 ± 10 μm in thickness cut at different depths between the defect surface and bottom. The results were compared with those obtained from empty defects and from defects filled with the control material. Histological sections of the untreated and treated defects at 4 weeks are shown in Fig. 7.3. The formation of new bone (NB) in the untreated cavities of control specimens re-mained restricted to the edge of the defects (Fig. 7.3A). Newly formed bone was generated radially inward from the defect surfaces in both the synthetic polymer-treated control (Fig. 7.3B) and the fibroin hydrogel-treated experimental animals (Figs. 7.3C and D). Thin and dense new trabeculae (NB in Fig. 7.3) grew radially from the old bone (OB) surface of the synthetic polymer-treated defect, but with a noticeable interphase between the NB and the OB (Fig. 7.3B). Thin and dense new trabeculae (NB) also grew radially f rom the OB surface of the fibroin hydrogel-treated defect. In this case, however, there was no noticeable interface between the NB and the OB (Fig. 7.3C). One of the five rabbits treated with the fibroin hydrogel exhibited full recovery after 4 weeks (Fig. 7.3D). At 4 weeks, the regrown cancellous bone in the fibroin hydrogel-treated defects was significantly thicker and denser than either normal bone or bone grown in the synthetic polymer-treated defects. Twelve weeks after surgery, however, the bone grown in the fibroin hydrogel-treated defects had changed appearance completely. It appeared more similar to normal bone than bone in the synthetic polymer-treated defects. At 12 weeks all six rabbits (five fibroin-treated and one polyester-based control) showed full recovery, but the fibroin hydrogel accelerated bone remodeling. The distal femur areas were occupied by trabecular bone with the spatial orientation, shape, and size seen in healthy cancellous bone.

To function as an injectable scaffold, a hydrogel must be injected as a solution or dispersion that can be cross-linked in vivo. Alternatively, it must be injected as a reversible gel and restructured in the defect cavity. Either scaffold will have poor tensile and shear properties because of high water content. Where load bearing is needed during repair, an injectable, hydrogel-based scaffold may require external support or a composite scaffold structure that can support external stresses, yet retain the advantages of the osmotic environment provided by the hydrogel. A creative use of composite technology and an understanding of the viscoelastic properties of the material during injection and in the environment at the defect site are necessary to achieve this objective.

## 7.5 Outlook

Injectable scaffolds for the regeneration of bone fall into two main categories: flowable ceramic–water mixtures that set in situ as either compacted, porous scaffolds or porous particulate mixtures in a polymeric carrier; and natural or synthetic hydrogels with high water content that encapsulate and carry compounds and cells to the injection site.

Differences in composition and physical properties notwithstanding, these two categories of scaffolds can promote bone tissue regeneration and can replace invasive surgery in many situations. They do this, however, by two very different mechanisms. With ceramic-based scaffolds, solubilization and resorption of the scaffold furnish the appropriate mineral environment to guide the regeneration process. With hydrogel scaffolds, bone regeneration involves a self-regulating process that is guided by the cells and mineral components in the aqueous phase of the scaffold.

In materials that are calcium phosphate-based, osteoclasts adhere to the external surfaces and to accessible pore surfaces, i.e., those larger than 200 to 300 μm, generating an extracellular matrix that contains calcium ions from the ceramic. The osteoclasts initiate new bone growth at the ceramic surface and migrate at the receding bone–ceramic interfaces at a rate that is determined by the rate of dissolution of the ceramic. In the case of the highly swollen hydrogels, cells and dissolved mineral components from the host penetrate the dilute space of the scaffold, and it is the movement of cells, ionic mineral components, and polymer chains of the hydrogel skeletal network in the dilute environment of the body fluids that determines the rate of bone regeneration. Ideally, the freedom of movement of cells in the physiological environment of the hydrogel is much like that within an unmineralized extracellular matrix. If, in addition, the skeletal network has a specific, bioactive character, cells can proliferate and initiate a collagenous extracellular matrix throughout the entire volume of the hydrogel. Mineralized new bone trabeculae can then develop throughout, rather than only at scaffold surfaces. In this situation, degradation of the polymer skeletal network is less critical for the healing process than in ceramic-based materials.

It is not yet certain how the differences in mechanisms may affect the long-term properties of the regenerated bone. It is likely, however, that the natural remodeling and mineralization processes of the body will minimize differences in the long run.

Although the type of scaffold used for tissue regeneration is determined primarily by clinical needs, the principles involved in the development of each of the two types of injectable scaffolds are relevant to both. Future research and development will greatly benefit from collaboration and interdisciplinary activity by engineers and clinicians.

## References


1. Alpelt D, Theiss F, El-Warrak AO, Zlinszky K, Bettschart-Wolfisberger R, Bohner M, Matter S, Auer JA, von Rechenberg B (2004) In vivo behavior of three different injectable hydraulic calcium phosphate cements. Biomaterials 25:1439–1451.
2. Altman GH, Diaz F, Jakuba C, Calabro T, Horan RL, Chen J, Lu H, Richmond J, Kaplan DL (2003) Silk-based biomaterials. Biomaterials 24:401–416.

2A. ASTM Standard F2150-02e1, Standard Guide for Characterization and Testing of BIomaterial Scaffolds Used in Tissue Engineered Medical Products (2006) Annual Book of ASTM Standards 2006, Volume 13.01, ASTM International, West Conshohocken PA, USA.

3. Atala A, Cima LG, Kim W, Paige KT, Vacante JP, Retik AB, Vacanti CA (1993) Injectable alginate seeded with chondrocytes as a potential treatment for vesicoureteral reflux. J Urol 150:745–747.
4. Balakrishnan B, Jayakrishnan A (2005) Self-cross-linking biopolymers as injectable in-situ forming biodegradable scaffolds. Biomaterials 26:3941–3951.
5. Band PA (1998) Hyaluranon derivatives: chemistry and clinical applications In: Laurent TC, ed. The Chemistry, Biology and Medical Applications of Hyaluranon and its Derivatives. London: Portland Press, pp 33–42.
6. Berger J, Reist M, Mayer JM, Felt O, Peppas NA, Gurny R (2004) Structure and interactions in covalently and ionically crosslinked chitosan hydrogels for biomedical applications. Eur J Pharm Biopharm 57:19–34.
7. Bhattarai N, Ramay HR, Gunn J, Matsen FA, Zhang M (2005) PEG-grafted chitosan as an injectable themosensitive hydrogel for sustained protein release. J Control Release 103:609–624.
8. Blokhuis TJ, Termaat MF, den Boer FC, Patka P, Bakker FC, Haarman HJ (2000) Properties of calcium phosphate ceramics in relation to their in-vivo behavior. J Trauma 48:179–186.
9. Bohner M, Lemaitre J, Van Landuyt P, Zambelli PY, Merkle HP, Gander B (1997) Gentamicin-loaded hydraulic calcium phosphate bone cement as an antibiotic delivery system. J Pharm Sci 86:565–572.
10. Bouhadir KH, Lee KY, Alsberg E, Damm KL, Anderson KW, Mooney DJ (2001) Degradation of partially oxidized alginate and its potential applications for tissue engineering. Biotechnol Prog 17:945–950.
11. Bromberg LE, Ron ES (1998) Temperature-responsive gels and thermogelling polymer matrices for protein and peptide delivery. Adv Drug Deliv Rev 31:197–221.
12. Brown RQ, Mount A, Burg KJL (2005) Evaluation of polymer scaffolds to be used in a composite injectable system for intervertebral disc tissue engineering. J Biomed Mater Res 74A:32–39.
13. Brown WE, Chow LC (1983) A new calcium phosphate cement. J Dent Res 62(Abstract 207):672.
14. Burg KJL, Porter S, Kellam JF (2000) Biomaterial developments for bone tissue engineering. Biomaterials 21:2347–2359.
15. Cabana A, Ait-Kadi A, Juhasz J (1997) Study of the gelation process of poly(ethylene oxide)-poly(propylene oxide)-poly(ethylene oxide) copolymer (poloxamer 407) aqueous solutions. J Colloid Interface Sci 190:307–312.
16. Chenite A, Buschmann M, Wang D, Chaput C, Kandani N (2001) Rheological characterization of thermogelling chitosan/glycerol-phosphate solutions. Carbohydrate Polymers 46:39–47.
17. Chenite A, Chaput C, Wang D, Combes D, Buschmann M, Hoemann CD, Leroux J-C, Atkinson BL, Binette F, Selmani A (2000) Novel injectable neutral solutions form chitosan biodegradable gels in situ. Biomaterials 21:2155–2161.
18. Constanz BR, Barr BM, Ison IC, Fulmer MT, Baker J, McKinney L, Goodman SB, Gunasekaren S, Delaney DC, Ross J, Pose RD (1998) Histological, chemical and crystallographic analysis of four calcium phosphate cements in different rabbit osseous sites. J Biomed Mater Res B Appl Biomater 43:451–461.
19. Constantz BR, Ison IC, Fulme MT, Poser RD, Smith ST, VanWagoner M, Ross J, Goldstein SA (1995) Skeletal repair by in situ formation of the mineral phase of bone. Science 267:1796–1799.
20. Daculsi G (1998) Biphasic calcium phosphate concepts applied to artificial bone, implant coating and injectable bone substitutes. Biomaterials 19:1473–1478.
21. Daculsi G, Laboux O, Malard O, Weiss P (2003) Current state of the art of biphasic calcium phosphate bioceramics. J Mater Sci Mater Med 14:195–200.
22. Daculsi G, Legeros RZ, Nery E, Lynch K, Kerebel B (1989) Transformation of biphasic calcium phosphate ceramics in-vivo. Ultrastructural and physio-chemical characterization. J Biomed Mater Res 23:883–894.

23. Daculsi G, Passuti N (1990) Effect of the macroporosity for osseous substitution of calcium phosphate ceramics. Biomaterials 11:86–87.
24. Daculsi G, Passuti N, Martin S, Deudon C, Legeros RZ, Raher S (1990) Macroporous calcium phosphate ceramic for long bone surgery in humans and dogs. J Biomed Mater Res 24:379–396.
25. DiBenedetto AT, Huang SJ, Migliaresi C, Motta A (2003) Injectable bioactive gels and gel composites and method of use thereof. US Patent Application 39341.
26. Driskell TD, O'Hara MJ, Sheets HDJ, Greene GW Jr., Natiella JR, Armitage J (1972) Development of ceramic and ceramic composite devices for maxillofacial applications. J Biomater Res 6:345–361.
27. Elisseeff J (2005) Repairing knee joints by growing new cartilage using an injectable hydrogel http://www.birchbob.com/MarketingPieces/jhu/printversion2Elisseeff.PDF
28. Fini M, Motta A, Torricelli P, Giavaresi G, Nicoli Aldini N, Tschon M, Giardino R, Migliaresi C (2005) The healing of confined critical cancellous defects in the presence of silk fibroin hydrogel. Biomaterials 26:3527–3536.
29. Fujita M, Ishihara M, Simizu M, Obara K, Ishizuka T, Saito Y, Yura H, Morimoto Y, Takase B, Matsui T, Kikuchi M, Maehara T (2004) Vascularization in vivo caused by the controlled release of fibroblast growth factor-2 from an injectable chitosan/non-anticoagulant heparin hydrogel. Biomaterials 25: 699–706.
30. Gauthier O, Khairoun I, Bosco J, Obedia L, Bourges X, Rau C, Magne D, Bouler JM, Aguado E, Daculsi G, Weiss P (2003) Noninvasive bone replacement with new injectable calcium phosphate biomaterial. J Biomed Mater Res 66A:47–54.
31. Gauthier G, Muller R, von Stechow D, Lamy B, Weiss P, Bouler J-M, Aguado E, Daculsi G (2005) In-vivo bone regeneration with injectable calcium phosphate biomaterial: a three-dimensional micro-computed tomographic, biomechanical and SEM study. Biomaterials 26:5444–5453.
32. Gomes ME, Reis RL (2004) Biodegradable polymers and composites in biomedical applications: from catgut to tissue engineering. Part 2: Systems for temporary replacement and advanced tissue regeneration. Int Mater Rev 49:274–285.
33. Goodman SB, Bauer TW, Carter D, Casteleyn PP, Goldstein SA, Kyle RF, Larsson S, Stakewich CJ, Swiontkowski MF, Tencer AF, Yetkinler DN, Poser RD (1998) Norial SRS cement augmentation in hip fracture treatment. Clin Orthop Rel Res 348:42–50.
34. Graham NB (1998) Hydrogels: their future, Part 1. Med Device Technol 9:18–22.
35. Graham NB (1998) Hydrogels: their future, Part 2. Med Device Technol 9:22–25.
36. Grimandi G, Weiss P, Millot F, Daculsi G (1998) In vitro evaluation of a new injectable calcium phosphate material. J Biomed Mater Res 39:660–666.
37. Gutowska A, Jeong B, Jasionowski M (2001) Injectable gels for tissue engineering. Anat Rec 263: 342–349.
38. Gutowska A, Song L, Armstrong BL, Campbell AA (1998) Injectable stimuli-sensitive polymer ceramic composites for bone tissue regeneration. Trans Soc Biomater 21:450.
39. Hench LH (1998) Bioceramic. J Am Ceram Soc 81:1705–1728.
40. Heymann D, Guicheux J, Rouselle AV (2001) Ultrastructural evidence in-vitro of osteoclastic-induced degradation of calcium phosphate ceramic by simultaneous resorption and phagocytosis mechanisms. Histol Histopathol 16:37–44.
41. Holland TA, Tessmar JKV, Tabata Y, Mikos AG (2003) Transforming growth factor-β1 release from oligo(poly(ethylene glycol) fumarate hydrogels in conditions that model the cartilage wound healing environment. J Control Release 94:101–114.
42. Hutmacher DW, Sittinger M (2003) Periosteal cells in bone tissue engineering. Tissue Eng 9:45–64.
43. Ikenaga M, Hardouin P, Lemaitre J, Andrianjatovo H, Flautre B (1998) Biomedical characterization of a biodegradable calcium phosphate hydraulic cement: a comparison with porous biphasic calcium phosphate ceramics. J Biomed Mater Res A Appl Biomater 40:139–144.
44. Ishihara M, Obara K, Ishizuka T, Fujita M, Sato M, Masuoka K, Saito Y, Yura H, Matsui T, Hattori H, Kikuchi M, Kurita A (2003) Controlled release of fibroblast growth factors and heparin from photocrosslinked chitosan hydrogels and subsequent effect on in vivo vascularization. J Biomed Mater Res 64:551–559.
45. Jaisionowski M, Krzyminski K, Chrisler W, Markille LM, Morris J, Gutowska A (2004) Thermally-reversible gel for 3-D cell culture of chondrocytes. J Mater Sci Mater Med 15:575–582.
46. Jeong B, Bae YH, Kim SW (1999) Biodegradable thermosensitive micelles of PEG-PLGA-PEG triblock copolymers. Colloids and Surfaces, B: Biointerfaces 16:185–193.
47. Jeong B, Bae YH, Kim SW (1999) Thermoreversible gelation of PEG-PLGA-PEG triblock copolymer aqueous solutions. Macromolecules 32:7064–7069.
48. Jeong B, Bae YH, Kim SW (2000). In situ gelation of PEG-PLGA-PEG triblock copolymer aqueous solutions and degradation thereof. J Biomed Mater Res 50:171–177.
49. Jeong B, Bae YH, Lee DS, Kim SW (1997) Biodegradable block copolymers as injectable drug-delivery systems. Nature 388:860–862.
50. Jeong B, Kibbey MR, Birnbaum JC, Won Y-Y, Gutowska A (2000) Thermogelling biodegradable polymers with hydrophilic backbones: PEG-PLGA. Macromolecules 33:8317–8322.
51. Jeong JH, Lim DW, Han DK, Park TG (2000) Synthesis, characterization and protein adsorption behaviors of PLGA/PEG di-block co-polymer blend films. Colloids and Surfaces, B: Biointerfaces 18:371–379.
52. Jin H-J, Kaplin DL (2003) Mechanisms of silk processing in insects and spiders. Nature 424:1057–1061.
53. Kemal K, Motta A, Fambri L, Migliaresi C (2001) Poly(ε-Caprolactone-co-D,L-lactide)/silk fibroin particles composite materials: preparation and characterization. J Biomater Sci Polym Ed 12:337–351.
54. Kisiday J, Jin M, Kurz B, Hung H, Semino C, Zhang S, Grodzinsky AJ (2002) Self assembling peptide hydrogel fosters chondrocyte extracellular matrix production and cell division: implications for cartilage tissue repair. Proc Natl Acad Sci USA 99:9996–10001.

55. Knaack D, Goad ME, Aiolova M, Rey C, Tofighi A, Chakravarthy P, Lee DD (1998) A resorbable calcium phosphate bone substitute. J Biomed Mater Res B Appl Biomater 43:399–409.
56. Komath M, Varma HK (2003) Development of a fully injectable calcium phosphate cement for orthopedic and dental applications. Bull Mater Sci 26:415–422.
57. Kuo CK, Ma PX (2001) Ionically crosslinked alginate hydrogels as scaffolds for tissue engineering: Part 1. Structure, gelation rate and mechanical properties. Biomaterials 22:511–521.
58. Larsson S, Bauer TW (2002). Use of injectable calcium phosphate cement for fracture fixation: a review. Clin Orthop Relat Res 395:23–32.
59. Laurencin CT, Khan Y (2005) Bone graft substitute materials. www.emedicine.com
60. Leach JB, Bivens KA, Patrick Jr CW, Schmidt CE (2003) Photocrosslinked hyaluronic acid hydrogels: natural, biodegradable tissue engineering scaffolds. Biotechnol Bioeng 82:578–589.
61. Lee DS, Shim MS, Kim SW, Lee H, Park I, Chang T (2001) Novel thermoreversible gelation of biodegradable PLGA-block-PEO-block-PLGA triblock copolymers in aqueous solution. Macromol Rapid Commun 22:587–592.
62. Lee KY, Mooney DJ (2001) Hydrogels for tissue engineering. Chem Rev 101:1869–1879.
63. LeGeros RZ (1988) Calcium phosphate materials in restorative dentistry: a review. Adv Dent Res 2:164–183.
64. Luginbuehl V, Wenk E, Koch A, Gander B, Merkle HP, Meindel L (2005) Insulin-like growth factor I-releasing alginate-tricalcium phosphate composites for bone regeneration. Pharm Res 22:940–950.
65. Luo Y, Kirker KR, Prestwich GD (2000) Cross-linked hyaluronic acid hydrogel films: new biomaterials for drug delivery. J Control Release 69:169–184.
66. Ma PX (2004) Scaffolds for tissue fabrication. Materials Today, May, pp 30–40.
67. Mahr M, Bartle GB, Bite U, Clay R, Kasperbauer JL, Holmes JM (2000) Norian craniofacial repair system bone cement for the repair of craniofacial skeletal defects. Ophthal Plast Reconstr Surg 16:393–398.
68. Mallapragada KS, Narasimhan B, eds (2002) Injectable polymeric biomaterials. Special issue. Biomaterials 23:4305–4333.
69. Mano JF, Sousa RA, Boesel LF, Neves NM, Reis RL (2004) Bioinert, biodegradable and injectable polymeric matrix composites for hard tissue replacement: state of the art and recent developments. Composites Sci and Technol 64:789–817.
70. Minoura N, Aiba S-I, Gotoh Y, Tsukada M, Imai Y (1995) Attachment and growth of cultured fibroblast cells on silk protein matrices. J Biomed Mater Res 29:1215–1221.
71. Mortensen K, Pedersen JS (1993) Structural study on the micelle formation of poly(ethylene oxide)-poly(propylene oxide)-poly(ethylene oxide) triblock copolymer in aqueous solution. Macromolecules 26:805–812.
72. Motta A, Fambri L, Migliaresi C (2002) Regenerated silk fibroin films: thermal and dynamic mechanical analysis. Macromol Chem Physics 203:1658–1665.
73. Motta A, Migliaresi C, Faccioni F, Torricelli P, Fini M, Giardino R (2004) Fibroin hydrogels for biomedical applications: preparation, characterization and in vitro cell culture studies. J Biomater Sci Polym Ed 15:851–864.
74. Motta A, Migliaresi C, Lloyd AW, Denyer SP, Santin M (2002) Serum protein adsorption on silk fibroin fibres and membranes: surface opsonization and binding strength. J Bioact Compact Polym 17:23–35.
75. Muzarelli R (1973) Chitosan. In: Muzarelli R, ed. Natural Chelating Polymers. Oxford: Pergamon Press, pp 144–176.
76. Nakayama OS, Matsuda T (2001) Thermoresponsive artificial extracellular matrix for tissue engineering: hyaluronic acid biconjugated with poly(N-isoproylacrylamide) grafts. Biomacromolecules 2: 856–863.
77. Nguyen H, Qian JJ, Bhatnagar RS, Li S (2003) Enhanced cell attachment and osteoblastic activity by P-15 peptide coated matrix in hydrogels. Biochem Biophys Res Commun 311:179–186.
78. Nguyen KT, West JL (2002) Photopolymerizable hydrogels for tissue engineering applications. Biomaterials 23:4307–4314.
79. Osborne JF, Newesely H (1980) The materials science of calcium phosphate ceramics. Biomaterials 1:108–111.
80. Otsuka M, Nakahigashi Y, Matsuda Y, Fox JL, Higuchi WI (1994) A novel skeletal drug delivery system using self-setting calcium phosphate cement 7. Effect of biological factors on indomethacin release from the cement loaded on bovine bone. J Pharm Sci 83:1569–1573.
81. Park H, Temenoff JS, Holland TA, Tabata Y, Mikos AG (2005) Delivery of TGF-β1 and chondrocytes via injectable, biodegradable hydrogels for cartilage tissue engineering applications. Biomaterials 26:7095–7103.
82. Payne RG, McGonigle S, Yaszemski MJ, Yasko AW, Mikos AG (2002) Development of an injectable in situ crosslinkable, degradable polymer carrier for osteogenic cell populations. Part 3. Proliferation and differentiation of encapsulated marrow stromal osteoblasts cultured on crosslinking poly(propylene fumarate). Biomaterials 23:4381–4387.
83. Quellec P, Gref R, Perrin L, Dellacherie E, Sommer F, Verbavatz JM, Alonso MJ (1998) Protein encapsulation within polyethylene glycol-coated nanospheres. I. Physiochemical characterization. J Biomed Mater Res 42:45–54.
84. Ratier A, Freche M, Lacout JL, Rodriguez F (2004) Behavior of an injectable calcium phosphate cement with added tetracycline. Int J Pharm 274:261–268.
85. Ratner BD, Bryant SJ (2004) Biomaterials: where we have been and where we are going. Annu Rev Biomed Eng 6:41–75.
86. Santin M, Denyer SP, Lloyd AW, Motta A (2002) Domain-driven binding of fibrin(ogen) onto silk fibroin biomaterials. J Bioact Compat Polym 17:195–208.
87. Santin M, Motta A, Freddi G, Cannas M (1999) In vitro evaluation of the inflammatory potential of the silk fibroin. J Biomed Mater Res 46:382–389.
88. Schnettler R, Stahl JP, Alt V, Pavlidis T, Dingeldein E, Wenisch S (2004) Calcium phosphate-based bone substitutes. Eur J Trauma 30:219–229.
89. Seong J-Y, Jun YJ, Jeong B, Sohn YS (2005) New thermogelling poly(organophosphazines) with methoxypoly(ethylene glycol) and oligopeptide as side groups. Polymer 46:5075–5081.

90. Shin H, Quinten Ruhé P, Mikos AG, Jansen JA (2003) In-vivo bone and soft tissue response to injectable biodegradable oligo(poly(ethylene glycol) fumarate) hydrogels. Biomaterials 24:3201–3211.
91. Shu XZ, Liu Y, Palumbo FS, Luo Y, Prestwich GD (2004) In situ crosslinkable hyaluranon hydrogels for tissue engineering. Biomaterials 25:1339–1348.
92. Silberberg A (1989) Gelled aqueous systems In: Glass JE, ed. Polymers in Aqueous Media. Advances in Chemistry Series 223. Washington, DC: American Chemical Society, pp 1–13.
93. Song JS, Such CH, Park YB, Lee SH, Yoo NC, Lee JD, Kim KH, Lee SK (2001) A phase I/IIa study on intra-articular injection of holmium-166 chitosan complex for the treatment of knee synitis of rheumatoid arthritis. Eur J Nucl Med 28:489–497.
94. Sosnik A, Cohn D (2004) Ethoxysilane-capped PEO-PPO-PEO triblocks: a new family of reverse thermo-reponsive polymers. Biomaterials 25:2851–2858.
95. Takagi S, Chow LC, Ishikawa K (1998) Formation of hydroxyapatite in new calcium phosphate cements. Biomaterials 19:1593–1599.
96. Temenoff JS, Mikos AG (2000) Injectable biodegradable materials for orthopedic tissue engineering. Biomaterials 21:2405–2412.
97. Temenoff JS, Shin H, Conway DE, Engel PS, Mikos AG (2003) In vitro cytotoxicity of redox radical initiators for cross-linking of oligo(poly(ethylene glycol) fumarate) macromonomers. Biomacromolecules 4:1605–1613.
98. Unger RE, Wolf M, Peters K, Motta A, Migliaresi C, Kirkpatrick J (2004) Growth of human cells on a novel non-woven silk fibroin net: a potential use for tissue engineering. Biomaterials 25:1069–1075.
99. Unger RE, Peters K, Wolf M, Motta A, Migliaresi C, Kirkpatrick CJ (2004) Endothelialization of a non-woven silk fibroin net for use in tissue engineering: growth and gene regulation of human endothelial cells. Biomaterials 25:5137–5146.
100. Williams DF (1991) Concise Encyclopedia of Medical and Dental Materials. Oxford: Pergamon Press.
101. Williams DF (2003) Revisiting the definition of biocompatibility. Med Device Technol 14:10–13.
102. Yamada S, Heymann D, Bouler JM, Daculsi G (1997) Osteoclastic resorption of calcium phosphate ceramics with different hydroxyapatite/beta-tricalcium phosphate ratios. Biomaterials 18:1037–1041.
103. Zahraoui C, Sharrock P (1999) Influence of sterilization on injectable bone biomaterials. Bone 25(2 Suppl):1037–1041.

# 8. Motion and Bone Regeneration

*Ching-Chang Ko, Martha J. Somerman, and Kai-Nan An*

## 8.1 Introduction

Bone is a living material composed of cells and an extracellular matrix (ECM) that has a multicomponent structure [4]. The ECM of bone is composed of three phases: an inorganic mineral phase, an organic phase, and an aqueous phase. The inorganic phase of bone is calcium hydroxyapatite, $Ca_{10}(PO_4)_6(OH)_2$. The organic phase consists primarily of collagen fibers and associated noncollagenous ECM proteins. The molecular configuration of collagen provides binding sites for hydroxyapatite crystal nucleation and growth. The ECM is created and maintained by active bone cells: osteoblasts, osteoclasts, and osteocytes. Osteoblasts and osteocytes are involved in bone formation and maintenance, respectively, whereas osteoclasts promote resorption of bone [2, 99]. Bone is, in general, dynamic and constantly being remodeled by the action of these cells, and thus can regenerate itself.

Bone regeneration is an important function of the living organism. It provides reparative power to the vertebrae organism, including the ability to unite broken bones and to refill defects [110]. A complicated bony fracture will require a healing construct (callus) to glue the fragmented bone together. This callus sets the foundation for regeneration to occur. Bone regeneration is defined as a dynamic process that consists of episodes of cell recruitment, cell differentiation, mineralization, and reorganization of mineral structures [22, 110]. These episodes are similar to the biological cascades that occur during wound healing. Therefore, the first portion of this chapter (Section 8.2) will review the biological aspects of bone healing.

One of the most interesting aspects of the regeneration process is that it allows alteration of its cellular activities through physical means, such as exercise or motion [19, 20, 21, 55, 62]. This characteristic conveys information of structure-functional relationships (adaptation) during the early stages of regeneration. The relationship between motion and bone regeneration is thought to be preprogrammed in cells. Physical deformation—the distortion of tissue by movement—can be transmitted into the cell cytoskeleton and converted into biochemical signals for promotion of osteogenesis [18, 60]. It is generally accepted that precise engineering motion parameters (magnitude, frequency, direction, etc.) can provide useful tools for designing therapies to regenerate bone. Specific clinical movements, such as dynamization [28], osteogenic distraction applied to healing callus [57, 58, 59], and orthodontic tooth movement [95], have been used in efforts to increase bone formation. Bone regeneration includes bone healing, osteogenesis, and osseointegration. In this order, the second portion of this chapter (8.3 and 8.4) is devoted to a review of adaptation theory and mechanobiology as they relate to bone regeneration. Examples including load-enhanced implant osseointegration and mandibular distraction osteogenesis will be used to illustrate the relationship between physical stimuli and tissue regeneration.

# 8.2 Bone Healing

Bone regeneration is often associated with wound repair, which involves complex biochemical interactions among cells and associated factors. Research targeted at understanding modulation of tissue development and early embryogenesis has contributed to the field of wound healing, because similar molecules, common cell types including "stemlike" cells, and parallel processes are involved [13, 22, 32, 33, 34, 52]. In both cases, mesenchymal "stemlike" cells migrate to and aggregate within the matrix core and begin to proliferate and differentiate in order to form the required tissues. Injured bone attracts platelets, growth factors, and blood capillaries to the local site, and this process allows for recruitment of mesenchymal stem cells (MSCs) to form the matrix. When the MSCs differentiate, they produce cartilaginous or osseous tissues, depending on whether they are forming endochondral bone (cartilage to bone, e.g., long bone) or intramembranous bone (does not form cartilage first, e.g., calvarium, periodontal wound healing). Bone healing is functionally divided into three phases: the inflammatory phase, the reparative phase, and the remodeling phase [79].

## 8.2.1. Three Phases Of Bone Healing

### 8.2.1.1 Inflammatory Phase

The inflammatory phase begins at the onset of bone damage. As blood vessels in the damaged region rupture and clot formation is initiated, they signal the body to dispatch macrophages to the wounded area [108]. These macrophages absorb and break down necrotic or damaged tissues, and in turn lure osteoclasts, which debride broken bone fragments. Whether the next two phases of bone healing are activated is contingent upon whether both the macrophages and the osteoclasts can "clean up" the field of biological debris. In the meantime, granuloma tissue, a repair blastema, is formed to provide structural support in this vulnerable area, while endothelial cells form capillaries that provide basic nutrition and also deliver cells to the healing site. Growth factors are released from the local environment; these include insulin-like growth factor (IGF), vascular endothelial growth factor (VEGF), platelet-derived growth factor (PDGF), transforming growth factor β (TGF-β), fibroblast growth factor (FGF), and epidermal growth factor (EGF) [22]. These secreted growth factors play important roles in the proliferation and differentiation of "stem-like" cells at the healing site.

### 8.2.1.2 Reparative Phase

As the MSCs begin to differentiate, a loose, unorganized callus is formed. This phase requires the presence of collagens and many noncollagenous proteins, including bone morphogenetic protein (BMP), osteopontin (OPN), osteocalcin (OCN), alkaline phosphatase (ALP), and bone sialoprotein (BSP), along with several others. Collagen provides a protein bed (osteoid) for biomineralization to occur. Some noncollagenous proteins serve as adhesive molecules that immobilize cells. Others are enzymes that activate the binding sites in the collagen through phosphorylation/dephosphorylation. Still others play a role as carriers to deliver calcium and phosphate ions to supersaturated loci for apatite nucleation. Table 8.1 lists the putative roles of selected proteins involved in regulating osteoblast activity and biomineralization.

Linked with expression of these proteins are transcriptional factors, including osterix and Runx-2/Cbfa-1, which have been identified as key factors required for osteoblast differentiation and skeletal development. The *cis*-binding element (OSE2) for Runx2/Cbfa-1 has been identified in various osteoblast-specific genes, including type I collagen, BSP, OPN, OCN, and ALP. Binding of Runx2/Cbfa-1 to the OSE2 site regulates the expression of transcripts encoding these proteins [128]. A key result defining the critical role of Runx2/Cbfa-1 expression during skeletal development was the observation of the absence of mineralized bone in mice with homogyzous deletion of the Runx2/Cbfa-1 gene [119, 120]. Therefore, any factors that influence Run2/Cbfa-1 and osterix expression of these transcription factors may affect this phase of healing.

There are two types of reparative cascades: one involves a phase of chondrogenesis (cartilage formation) first and then converts to osteogenesis (bone formation), a process that occurs primarily in appendicular bone and vertebrae (endochondral bone formation). In the second type, intramembranous bone formation, bone is formed directly without a cartilage template.

**Table 8.1.** Major proteins associated with the osteoblast phenotype

| Protein | Function [reference] |
|---|---|
| Type I collagen | Provides the organic matrix for mineralization |
| Alkaline phosphatase | Marker for osteoblast differentiation; thought to be critical for regulating Pi/PPi and subsequently biomineralization [51] |
| Osteopontin (OPN) | Present in many tissues, with high concentration in bone. Various roles assigned to OPN include regulation of crystal growth, protection against cell death, regulation of inflammation, and promotion of osteoclast adhesion [12] |
| Osteocalcin | A late marker of the osteoblast phenotype. A modulator of crystal growth [26, 126] |
| Osteonectin/SPARC | Found in many tissues. In bone, rises during the increased mineralization (reparative) phase; may mediate deposition of hydroxyapatite; considered to have a role in angiogenesis [80] |
| Bone sialoprotein | Thought to enhance mineralization and support osteoblast cell attachment [9, 35, 56, 66] |
| Wnts | Bind to their receptors and then regulate LEF1/TCF; promote osteoblast maturation and may play a role in lineage commitment of mesenchymal precursor cells [87, 114, 125] |
| Transforming growth factor β (TGF-β) superfamily (bone morphogentic proteins, BMPs) | Regulate a myriad of cellular processes based on their extracellular concentration. At low concentrations, promote chemotaxis and cellular proliferation; at high concentrations, facilitate cellular differentiation and bone formation (e.g., BMP-2, -4, -7) [115, 116, 127] |
| Parathyroid hormone-related peptide (PTHrP) | Proven to act in many tissues to regulate both development and function; inhibits bone resorption; thought to be a signaling molecule in epithelial–mesenchymal interactions [11, 49, 122] |
| Fibroblast growth factors (FGFs) | Modulate cell migration, angiogenesis, bone development and repair, and epithelial–mesenchymal interactions; e.g., FGF-2 stimulates osteoblast proliferation and enhances bone formation [42, 83, 89, 90] |

Healing in craniofacial bone, ileum, scapula, and clavicle mainly involves intramembranous bone formation. Some examples of factors and proteins that affect the differentiation of MSCs into cartilage and bone during postnatal bone growth include Indian hedgehog (*ihh*), BMPs, Wnts, Sox, parathyroid hormone-related peptide (PTHrP), and transcription factor *gli3* (Table 8.2). During this stage, collagen fibers are not perfectly aligned, and as a result a loose, unorganized woven tissue is produced.

### 8.2.1.3 Remodeling Phase

After primary formation, healing skeletal tissue reaches phase 3, the remodeling phase. In this phase, the unorganized bone woven produced in phase 2 is replaced with a more organized structure, signaling the complete restoration of damaged bone. The remodeling process, known as activation-resorption-formation (ARF), is guided by expression of specific genes and associated proteins, protein synthesis and secretion, and physical activity.

Numerous factors, including macrophage colony-stimulating factor (M-CSF), tumor necrosis factor α (TNF-α), receptor activator of nuclear factor κB (RANK) and its ligand RANKL, and osteoprotegerin (OPG) [72], have been shown to play critical roles in balancing osteoblast–osteoclast homeostasis. M-CSF promotes osteoclast maturation, whereas RANKL is required for activating the osteoclast to resorb bone. RANK, which is expressed by osteoclast progenitors and mature osteoclasts, binds to its ligand, RANKL, which is expressed on osteoblasts and stromal cells. For example, osteoblasts activated by signaling factors such as parathyroid hormone (PTH) and lipopolysaccharide (LPS) enhance their secretion of OPG and/or RANKL. RANKL binds to RANK receptors and activates osteoclasts, whereas OPG acts as a delay and blocks RANKL–RANK-mediated osteoclast activation. When osteoclasts become stimulated, they home to osteoblast-vacant zones, attach at these sites, and resorb mineralized tissues.

**Table 8.2.** Transcription factors associated with bone metabolism

| Factors | Functions [reference] |
|---|---|
| Runx2/Cbfa1 | A runt domain containing transcription factor essential for osteoblast and hypertrophic chondrocyte differentiation and bone formation during embryogenesis and postnatal life [27, 76, 88] |
| Osterix (Osx; SP7) | Transcription factor containing a zinc-finger motif and essential function for osteoblast differentiation; may prevent chondrocyte differentiation [17, 40, 91] |
| ATF4 (CREB2; cAMP response-element binding protein 2) | A basic leucine-zipper transcription factor and a member of the ATF/CREB protein family. ATF4 is involved in regulation of osteoblast differentiation and bone formation and exhibits cooperative interactions with Runx2/Cbfa1 [123, 124] |
| NFAT (nuclear factor of activated T cells) | NFAT forms a complex with osterix that binds to DNA. This interaction appears to be important for the transcriptional activity of osterix [73] |
| β-Catenin/TCF/LEF (TCF, T-cell factor; LEF, lymphoid enhancer factor) | Transcription regulatory DNA binding complex considered to play multiple critical roles in osteoblast differentiation [10, 41, 48] |
| Osteoclast transcription factors: PU.1; Fos/Fra1; NFATc1 (nuclear factor of activated T cells, cytoplasmic, calcineurin-dependent 1); NFκB (nuclear factor κB); MITFs (microphthalmia-associated transcription factors) | Several transcription factors are involved in promoting osteoclast differentiation and maturation from hematopoietic lineage cells. A few such factors are listed here [72] |

This is a very limited list of some of the key transcription factors associated with bone tissues. For more details, please refer to the references cited in the table and Bilezekian et al. [4]

Following activation, lysosomal enzymes, such as tartrate-resistant acid phosphatase (TRAP) and cathepsin K, are synthesized and secreted through the ruffled border into the extracellular bone-resorbing compartment of woven bone. The osteoclast also secretes various metalloproteinases, including collagenases. These enzymes dissolve and degrade the bone mineral and organic matrix. Resorption of woven bone releases noncollagenous proteins, such as BMP and TGF-β, which are thought to stimulate osteoblastic activity. In response to an as yet unidentified signal, osteoclasts cease resorbing and abandon their attachment to bone [52]. Osteoclastic resorptive pits (Howship's lacunae) are repopulated by osteoblasts to produce osteoid, which then mineralizes to restore bone.

It is now recognized that osteocytes, in addition to osteoblasts and osteoclasts, play an important role in remodeling [7]. Osteocytes can be modulated by environmental factors such as fluid flow. Mechanical deformation of osteocytes alters gene expression and leads to secretion of biochemical signals that regulate osteoblast and osteoclast activity. The group of cells (osteoblast, osteoclast, and osteocyte) responsible for the ARF process is called the *basic multicellular unit* (BMU) [36]. In humans, the ARF takes approximately 3 to 6 months to complete a remodeling cycle, and the events continue throughout adult life. Changes in estrogen levels, such as those occurring in postmenopausal women, can alter bone homeostasis; hence the increased susceptibility of women in comparison with men to bone fractures with age. This dynamic remodeling process generates microstructures, including lamellae, haversian canals, and the defined orientations of mature trabecular bone.

## 8.2.2 Mechanical Effects on Bone Healing

A century ago, Wolff [121] studied biomechanics using the femur and discovered that the orientation of trabecular bone coincided with stress patterns. This discovery came to be called Wolff's law: bony architecture aligns with the direction of principal stresses. Since that time, researchers have worked diligently to find a mathematical relationship to solve this observational enigma. A large number of studies, primarily performed during the 1980s, were undertaken to examine the local mechanical responses of mineralized tissues during fracture healing. These studies attempted to develop a physical rule to explain the link between callus formation and functional stresses at different stages of healing.

During the healing process, many factors, including the sex, genotype, and age of an individual and the presence of chemical and physical stimuli, influence the gene/protein expression of cells and thus affect the outcome of regeneration [15, 44, 46, 47]. The degree of influence depends on the local bone quality, which governs cell activities and determines the degree of maturation possible. The sequential expression of tissue-specific genes encoding collagens, proteoglycans, and other noncollagenous ECM proteins provides patterning for the development of new ECM.

An ordered sequence of cell differentiation and mRNA expression of bone-matrix proteins governs the evolving histological changes observed during fracture healing [6, 63, 102, 126]. The expression of markers of osteoblast activity during bone healing follows a temporal pattern that is similar to the adolescence growth curve. It contains a lag phase (prespurt) followed by a spurt and then a decline (postspurt). During the spurt stage, the cells secrete profoundly more ECM than does unwounded, normal bone. This overexpression of genes/proteins may facilitate a rapid mineralization during healing and was termed the regional acceleratory phenomenon (RAP) by Frost in 1986 [37]. Frost postulated that the RAP healing process is controlled predominantly by biological factors (cells and molecules) released at sites of injury, and that this phase is insensitive to other physical stimuli, such as stresses and strains. He implied that accelerated ossification is controlled by specific genes. In contrast to this idea, a significant body of histomorphometric research has shown that early weight bearing (loading 2 to 7 days after fracture) can enhance both endochondral and intramembranous ossification during fracture healing [24, 44, 81, 92, 104]. Distracted motion perpendicular to an osteotomized bone section has also been shown to stimulate profound bone formation. In addition, osseous implants show evidence of an improved osseointegration effect, similar to fracture healing and distraction osteogenesis, when early-loading protocols are used [112]. Taken together, these observations indicate that there is a therapeutic benefit of functional loading prior to the postspurt RAP, e.g., to stimulate healing tissue to produce bone faster and to maintain osteoblastic cells in highly active and mitogenic states.

In intact bone, the pathway for transducing mechanical signals in bone cells includes an increase of the mRNA level of the protooncogene *c-fos* and of the bone matrix proteins collagen and ALP [97]. In addition, several other genes/proteins have been shown to be altered by mechanical stimuli. Similar findings were reported in a healing scaffold as a result of loading the 8-week healed implant [86]. Further, Ziros et al. [128] identified Runx-2/Cbfa-1 as a target for mechanotransduction in human periodontal ligament osteoblastic fibroblast-cell culture. Following mechanical stretch, both Runx2/Cbfa-1 mRNA and proteins were up-regulated. The authors reported that MAPK could physically interact with and phosphorylate Runx2/Cbfa-1. Other physical stimuli, such as a shock wave [119, 120] and ultrasound [105], have also been shown to influence the phosphorylation of various genes, such as Runx2/Cbfa-1. Thus, the more recent studies have taken into account that mechanical signal transduction is part of the regenerative effects of late RAP.

During the accelerated regenerative stage (prespurt or spurt) of fracture healing, distraction osteogenesis, and endosteal implantation, the mechanical response of osteoblastic cells, has been shown to produce more ECM than that evoked during the subsided RAP (postspurt) [44, 68]. This finding contradicts Frost's hypothesis that RAP is not affected by physical stimulation. A decrease of cellular activity during the late healing stage may result in cells being less responsive to mechanical stimuli. If this were the case, the situation would be analogous to that in intact, disused bone in which bone cells resist gene regulation, and the result is decreased bone formation [23, 30, 50, 77, 109]. Questions exist as to whether the accelerated molecular expression during early healing elevates tissue sensitivity to mechanical stimulation, and whether the enhanced osteoblastic bioactivity suppresses osteoclastic resorption and callus shrinkage. Other questions concern the role of apoptosis of osteoblasts that occurs during the late healing stage [78, 103].

Each form of regenerated tissue at various healing stages possesses unique viscoelastic properties that depend on the maturity of the collagen network and the mineralization of the hydroxyapatite frames. In all cases, the tissue should be strong enough to sustain deformation but compliant enough to allow deformation. An exact deformation that is within the tolerance of the mechanical strength of the tissue and beyond the threshold of cellular sensitivity is important for the success of regenerative therapies.

## 8.3 Motion and Osteogenesis

The physical interaction between the various cells and their resulting ECM must be considered. In continuum mechanics, motion can create a relative movement in ECM to a selected reference point that needs to be considered in a three-dimensional space. Internal stresses and strains created by motion are transmitted in various directions to maintain equilibrium of cells and associated ECM within the local environment. From the mechanical theory, nine components of stresses and strains, as illustrated in Fig. 8.1, exist at any point in ECM. The components $\sigma_{xx}$, $\sigma_{yy}$, $\sigma_{zz}$ are considered normal stresses, and the remaining components $\sigma_{xy}$, $\sigma_{xz}$, $\sigma_{yx}$, $\sigma_{yz}$, $\sigma_{zx}$, $\sigma_{zy}$, etc. are considered shearing stresses. Each of these components is expressed in the dimensions of force per unit area [39]. These nine components of stress form a symmetrical mathematical matrix in a cartesian coordinate system. Each stress component is expressed by the orthogonal vectors in the cartesian system, which represents the viewpoint of the observer. The magnitude of each stress component changes when the system rotates. At one particular rotation angle, we can eliminate confounding dimensions, thus reducing the system to a simpler diagonal matrix with three nonzero stresses.

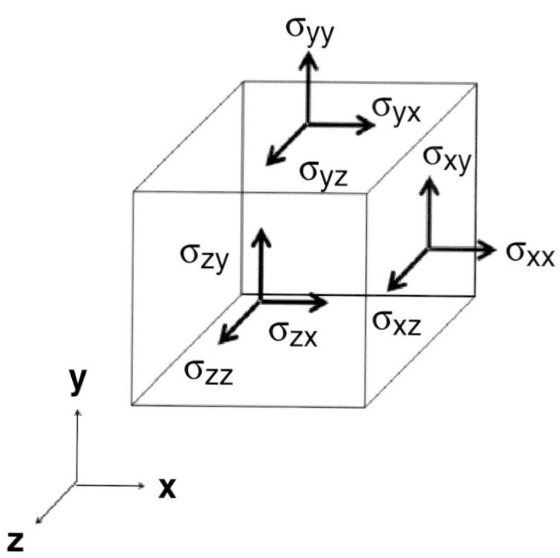


**Figure 8.1.** Schematic drawing of nine stress components at any point inside an object.

The coordinate axes of the diagonal matrix are called principal axes, and the corresponding stress components are called principal stresses. In engineering terms, the maximum of these principal stresses is referred to as the *tensile stress*, and the minimum is referred to as the *compressive stress*. In physics, tensile stresses pull the neighboring ECM molecules apart, which tends to unfold and straighten protein molecules. Compressive stresses, on the other hand, condense neighboring particles, which shortens the distance between hydroxyapatite crystals in mineralized bone and affects the growth of the minerals. In addition, the same cells in a given area will respond differently to tensile and compressive stress. When a force is applied to tissue, both of these stresses combine to alter bone homeostasis at the local site. Such a differential effect of stresses becomes critical when motion, such as distraction, is used to regenerate bone.

Functionally, skeletal regeneration is an extension of adaptive responses, which are controlled by genetically and environmentally determined factors. Mechanical adaptation, paralleling Wolff's law to a great degree, results in remodeling of the proteins and minerals within ECM to accommodate the applied stress patterns. Stresses can be carefully engineered along a specific direction, and thus designing mechanotherapies where stresses are controlled may assist in providing predictable regenerative approaches. Two clinical examples using such strategies are distraction osteogenesis for lengthening of limbs and jaw bones, and alveolar bone regeneration through control of tooth movement.

### 8.3.1 Distraction Osteogenesis

Distraction osteogenesis provides an attractive model for the study of mechanical forces and their effects on bone formation, because this technique produces a large volume of new bone in a controlled fashion [25, 98]. The technique of distraction osteogenesis has been used in the practice of orthopedics and oral maxillofacial surgery. The procedure, following osteotomy, includes a latency period of up to 6 days, a distraction period during which the osteotomized gap is lengthened by 0.50 to 0.75 mm per day for 14 days, and a consolidation period of 8 weeks. These procedures were designed on the basis of the compliance or stiffness of the ECM and result in increasing the length of bone by 1 to 10 cm, depending on the duration of distraction.

The latency period is the period from bone division to the onset of traction and represents the time required for reparative callus formation. Prolonging the latency period of bone healing may prevent distraction. If the bone matures to a point at which mineralization is significant, the distraction process will fracture bone instead of inducing growth. As bone matures, it accumulates hydroxyapatite and becomes brittle, giving it very little deformation range. Thus, the correct length of the latency period should be determined before proceeding to the distraction stage. Inadequate scheduling of each phase may result in relapse and failure to lengthen [1, 111].

Formation of soft callus is the key to successful distraction and vascularization. The soft callus includes collagen and progenitor cells. The distraction appears to be related to the movement of flexible, threadlike, long-chain molecules of collagen. Long collagen fibers along the direction of stretch have been reported in the distraction gap. When collagen undergoes changes in configuration, internal cohesive force (stress) is developed. This internal stress is then transferred into the cells, where collagen transcription rates are increased. Like a manufacturing plant, the cells produce large amounts of ECM until the gap is filled. It has been shown that the increased products in the distraction stage consist of primary fibrous and vascular tissues; little hydroxyapatite has been found. The strain decreases every day during distraction from infinite strain (day 1) to less than 0.1 strain by day 10 (Fig. 8.2). The strain level appears much higher than the levels that Frost and other researchers reported to occur during bone remodeling [38, 84, 100]. In addition, the direction of the tensile strain, rather than the direction of the compression or the shear strain, corresponds well with the direction of fibrous tissue formation.

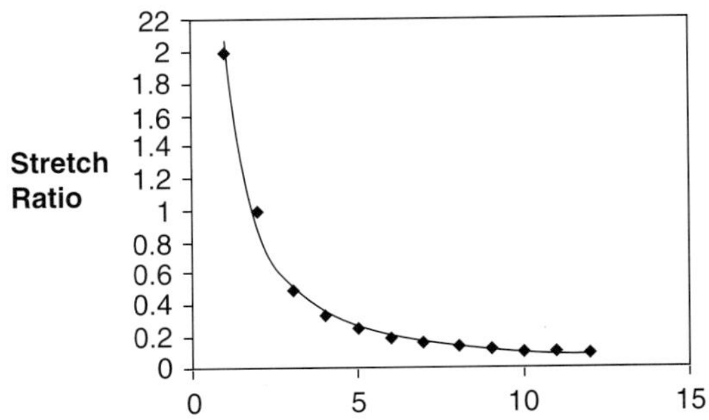


**Figure 8.2.** Stretch ratio, defined as deformation divided by the original length of the distracted gap, during distraction osteogenesis. The stretch ratio decreases as the distraction period increases.

The consolidation stage represents the time at which large amounts of hydroxyapatite are deposited in bone. The rate of bone formation appears to reach a maximum 2 to 4 weeks after the completion of distraction (Fig. 8.3). At the beginning of consolidation (end of distraction), the tissue is filled with fibrovascular tissue comprising 70% to 93% of the total regeneration area and and organized as parallel collagen bundles with interspersed vascular channels. Only 2% to 5% of the regenerated tissue consists of bony trabeculae; the remaining 4% to 27% is marrow space (Fig. 8.4A). At 2 weeks of consolidation, new bone formation occupies up to 30% of the distracted region (Fig. 8.4B). At 4 and 6 weeks, new bone occupies 40% to 45% of the area, with a small fibrous interzone remaining (Fig. 8.4C). At 8 weeks, the regenerated area is filled with trabecular bone and lacks a fibrous interzone. The trabeculae increase in both length and thickness during the time of consolidation and are oriented parallel to the direction of distraction. No tensile strain greater than 0.1 should be applied to the tissues during the consolidation stage, because excessive distraction inhibits crystal growth. However, Richards [98] has shown that adding small compressive strains (strains less than 0.003; 3000 $\mu\varepsilon$) might provide additional stimulation to cells and produce more bone than would be produced without any additional forces. In addition, pressure at the local site may produce a consolidation effect on the formation of hydroxyapatite crystals and hydroxyapatite–collagen complex.

**Figure 8.3.** Bone formation rates (A) percentage trabecular bone area and (B) mineral appositional rate, measured in the region of interest, reached a maximum 4 weeks after the end of distraction. Data adapted from Samchukov et al, 2001 [101].

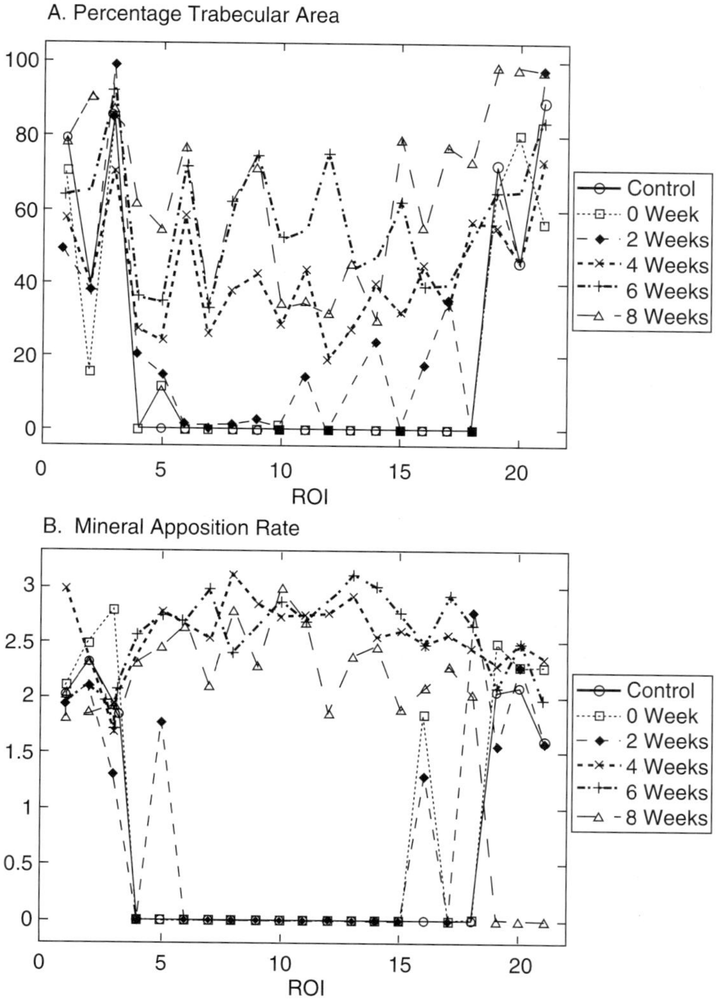


The effect of motion on distraction regeneration relies on two mechanisms: formation of fibrovascular matrix, and growth and condensation of hydroxyapatite crystals. We propose the use of a stress-strain diagram (Fig. 8.5) to predict differential bone phases analogous to the general chemistry phase diagram that is used to explain the relationship between ice, water, and water vapor. This diagram is a modification of Carter's previous work [16] and is based on histomorphometric observations. The three phases of bone formation (endochondral) are the fibrous, cartilage, and osseous phases. Neither the fibrous nor the cartilage phase contains hydroxyapatite, which means that $Ca^{2+}$ and $PO_4^{3-}$ ions in the tissue do not form apatite crystals. High tensile stress and strain cause fibrous tissue formation, while high compression produces osseous tissue (tissue mineralized with hydroxyapatite). In the area between low tensile and low compression, all threee phases (cartilage, fibrous cartilage, and osscous tissue) can be developed. Molecular studies have shown that during the early stage of healing, motion inhibits mesenchymal cell differentiation into osteoblasts by increasing expression of the *ihh* gene, which regulates chondrocyte maturation during fetal and early postnatal skeletogenesis [82]. Collagen, ECM proteins and hydroxyapatite crystals are associated with the osseous phase. Cytokines such as IGF-1, TGF-β, BMP-2, and BMP-4, which can modulate mineralization, are up-regulated in cells associated with distracted tissues [104]. The formation of these phases is controlled by the products of the cells (e.g., collagen, and ECM proteins) and physical-chemical reactions among the secreted cell products and calcium and phosphate ions. The stress/strain history of the tissues can affect both the cell products and the subsequent physical-chemical interactions.

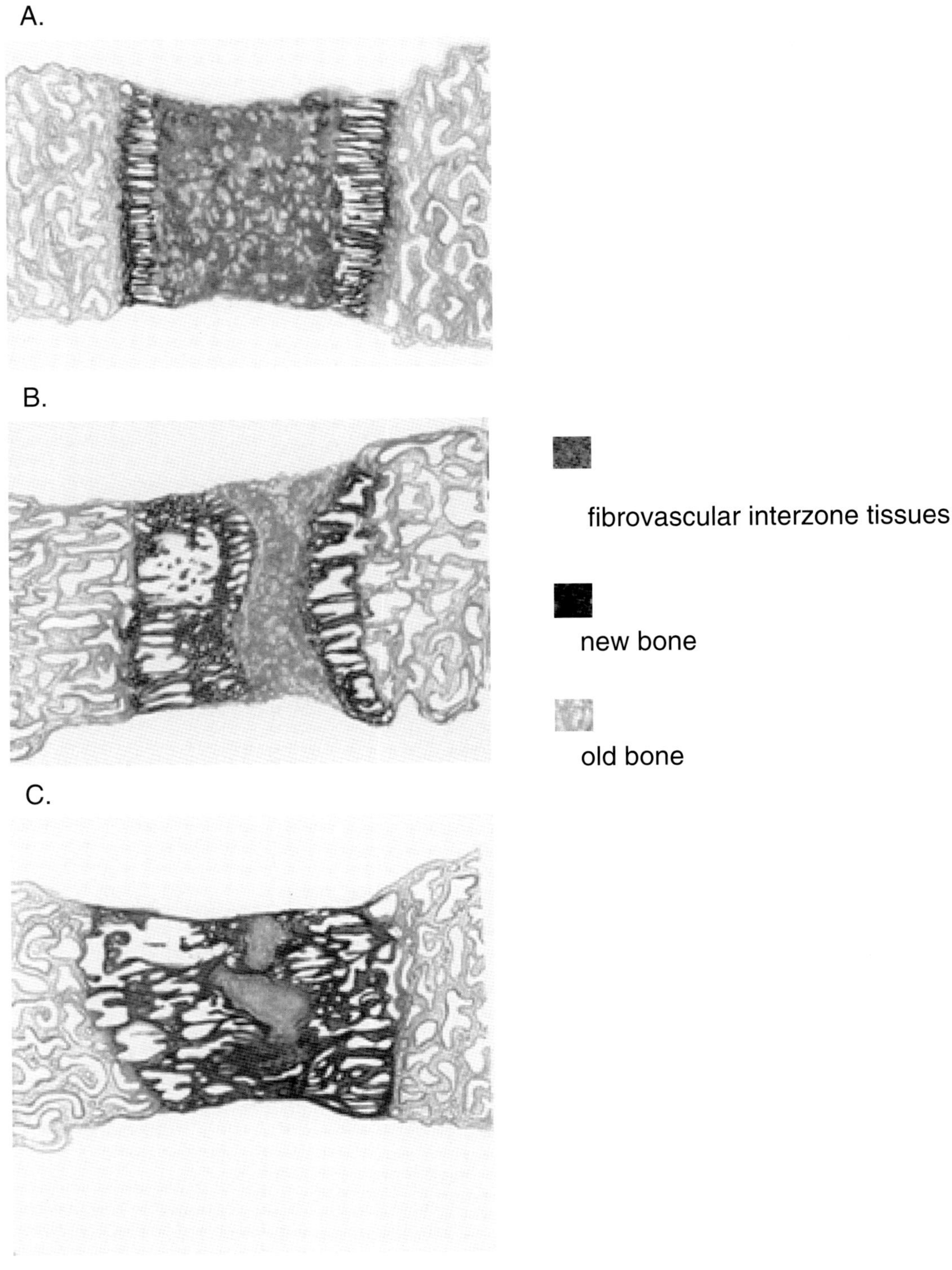


**Figure 8.4.** Schematics showing histomorphology of distraction regeneration at week 0 (A), week 2 (B), and week 4 (C) after distraction of rat mandible. The proportion of old bone, new bone, and fibrovascular interzone tissues varied at different stages. Diagram was adapted from the original histological findings of Samchukov et al [101].

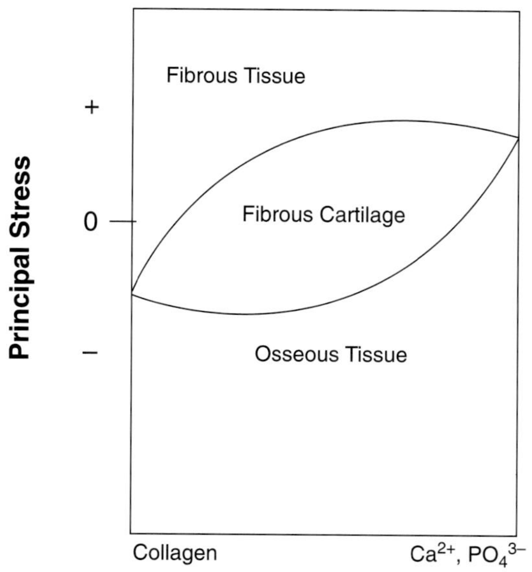


**Figure 8.5.** Phase diagram showing three tissue phases associated with tissue composition and stress received in a given tissue.

## 8.3.2 Tooth Movement

Some of the aforementioned principles are being practiced by a large percentage of orthodontists today. Moving a tooth on the tensile side of its alveolar socket can regenerate bone [74, 75, 95, 113]. In theory, tooth movement is equivalent to "distraction". The periodontal ligament is composed of fibrous tissue, and the cells of the periodontal ligament serve as an interzone tissue analogous to the osteotomized gap in distraction osteogenesis. The continuous forces generated during the orthodontic treatment, in general, are reactivated every 1 to 2 months by the use of a metallic wire. Within this 2-month (nonadjustment) period, the distraction and consolidation phases occur. At the beginning of loading (in this case, a new adjustment made by adjusting an orthodontic wire), the periodontal ligament is stretched for approximately 0.3 mm within a day and then left to consolidate for the rest of the nonadjustment period. The tensile strain level applied to the periodontal tissue in a central incisor is estimated to be 500 to 1000 $\mu\varepsilon$ for the periodontal ligament and 10 $\mu\varepsilon$ for the lamella dura. It has been postulated that these strains can increase blood flow [65] and oxygen levels in bone, as well as activate osteoclasts, a process that will stimulate cell signaling, differentiation, and, subsequently, remodeling and new bone formation. Conventionally, tooth movement is used to align tooth position by remodeling the bone surrounding the roots of the teeth. More recently, these principles have been used to regenerate large amounts of bone for replacement of periodontal defects and atrophied ridges [31].

## 8.3.3 Growth Modification

Dental functional appliances (e.g., the Herbst appliance) have been used as bite-jumping devices to modify the growth of the jaw bones. The movement of the jaw, which is guided by these devices, modifies the stress/strain fields of the mandibular condyle and glenoid fossa. As a result, the growth of the mandible and the maxilla can be redirected to correct abnormal bite patterns of individuals. The theory of growth relativity states that bone growth modifications occur relative to retrodiscal tissues, which are temporomandibular ligaments, and the transduction of the nonmuscular forces [64, 117]. The retrodiscal tissues are stretched like a large elastic band between the fossa and the displaced condyle. The transduction of these nonmuscular forces has been shown to be effective at a significant distance from the actual physical soft-tissue attachments. In a simplified two-dimensional finite element analysis, we found that tensile strains of 1700 to 3000 $\mu\varepsilon$ could be created by the temporomandibular ligament. This yielded a biomechanical effect on condylar growth. These values were estimated on the basis of a 1-mm forward movement of the mandible. This model did not include muscular system and occlusal forces; future modeling will include more detailed anatomy and three-dimensional dental-muscular structures. Nevertheless, the tensile strain vectors ran posteriorly and matched with the posterior growth direction observed in animal studies [64, 117, 118]. Our preliminary data support the growth relativity concept, which suggests that tensile strains and stresses due to the constraint of mandibular movement are related to modification of osteogenesis within the mandibular condyle.

## 8.4 Micromotion and Implant Osseointegration

### 8.4.1 Orthopedic Porous Implants

We have developed a series of two-dimensional finite element models to calculate interfacial tissue strains inside the pores of porous coated implants, based upon analyses of histological sections obtained from canine tibia [8, 67]. We hypothesized that long-term bone ingrowth to the porous interface is related to the state of local interface tissue strain. This hypothesis was tested by comparing predicted values of the amount of local bone (Ip) in the interfacial zone with actual ingrowth (Ia) measurements obtained from animal experiments [43], using a rule relating ingrowth distribution to calculated tissue strain magnitude. The rule is based on the assumption that bone adaptation of osseointegrated tissues obeys the minimum effective strain (MES) theory postulated by Frost [37]. If the local bone element is subjected to a maximum shear strain greater than a given threshold value (MES), it will undergo a maintenance process, resorption-formation equilibrium, in which the element retains the same properties as defined in the original model. Otherwise, the bone is resorbed and replaced with soft tissue, thus decreasing element stiffness. Here, the MES was computed locally in 10-με-thick regions of osseointegrated tissues around individual beads over a whole implant interface. Homogenization theory [45, 53, 67, 69] was used to calculate local strains by coupling the local microstructural model with the global model.

Parametrical analyses by varying loading directions, osseointegrated tissue modulus (E = 0.01, 0.1, 0.5, 1, 5, and 10 GPa), and levels of MES (500, 1000, 1500, 2000, 2500, and 3000 με) were used to investigate the relationship between motion and bone regeneration (ingrowth). Linear regression between Ip and the percentage of actual measured bone ingrowth, Ia, was calculated. Coefficients of determination, *r*, and the sum of squares difference, SSQD ($\sigma|Ia\text{-}Ip|^2$), the accumulated difference between Ia and Ip, were then compared by a three-way analysis of variancee (ANOVA) (Prophet 5.0, BBN Co., MA). The optimal range of MES for correlating tissue strain with the osseointegrated tissue was determined according to the statistical results.

The results (Table 8.3) showed that the best (highest *r*) angle for predicting Ia was 0. This result is consistent with the primarily vertical loads applied to the canine tibia during gait. The best bone modulus (E) for prediction was 0.5 GPa, followed by 0.1 and 1 GPa; however,

**Table 8.3.** Summary of the results of analyses with respect to correlation coefficient *r* (top) and SSQD (bottom)

| Source | DF | Sum of squares | *F* ratio | Prob > *F* |
|---|---|---|---|---|
| Loading angle | 6 | 1.852062 | 44.5449 | 0.0000 |
| Modulus (GPa) | 5 | 10.974497 | 316.7441 | 0.0000 |
| Loading angle* modulus | 30 | 2.039731 | 9.8117 | 0.0000 |
| MES | 5 | 0.123563 | 3.5662 | 0.0045 |
| Loading angle* MES | 30 | 0.084647 | 0.4072 | 0.9974 |
| Modulus * MES | 25 | 0.796406 | 4.5971 | 0.0000 |
| **Source** | **DF** | **Sum of squares** | ***F* ratio** | **Prob > *F*** |
| Loading angle | 6 | 9754948746 | 21.7949 | 0.0000 |
| Modulus (GPa) | 5 | $3.18913*10^{11}$ | 855.0325 | 0.0000 |
| Loading angle* modulus | 30 | 6909422103 | 3.0875 | 0.0000 |
| MES | 5 | $7.33802*10^{10}$ | 196.7383 | 0.0000 |
| Loading angle* MES | 30 | 1018415745 | 0.4551 | 0.9933 |
| Modulus * MES | 25 | $1.95474*10^{10}$ | 10.4816 | 0.0000 |

The relationship between Ia and Ip varied significantly ($p < 0.05$) with loading angle, modulus, and minimum effective strain (MES). Abbreviations: DF, degrees of freedom; SSQD, sum of squares difference.

MES had little effect on the prediction. The interaction between E and MES revealed that for the three middle E values (0.1, 0.5, 1 GPa), *r* showed an increasing trend in MES. The interaction between loading angle and E indicated that for all angles, 0.01, 5, and 10 GPa had poor *r* values. For E values of 0.1 through 1 GPa, the *r* value varied considerably in accordance with the loading angle, with 0° being superior to the other angle.

Similar to the results of the *r* values, the best (smallest SSQD) loading direction on average was 0; the best E and MES values were the high ones. When the interaction between E and MES was examined, the prediction improved for all E as MES increased. However, for the four highest E values (0.5, 1, 5, and 10 GPa), the prediction increased substantially as MES increased from 500 to 1500 με, but modestly or not all for larger MES values. The results indicated that 1500 με may be a reasonable cutoff strain for MES. From examination of the interaction between angle and E, loading angle had a similar performance for E ≤ 0.5 GPa, but these values separated for 1 through 10 Gpa, with 0 being the best prediction.

The top-10 list (Table 8.4) for the highest Ia-Ip correlation and smallest deviation showed that two criteria, *r* and SSQD, did not agree entirely, although the same combination (0–0.5 GPa–3000 με) yielded the best agreement between experiment and theory according to both criteria. However, some combinations with high *r* had poor SSQD (such as number 2 on the *r* list) and were ruled out of consideration as strain-mediated ingrowth parameters. In these cases, the data fell close to a straight line, but not the line Ia = Ip. Therefore, the results indicate that, by using a range of MES

**Table 8.4.** Top-10 list ranked by *r* and SSQD

| Ranked by *r* | Loading angle | E (GPa) | MES (με) | SSQD | *r* | A | B |
|---|---|---|---|---|---|---|---|
| 1 | 0° | 0.5 | 3000 | 6062 | 0.83 | −4.94 | 1.33 |
| 2 | 0° | 0.1 | 1500 | 122234 | 0.82 | 93.57 | 0.10 |
| 3 | 0° | 0.5 | 2500 | 13576 | 0.82 | 6.71 | 1.36 |
| 4 | 0° | 0.1 | 3000 | 82441 | 0.82 | 71.69 | 0.46 |
| 5 | 0° | 0.1 | 2000 | 109480 | 0.80 | 87.39 | 0.19 |
| 6 | 0° | 0.1 | 2500 | 95493 | 0.80 | 79.47 | 0.33 |
| 7 | 0° | 1 | 1500 | 9834 | 0.79 | 1.05 | 1.31 |
| 8 | 0° | 0.5 | 2000 | 29026 | 0.75 | 20.53 | 1.29 |
| 9 | 0° | 0.1 | 1000 | 132710 | 0.74 | 98.19 | 0.03 |
| 10 | 0° | 1 | 1000 | 37858 | 0.72 | 26.33 | 1.26 |
| **Ranked by SSQD** | **Loading angle** | **E (GPa)** | **MES (με)** | **SSQD** | ***r*** | **A** | **B** |
| 1 | 0° | 0.5 | 3000 | 6062 | 0.83 | −4.94 | 1.33 |
| 2 | 0° | 1 | 2000 | 7759 | 0.71 | 93.57 | 0.10 |
| 3 | 0° | 1 | 1500 | 9834 | 0.79 | 6.71 | 1.36 |
| 4 | 0° | 5 | 500 | 10073 | 0.49 | 71.69 | 0.46 |
| 5 | 0° | 1 | 2500 | 11505 | 0.68 | 87.39 | 0.19 |
| 6 | 0° | 0.5 | 2500 | 13576 | 0.82 | 79.47 | 0.33 |
| 7 | 0° | 1 | 3000 | 16047 | 0.66 | 1.05 | 1.31 |
| 8 | 0° | 10 | 500 | 21387 | −0.36 | 20.53 | 1.29 |
| 9 | 30° | 1 | 3000 | 22374 | 0.42 | 98.19 | 0.03 |
| 10 | 0° | 5 | 1000 | 22507 | 0.14 | 26.33 | 1.26 |

A and B are the coefficients in the regression equation of Ia = A + B * Ip. The three-factor combined effects indicate that 0°–0.5 Gpa–3000 με and 0°–1 Gpa–1500 με were the best combinations matching both criteria to establish strain-ingrowth relationships. The cases with 0.1 GPa were ruled out by two-factor interaction analyses. Abbreviations: E, elastic modulus; MES, minimum effective strain; Ia, amount of bone ingrowth measured by histological section; Ip, amount of bone ingrowth predicted by computer modeling.

(1500 to 3000 με), combined with an interfacial bone modulus of 0.5 and 1 GPa, the Frost theory can best predict bone ingrowth corresponding to actual experimental ingrowth ($r \approx 0.8$).

This local strain-ingrowth relation has not been previously described for osseointegration with the use of either isolated local models [71, 93] or global models [3, 15, 54, 55]. Our data showed that such a local correlation depends upon the synergy between global boundary conditions and local bone properties. If off-axis loading occurs or if bone is either too compliant or too stiff (for example outside of the physiological range), the strain-ingrowth relation, even with use of the same MES, becomes statistically weaker. A vertical load yields the best association for the strain-ingrowth relation. The effect of the other anterior-posterior loading components on the strain-ingrowth relation was poor, confirming previous findings that a vertical load is the dominant mode of motion on the canine tibia [8], and that other loading conditions in an optimization simulation were not significantly correlated with the actual ingrowth.

In the study described above, the binary rule that was used to relate strain state to the amount of bone is limited to two hypothetical strain regions for bone adaptation: below MES (a single value) for resorption and above MES for maintenance. However, according to Frost's theory, MES is a physiological strain window (200 to 2500 με), not a single value, within which bone retains its mineral (maintenance). Bone resorbs if strain is below the window. If the strain is above the window, the bone will either deposit more mass (formation) to strengthen the structure or undergo tissue damage or necrotic resorption. Our binary rule simplifies interfacial bone remodeling to two functions, resorption (<MES) and maintenance (≥MES) only, by seeking a strain threshold for the lower bound of the strain window. The predicted areas of bone ingrowth (strain ≥ MES) from the rule eventually include homeostasis of maintenance, formation, and resorption. Potential necrotic resorption was not separated in the present study, because only a small portion (<4%) of the tissue area was predicted to be under a high-strain condition (>4000 με). Further studies are needed to include the upper bound of the strain window for more accurate predictions of the amount of bone growth that can be achieved.

Despite the limitation, the overall results suggest that mechanically driven osseointegration adaptation is possible. The results support the hypothesis that long-term bone distribution in the implant interface is related to the local interface tissue strain state. Thus, mechanical adaptation for long-term osseointegration of porous coated implants seems to occur at a local level of micromotion.

## 8.4.2 Dental Implants

Endosteal implants have been among the most significant developments in dentistry over the past 20 years [5, 29, 85, 94]. The use of implants for edentulous patients and for single tooth replacement has grown exponentially. Previously we found that alveolar ridges, which contain distinct porosity with active synthesis of trabecular bone (intramembrane ossification [61, 107]), may respond to mechanical stimuli differently than long bone. This difference relates implant osseointegration to alveolar crest bone adaptation and emphasizes the importance of early loading on a mandibular microenvironment. This finding is particularly acute in dental clinical practice where immediate functional loading, i.e. no waiting or healing period, is preferred for newly inserted implants.

Dental implants have a specific thread design to lock into the jawbone, minimizing interfacial motion. In addition, various adjuvant treatments, such as food selection and implant splinting (connecting implants together), can be applied to reduce bite forces and minimize interfacial movements. It has been shown that with careful clinical designation, early loading does not cause excessive relative movement that would result in failure of osseointegration between the implant and bone. For these reasons, dental implants have started a trial in immediately loading in the past two decades [111].

We tested alveolar bone regeneration adaptation using an animal model [68, 71]. In this study, Sinclair mini-pigs (Sinclair Research Center, Columbia, MO) and in vivo controlled loads were used. Titanium threaded dental implants (Walter Lorenz Co., Jacksonville, FL) and intraoral hydraulic devices were placed unilaterally in the premolar alveolar ridges of mandibles. Both the implant and the device were protected from any bite forces. A controlled load consisting of daily loading for a 5-month period was administered to the implant through the device. The implant was loaded with a 6.5 N force, with a cyclic square wave at 1 Hz (600 cycles/day). The implant was immobilized and allowed to heal for 1 month prior to loading. Microcomputer tomographic (μCT) scans were used to determine the peri-implant bone density of the experimental implants.

The three-dimensional osseous architecture of the μCT images showed qualitatively higher bone density, thicker trabeculae, and fewer intertrabecular spaces surrounding the 5-month-loaded implants [68]. The trabeculae appeared to orient in a specific apical direction running from the cortical shell to the implant; this suggests that there was an adaptation response to loading. This adaptation reached remodeling equilibrium at sites where tissues received daily attractor stress $\Psi_{bas}$, the stress value of which provides adequate stimuli to bone cells for maintaining a balance between formation and resorption [14, 15]. Based on the μCT image, a two-dimensional finite element model was constructed and used to determine the daily attractor stress value (Fig. 8.6). The model describes the mathematical relationship

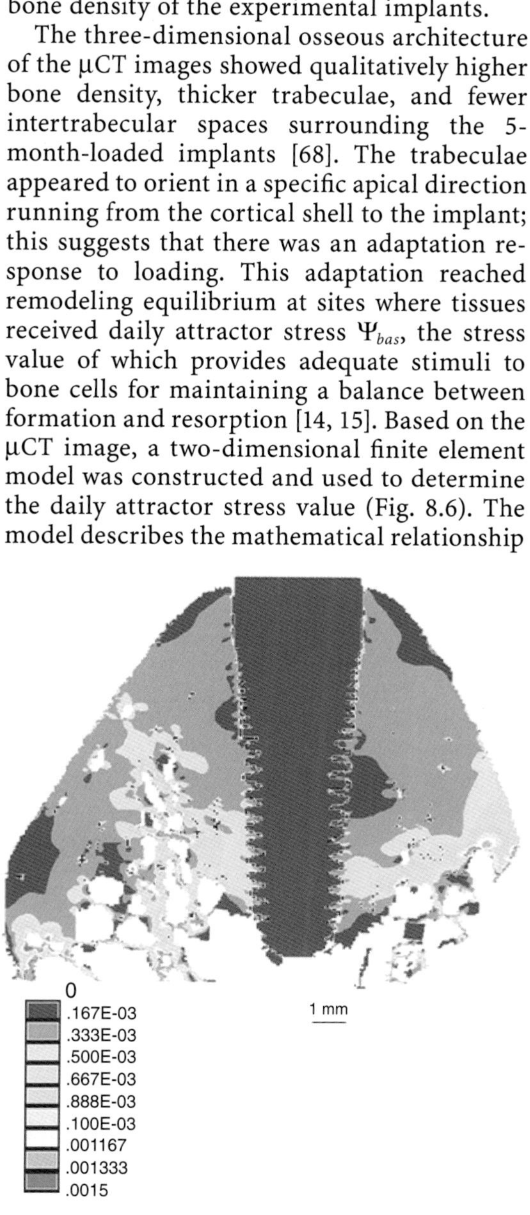


**Figure 8.6.** Strain distributions of the implant–alveolar bone complex were computed based on the outcome of our previous study [68]. With the use of this model, an equilibrium stage was reached after a 5-month loading. The tissue strains around the coronal and middle third of the implant appeared very uniform. This stress value was taken as the attractor stress state.

$\psi_{bas} = (\sum_{day} n_i \sigma_{basi})$, in which $\sigma_{basi}$ is the tissue stress generated by a single load cycle at the tissue equilibrium stage, and $n_i$ is the number of cycles of load type $i$ per day.

An idealized finite element model was further constructed to predict bony patterns by using adaptive methods. It was assumed that during healing, cells continuously modify the mineral density of the surrounding bone, according to the equation $r = c^*(\psi_b - \psi_{bas})$, in which $\Psi_{bas}$ is the daily stress stimulus $\psi_b = (\sum_{day} n_i \sigma_{bi})$ created by the loading device. In this equation, the difference between a daily tissue-level stress stimulus and the attractor state stress stimulus is named the tissue-remodeling criterion. In our study, the increase (positive) of the remodeling criteria (MPa/day) would raise the elastic modulus of bone proportionally to the increase of bone density, and vica versa. The variable $c$ is an empirically determined value and was set to equal one [(MPa/day)/(MPa/day)] in our case. This remodeling equation was adapted from Beaupre et al. [3] for long-bone studies.

Results showed that with the use of tensile stress criteria, the predicted bony pattern matched that of experimental μCT data (Fig. 8.7). Other stress components (e.g., Von Mises and compressive stress) cannot provide similar predictions to relate motion-derived stresses with regenerated bony architecture, thus indicating that the cells in alveolar bone are more prone to tension stimuli than the cells in long bone. As described in Section 8.4.1 and in other orthopedic literature, compression and shear stresses provide greater stimulation to long-bone adaptation than tensile stress. This type of tensile stress, which stimulates alveolar bone osseointegration, is consistent with the forces used for tooth movement. Nevertheless, a peri-implant ligament analog to the periodontal ligament does not form, and thus a tension zone, as seen in normal orthodontic procedures, does not occur. The similarity of tension-stress effects may be due to the prevalence of soft callus and progenitor cells in the early healing stage. The soft callus allows a large stretch range of the tissues. Fiber extrusion along the direction of tensile stress, similar to that observed in distraction osteogenesis, can occur in the interfacial tissue (Fig. 8.8). This explains why trabecular bone formation aligns with the principal direction of tensile stress. The estimated maximum tissue strain (70 με) was much lower than the MES values

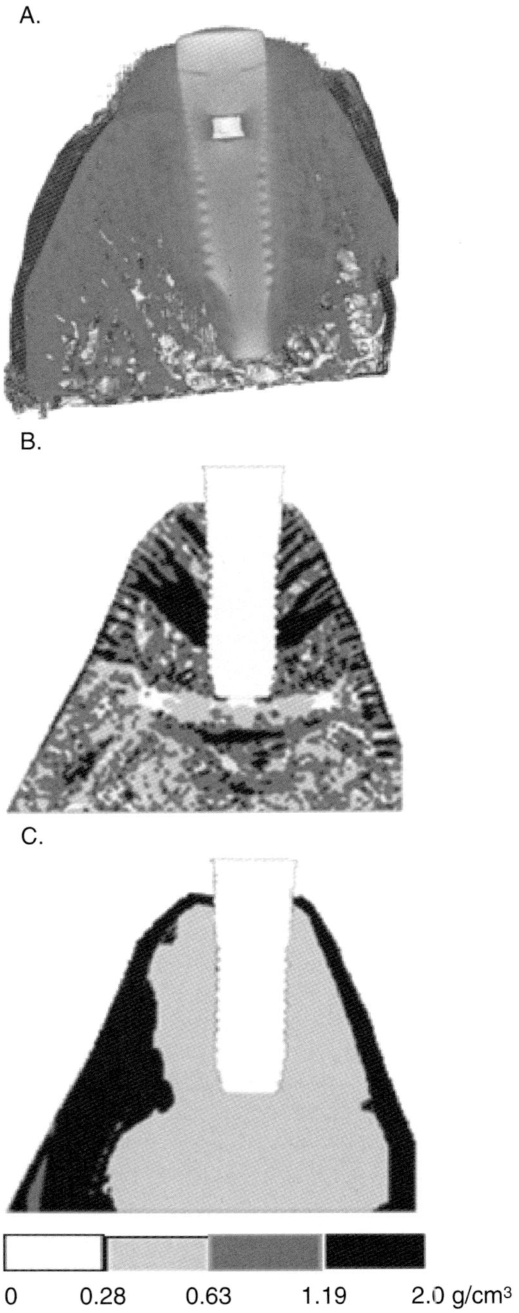


**Figure 8.7.** (A) Microcomputer tomographic image of tissue architecture produced from one of our animal studies. This was considered to represent an equilibrium stage (attractor state). (B) The bony density and architecture predicted by tensile stress criteria appeared to match the patterns shown in (A). (C) The bony pattern predicted by Von Mises criteria did not match that of experimental data. This result indicated that cells surrounding an osseointegrated dental implant are susceptible to tensile force stimulation, which differs from implants placed in long bone.

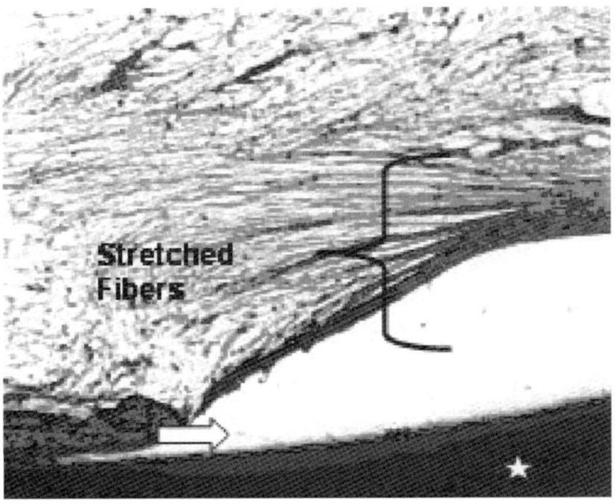


**Figure 8.8.** Selected region of the interfacial tissue surrounding a dental implant showing stretched collagen fibers, similar to those observed during distraction osteogenesis. The section along the long axis of the dental implant was stained with Stevenel's Blue and van Gieson's Picro-Fuchsin and was observed under 30× magnification. Arrow shows bone; star shows the implant.

(700 and 1000 $\mu\varepsilon$) reported by McLeod and Rubin [84] and by Rubin and Lanyon [100], respectively. Nonetheless, the effect on bone formation was clear. Qin et al. [96] reported that a threshold near 70 $\mu\varepsilon$ can induce antiresorptive bone formation, and our findings agree with these conclusions. It is possible that progenitor cells within alveolar or healing bone are more susceptible to low strain thresholds, so that even a small strain can induce substantial bone regeneration.

# 8.5 Summary

Mineralized biological tissues are essentially composite materials with dynamic structures that can change because of increases or decreases in mineralization via apatite nucleation and growth or dissolution processes. This balance gives rise to structures not seen in engineering composites. Biological and biomechanical factors are two of the most important factors influencing the dynamics of regenerated bone. Healing can regenerate bone through biological cascades by altering gene expression in bone cells. This natural process involves soluble factors delivered from blood or released at the local injured site. Modification of healing regeneration uses physical movement, including distraction osteogenesis, orthodontic tooth movement, and implant osseointegration. The relationship between motion and these three types of bone regeneration have been discussed in this chapter. At the beginning of distraction osteogenesis, high strain (>0.1) is required to regenerate large amounts of collagen substrate. Once the ECM induces mineralization, the effective strain level decreases to the range of 500 to 3000 $\mu\varepsilon$. Extremely low effective strain (70 $\mu\varepsilon$) was identified for dental implants at early stages of healing. The alveolar ridge is more susceptible to tensile stress stimuli, whereas long bone responds to shear stress.

# References


1. Altuna GW, Walker DA, Freeman E (1995) Surgically assisted-rapid orthopedic lengthening of the maxilla in primates: relapse following distraction osteogenesis. Int J Adult Orthod Orthog Surg 10:269–275.
2. Baron RR, Ravesloot JH, Neff L, Chakraborty M, Chatterjee D, Lomri A, Horne WC (1993) Cellular and molecular biology of the osteoclast. In: Noda M, ed. Cellular and Molecular Biology of Bone. Academic Press, San Diego, pp 446–495.
3. Beaupré GS, Orr TE, Carter DR (1990) An approach for time-dependent bone modeling and remodeling—theoretical development. J Orthop Res 8: 651–661.
4. Bilezikian JP, Raisz L, Rodan GA (2002) Principles of Bone Biology, 2nd ed. Academic Press, San Diego.
5. Block MS, Kent J, Guerra LR (1997) Implants in Dentistry W.B. Saunders Co, Philadelphia.
6. Bolander M (1994) Regulation of fracture repair and synthesis of matrix macromolecules. In: Brighton CT, Friedlaender GE, Lane JM. eds. Bone Formation and Repair. American Academy of Orthopaedic Surgeons, Rosemont, IL. pp 185–196.
7. Bonewald L (2002) Osteocytes: a proposed multifunctional bone cell. J Musculoskelet Neuronal Interact 2:239–241.
8. Borodkin JL, Eadie JS, Choi K, Hollister SJ, Goldstein SA (1994) The effect of mechanical stimuli on bone ingrowth into porous coated implants. In: 40th Meeting ORS, New Orleans, p 582.
9. Boskey A (1989) Noncollagenous matrix proteins and their role in mineralization. Bone Miner 6:111–123.
10. Brault VM, Moore R, Kutsch S, Ishibashi M, Rowitch DH, McMahon AP, Sommer L, Boussadia O, Kemler R (2001) Inactivation of the β-catenin gene by Wnt1-Cre-mediated deletion results in dramatic brain malformation and failure of craniofacial development. Development 128:1265–1273.
11. Broadus AE, Stewart AF (1996) Parathyroid hormone-related protein: structure, processing, and physiological actions. In: Bilezikian JP, Levine AM, Marcus R, eds. The Parathyroids. Raven Press, New York, pp 259–294.
12. Butler WT, Ridall AL, McKee MD (1996) Osteopontin. In: Bilezikian JP, Raisz L, Rodan GA, eds. Principles of Bone Biology. Academic Press, San Diego, pp 167–182.
13. Caplan A (1991) Mesenchymal stem cells. J Orthop Res 9:641–650.
14. Carter D (1987) Mechanical loading history and skeletal biology. J Biomech 20:1095–1109.
15. Carter DR, Beaupre GS, Giori NJ, Helms JA (1998) Mechanobiology of skeletal regeneration. Clin Orthop Relat Res 355:S41–S55.
16. Carter DG, Giori NJ (1990) Effect of mechanical stress on tissue differentiation in bony implant bed. In: Davies J, ed. The Bone-Biomaterial Interface. University of Toronto Press, Toronto, pp 367–379.
17. Celil AH, Hollinger JO, Campbell PG (2005) Osx transcriptional regulation is mediated by additional pathways to BMP2/Smad signaling. J Cell Biochem 95:518–528.
18. Chien SSJ (1998) Effects of hemodynamic forces on gene expression and signal transduction in endothelial cells. Biol Bull 194:390–393.
19. Chow JW, Fox SW, Lean JM, Chambers TJ (1998) Role of nitric oxide and prostaglandins in mechanically induced bone formation. J Bone Miner Res 13: 1039–1044.
20. Chow JW, Jagger CJ, Chambers TJ (1993) Characterization of osteogenic response to mechanical stimulation in cancellous bone of rat caudal vertebrae. Am J Physiol 265:E340–E347.
21. Claes L, Eckert-Hubner K, Augat P (2002) The effect of mechanical stability on local vascularization and tissue differentiation in callus healing. J Orthop Res 20:1099–1105.
22. Clark RAF (1988) The Molecular and Cellular Biology of Wound Repair, 2nd ed. Plenum Press, New York.
23. Colleran PN, Wilkerson MK, Bloomfield SA, Suva LJ, Turner RT, Delp MD (2000) Alternatives in skeletal perfusion with simulated microgravity: a possible mechanism for bone remodeling. J Appl Physiol 89: 1046–1054.
24. Connolly JF, Hahn H, Davy D (1978) Fracture healing in weight-bearing and nonweight-bearing bones. J Trauma 18:766–770.
25. Cope JBS, Mikhail L (2000) Regenerate bone formation and remodeling during mandibular osteodistraction. Angle Ortho 70:99–111.
26. Ducy P, Desbois C, Boyce B, Pinero G, Story B, Dunstan C, Smith E, Bonadio J, Goldstein S, Gundberg C, Bradley A, Karsenty G (1996) Increased bone formation in osteocalcin-deficient mice. Nature 382: 448–452.
27. Ducy P, Zhang R, Geoffroy V, Ridall AL, Karsenty G (1997) Osf2/Cbfa1: a transcriptional activator of osteoblast differentiation. Cell 89:747–754.
28. Egger EL, Gottsauner-Wolf F, Palmer J, Aro HT, Chao EY (1993) Effects of axial dynamization on bone healing. J Trauma 34:185–192.
29. Escobar V, Epker BN (1998) Alveolar bone growth in response to endosteal implants in two patients with ectodermal dysplasia. Int J Oral Maxillofac Surg 27:445–447.
30. Evans GL, Morey-Holton E, Turner RT (1998) Spaceflight has compartment- and gene-specific effects on mRNA levels for bone matrix proteins in rat femur. J Appl Physiol 84:2132–2137.

31. Faber J, Azevedo RB, Bao SN (2005) Distraction osteogenesis may promote periodontal bone regeneration. J Dent Res 84:757–761.
32. Ferguson CM, Alpern E, Miclau T, Helms JA (1999) Does adult fracture repair recapitulate embryonic skeletal formation? Mech Dev 87:57–66.
33. Ferguson CM, Miclau T, Hu D, Alpern E, Helms JA (1998) Common molecular pathways in skeletal morphogenesis and repair. Ann NY Acad Sci 23:33–42.
34. Franceschi R (1999) The developmental control of osteoblast-specific gene expression: role of specific transcription factors and the extracellular matrix environment. Crit Rev Oral Biol Med 10:40–57.
35. Franzen AH, Heinegard D (1985) Isolation and characterization of two sialoproteins present only in bone calcified matrix. Biochem J 232:715–724.
36. Frost H (1987) Bone "mass" and the "mechanostat": a proposal. Anat Rec 219:1–9.
37. Frost H (1986) Intermediary Organization of the Skeleton. Vol 1. CRC Press, Boca Raton, FL.
38. Frost H (1990) Skeletal structural adaptations to mechanical usage (SATMU): 1. Redefining Wolff's law: the remodeling problem. Anat Rec 226: 414–422.
39. Fung Y (1977) A First Course In Continuum Mechanics, 2nd ed. Prentice-Hall, Englewood Cliffs, NJ.
40. Gao Y, Jheon A, Nourkeyhani H, Kobayashi H, Ganss B (2004) Molecular cloning, structure, expression, and chromosomal localization of the human Osterix (SP7) gene. Gene 341:101–110.
41. Gaur TL, Lengner CJ, Hovhannisyan H, Bhat RA, Bodine PV, Komm BS, Javed A, van Wijnen AJ, Stein JL, Stein GS, Lian JB. (2005) Canonical WNT signaling promotes osteogenesis by directly stimulating RUNX2 gene expression. J Biol Chem 280:33132–33140.
42. Globus RK, Patterson-Buckendahl P, Gospodarowicz D (1988) Regulation of bovine bone cell proliferation by fibroblast growth factor and transforming growth factor-beta. Endocrinology 123:98–105.
43. Goldstein SA, Matthews LS, Kuhn J, Hollister SJ (1991) Trabecular bone remodeling: an experimental model. J Biomech 24(Suppl):135–150.
44. Goodship AE, Cunningham JL, Kenwright J (1998) Strain rate and timing of stimulation in mechanical modulation of fracture healing. Clin Orthop Rel Res 355S:S105–S115.
45. Guedes JM, Kikuchi N (1990) Preprocessing and postprocessing for materials based on the homogenization method with adaptive finite element methods. Comput Meth Appl Mech Eng 83:143–198.
46. Guldberg RE, Caldwell NJ, Guo XE, Goulet RW, Hollister SJ, Goldstein SA (1997) Mechanical stimulation of tissue repair in the hydraulic bone chamber. J Bone Miner Res 12:1295–1302.
47. Guldberg RE, Richards M, Caldwell NJ, Kuelske CL, Goldstein SA (1997) Trabecular bone adaptation to variations in porous-coated implant topology. J Biomech 30:147–153.
48. Haegel HL, Larue L, Ohsugi M, Fedorov L, Herrenknecht K, Kemler R (1995) Lack of beta-catenin affects mouse development at gastrulation. Development 121:3529–3537.
49. Halloran BP, Nissenson RA (1992) Parathyroid Hormone-Related Protein: Normal Physiology and its Role in Cancer. CRC Press, Boca Raton, FL.
50. Harris SA, Zhang M, Kidder LS, Evans GL, Spelsberg TC, Turner RT (2000) Effects of orbital spaceflight on human osteoblastic cell physiology and gene expression. Bone 26:325–331.
51. Henthorn P (1996) Alkaline phosphatase. In: Bilezikian JP, Raisz L, Rodan GA, eds. Principles of Bone Biology. Academic Press, San Diego, pp 197–206.
52. Hollinger JO, Buck DC, Bruder SP (1999) Biology of bone healing: its impact on clinical therapy. In: Lynch SE, Genco R, Marx RE, eds. Tissue Engineering: Applications in Maxillofacial Surgery and Periodontics. Quintessence Publishing Co, Chicago, pp 17–54.
53. Hollister SJ, Fyhrie DP, Jepsen KJ, Goldstein SA (1991) Application of homogenization theory to the study of trabecular bone mechanics. J Biomech 24:825–839.
54. Hollister SJ, Ko CC, Kohn DH (1993) Bone density around screw thread dental implants predicted using topology optimization. In: Advances in Bioengineering. Vol 24. ASME, New York, pp 339–342.
55. Huiskes RR, Ruimerman R, van Lenthe GH, Jansen JD (2000) Effects of mechanical forces on maintenance and adaptation of form in trabecular bone. Nature 405:704–706.
56. Hunter GK, Goldberg HA (1993) Nucleation of hydroxyapatite by bone sialoprotein. Proc Natl Acad Sci USA 90:8562–8565.
57. Ilizarov GA (1990) Clinical application of the tension-stress effect for limb lengthening. Clin Orthop Relat Res 250:8–26.
58. Ilizarov GA (1989) The tension-stress effect on the genesis and growth of tissues: Part I. The influence of stability of fixation and soft-tissue preservation. Clin Orthop Relat Res 238:249–281.
59. Ilizarov GA (1989) The tension-stress effect on the genesis and growth of tissues: Part II. The influence of the rate and frequency of distraction. Clin Orthop Relat Res 239:263–285.
60. Ingber D (1998) Cellular basis of mechanotransduction. Biol Bull 194:323–327.
61. Jahangiri L, Devlin H, Ting K, Nishimura I (1998) Current perspectives in residual ridge remodeling and its clinical implications: a review. J Prosthet Dent 80:224–237.
62. Jarvinen MJ, Lehto MU (1993) The effects of early mobilisation and immobilisation on the healing process following muscle injuries. Sports Med 15:78–79.
63. Jingushi S, Bolander ME (1991) Biological cascades of fracture healing as models for bone-biomaterial interfacial reactions. In: Davies JE, ed. The Bone Biomaterial Interface. University of Toronto Press, Toronto, pp 250–262.
64. Johnston LJ (1996) Functional appliances: a mortgage on mandibular position. Austr Orthod 14: 154–157.
65. Khouw FE, Goldhaber P (1970) Changes in vasculature of the periodontium associated with tooth movement. Arch Oral Biol 15:1125–1132.
66. Kinne RW, Fisher LW (1987) Keratan sulfate proteoglycan in rabbit compact bone is sialoprotein II. J Biol Chem 262:10206–10211.
67. Ko CC (1994) Mechanical Characterization of Implant/Tissue Interfaces. University of Michigan Press, Ann Arbor.

68. Ko CC, Douglas WH, DeLong R, Rohrer MD, Swift JQ, Hodges JS, An K-N, Ritman EL (2003) Effects of implant healing time on crestal bone loss of a controlled-load dental implant. J Dent Res 82:585–591.
69. Ko CC, Kohn DH, Hollister SJ (1996) Effective anisotropic elastic constants of bimaterial interphases: comparison between experimental and analytical techniques. J Mater Sci Mater Med 7:109–117.
70. Ko CC, Kohn DH, Hollister SJ (1992) Micromechanics of implant/tissue interfaces. J Oral Implantol 18: 220–230.
71. Ko CC, Swift JQ, DeLong R, Douglas WH, Kim YI, An KN, Chang CH, Huang HL (2002) An intra-oral hydraulic system for controlled loading of dental implants. J Biomech 35:863–869.
72. Kobayashi T, Kronenberg H (2005) Minireview: transcriptional regulation in development of bone. Endocrinology 46:1012–1017.
73. Koga T, Matsui Y, Asagiri M, Kodama T, de Crombrugghe B, Nakashima K, Takayanagi H (2005) NFAT and Osterix cooperatively regulate bone formation. Nat Med 8:880–885.
74. Kokich V (1996) Esthetics: the orthodontic-periodontic restorative connection. Semin Orthod 2:21–30.
75. Kokich VJ (2002) Congenitally missing teeth: orthodontic management in the adolescent patient. Am J Orthod Dentofacial Orthop 121:594–595.
76. Komori T, Yagi H, Nomura S, Yamaguchi A, Sasaki K, Deguchi K, Shimizu Y, Bronson RT, Gao YH, Inada M, Sato M, Okamoto R, Kitamura Y, Yoshiki S, Kishimoto T (1997) Targeted disruption of Cbfal results in a complete lack of bone formation owing to maturational arrest of osteoblasts. Cell 89:755–764.
77. Kostenuik PJ, Harris J, Halloran BP, Turner RT, Morey-Holton ER, Bikle DD (1999) Skeletal unloading causes resistance of osteoprogenitor cells to parathyroid hormone and to insulin-like growth factor-I. J Bone Miner Res 14:21–31.
78. Landry P, Sadasivan K, Marino A, Albright J (1997) Apoptosis is coordinately regulated with osteoblast formation during bone healing. Tissue Cell 29: 413–419.
79. Landry PS, Marino AA, Sadasivan KK, Albright JA (1996) Bone injury response. An animal model for testing theories of regulation. Clin Orthop Rel Res 332:260–273.
80. Lane TF, Sage EH (1994) The biology of SPARC, a protein that modulates cell-matrix interactions. FASEB J 8:163–173.
81. Latta LL, Sarmiento A, Tarr RR (1980) The rationale of functional bracing of fractures. Clin Orthop Relat Res 146:28–36.
82. Le AX, Miclau T, Hu D, Helms JA (2001) Molecular aspects of healing in stabilized and non-stabilized fractures. J Orthop Res 19:78–84.
83. Mayahara H, Ito T, Nagai H, Miyajima H, Tsukuda R, Taketomi S, Mizoguchi J, Kato K (1993) In vivo stimulation of endosteal bone formation by basic fibroblast growth factor in rats. Growth Factors 9:73–80.
84. McLeod KJ, Rubin C (1992) Sensitivity of the bone remodeling response to the frequency of applied strain. Trans Orthop Res Soc 17:533.
85. Misch CE (1999) Contemporary Implant Dentistry, 2nd ed. Mosby, St. Louis.
86. Moalli MR, Caldwell NJ, Patil PV, Goldstein SA (2000) An in vivo model for investigations of mechanical signal transduction in trabecular bone. J Bone Miner Res 15:1346–1353.
87. Moon RT, Bowerman B, Boutros M, Perrimon N (2002) The promise and perils of Wnt signaling through beta-catenin. Science 296:1644–1646.
88. Mundlos S, Otto F, Mundlos C, Mulliken JB, Aylsworth AS, Albright S, Lindhout D, Cole WG, Henn W, Knoll JHM, Owen MJ, Mertelsmann R, Zabel BU, Olsen BR (1997) Mutations involving the transcription factor CBFA1 cause cleidocranial dysplasia. Cell 89:773–779.
89. Nakamura K, Kawaguchi H, Aoyama I, Hanada K, Hiyama Y, Awa T, Tamura M, Kurokawa T (1997) Stimulation of bone formation by intraosseous application of recombinant basic fibroblast growth factor in normal and ovariectomized rabbits. J Orthop Res 15:307–313.
90. Nakamura K, Kurokawa T, Kawaguchi H, Kato T, Hanada K, Hiyama Y, Aoyama I, Nakamura T, Tamura M (1997) Stimulation of endosteal bone formation by local intraosseous application of basic fibroblast growth factor in rats. Rev Rheum Engl Ed, 64:101–105.
91. Nakashima K, Zhou X, Kunkel G, Zhang Z, Deng JM, Behringer RR, de Crombrugghe B (2002) The novel zinc finger-containing transcription factor osterix is required for osteoblast differentiation and bone formation. Cell 108:17–29.
92. Park SH, O'Connor K, McKellop H, Sarmiento A (1998) The influence of active shear or compressive motion on fracture-healing. J Bone Joint Surg 80A: 868–878.
93. Pedersen DR, Brown TD, Brand RA (1991) Interstitial bone stress distributions accompanying ingrowth of a screen-like prosthesis anchorage layer. J Biomech 24:1131–1142.
94. Percinoto C, Vieira AE, Barbieri CM, Melhado FL, Moreira KS (2001) Use of dental implants in children: a literature review. Quintessence Int 32: 381–383.
95. Proffit WR (2000) Contemporary Orthodontics, 3rd ed. Mosby, St. Louis.
96. Qin YX, Rubin CT, McLeod KJ (1998) Nonlinear dependence of loading intensity and cycle number in the maintenance of bone mass and morphology. J Orthop Res 16:482–489.
97. Raab-Cullen DM, Thiede MA, Petersen DN, Kimmel DB, Recker RR (1994) Mechanical loading stimulates rapid changes in periosteal gene expression. Calcif Tissue Int 55:473–478.
98. Richards MG, James A, Weiss JA, Waanders NA, Schaffler MB, Goldstein SA (1998) Bone regeneration and fracture healing: experience with distraction osteogenesis model. Clin Orthop Relat Res 355S: S191–S204.
99. Roodman G (1996) Advances in bone biology: the osteoclast. Endocr Rev 17:308–332.
100. Rubin CT, Lanyon LE (1985) Regulation of bone mass by mechanical strain magnitude. Calcif Tissue Int 37:411–417.
101. Samchukov ML, Cope JB, Cherkashin AM (2001) Biologic basis of new bone formation under the influence of tension stress, in Samchukov ML, Cope JB, Cherkashin AM, eds. Craniofacial Distraction Osteogenesis, Mosby, Louis pp 21–41.

102. Sandberg MM, Aro HT, Vuorio EI (1993) Gene expression during bone repair. Clin Orthop Relat Res 289: 292–312.
103. Sarmiento A, Latta LL, Tarr RR (1984) The effects of function in fracture healing and stability. Instr Course Lect 33:83–106.
104. Sarmiento A, Schaeffer JF, Beckerman L, Latta LL, Enis JE (1977) Fracture healing in rat femora as affected by functional weight-bearing. J Bone Joint Surg 59A:369–375.
105. Sato M, Ochi T, Nakase T, Hirota S, Kitamura Y, Nomura S, Yasui N (1999) Mechanical tension-stress induces expression of bone morphogenetic protein (BMP)-2 and BMP-4, but not BMP-6, BMP-7, and GDF-5 mRNA, during distraction osteogenesis. J Bone Miner Res 14:1084–1095.
106. Sato W, Matsushita T, Nakamura K (1999) Acceleration of increase in bone mineral content by low-intensity ultrasound energy in leg lengthening. J Ultrasound Med 18:699–702.
107. Schenk RK, Hunziker EB (1994) Histologic and ultrastructural features of fracture healing. In: Brighton CT, Friedlaender G, Lane JM, eds. Bone Formation and Repair. American Academy of Orthopaedic Surgeons, Rosemont, IL, pp 117–146.
108. Seyfer AE, Hollinger JO (1994) Bone Repair and Regeneration. WB Saunders, Philadelphia.
109. Sibonga JD, Zhang M, Evans GL, Westerlind KC, Cavolina JM, Morey-Holton E, Turner RT (2000) Effects of spaceflight and simulated weightlessness on longitudinal bone growth. Bone 27:535–540.
110. Stocum D (1995) Wound Repair, Regeneration and Artificial Tissues. RG Landes, Austin, TX.
111. Stucki-McCormick S, Drew S, Mizrahi RD (2001) Distraction osteogenesis: overcoming the challenges of a new technique. In: Samchukov MC, Cope J, Cherkashin A, eds. Craniofacial Distraction Osteogenesis. Mosby, St. Louis, pp 595–603.
112. Szmukler-Moncler S, Salama H, Reingewirtz Y, Dubruille JH (1998) Timing of loading and effect of micromotion on bone-dental implant interface: review of experimental literature. J Biomed Mater Res Appl Biomater 43:192–203.
113. Thilander B, Nyman S, Karring T, Magnusson I (1983) Bone regeneration in alveolar bone dehiscences related to orthodontic tooth movements. Eur J Orthod 5:105–114.
114. Tuan R (2003) Cellular signaling in developmental chondrogenesis: N-cadherin, Wnts, and BMP-2. J Bone Joint Surg Am 85A(Suppl 2):137–141.
115. Urist M (1997) Bone morphogenetic protein: the molecularization of skeletal system development. J Bone Miner Res 12:343–346.
116. Urist MR, Mikulski AJ, Nakagawa M, Yen K (1977) A bone matrix calcification-initiator noncollagenous protein. Am J Physiol 232:C115–127.
117. Voudouris JC, Woodside DG, Altuna G, Angelopoulos G, Bourque PJ, Lacouture CY, Kuftinec MM (2003) Condyle-fossa modifications and muscle interactions during Herbst treatment, Part 2. Results and conclusions. Am J Orthod Dentofacial Orthop 124:13–29.
118. Voudouris JC, Woodside DG, Altuna G, Kuftinec MM, Angelopoulos G, Bourque PJ (2003) Condyle-fossa modifications and muscle interactions during Herbst treatment, Part 1. New technological methods. Am J Orthod Dentofacial Orthop 123:604–613.
119. Wang FS, Wang CJ, Huang HJ, Chung H, Chen RF, Yang KD (2001) Physical shock wave mediates membrane hyperpolarization and Ras activation for osteogenesis in human bone marrow stromal cells. Biochem Biophys Res Commun 287:648–655.
120. Wang FS, Wang CJ, Sheen-Chen SM, Kuo YR, Chen RF, Yang KD (2002) Superoxide mediates shock wave induction of ERK-dependent osteogenic transcription factor (Cbfa1) and mesenchymal cell differentiation toward osteoprogenitors. J Biol Chem 277: 10931–10937.
121. Wolff J, Das Gesetz der Transformation der Knochen (1892) Hirschwald, Berlin.
122. Wysolmerski JJ, Stewart AF (1998) The physiology of parathyroid hormone-related protein: an emerging role as a developmental factor. Annu Rev Physiol 60: 431–460.
123. Xiao G, Jiang D, Ge C, Zhao Z, Lai Y, Boules H, Phimphilai M, Yang X, Karsenty G, Franceschi RT (2005) Cooperative interactions between ATF4 and Runx2/Cbfa1 stimulate osteoblast-specific osteocalcin gene expression. J Biol Chem 280:30689–30696.
124. Yang X, Matsuda K, Bialek P, Jacquot S, Masuoka HC, Schinke T, Li L, Brancorsini S, Sassone-Corsi P, Townes TM, Hanauer A, Karsenty G (2004) ATF4 is a substrate of rsk2 and an essential regulator of osteoblast biology: implication for Coffin-Lowry syndrome. Cell 117:387–398.
125. Yang Y (2003) Wnts and wing: Wnt signaling in vertebrate limb development and musculoskeletal morphogenesis. Birth Defects Res, Part C, Embryo Today 69:305–317.
126. Zhou H, Choong P, McCarthy R, Chou ST, Martin TJ, Ng KW (1994) In situ hybridization to show sequential expression of osteoblast gene markers during bone formation in vivo. J Bone Miner Res 9: 1489–1499.
127. Zimmerman LB, DeJesus-Escobar JM, Harland RM (1996) The Spemann organizer signal noggin binds and inactivates bone morphogenetic protein 4. Cell 86:599–606.
128. Ziros PG, Gil AP, Georgakopoulos T, Habeos I, Kletsas D, Basdra EK, Papavassiliou AG (2002) The bone-specific transcriptional regulator Cbfa1 is a target of mechanical signals in osteoblastic cells. J Biol Chem 277:23934–23941.

# 9. Dental Applications of Bone Biology

*Thomas W. Oates and David L. Cochran*

## 9.1 Introduction

The teeth are implanted in depressions within alveolar bone and are surrounded by the periodontium which consists of bone, a suspensory ligament (the periodontal ligament), cementum on the root surface, and gingiva. In health, the bone tissue is located approximately 2 mm below the cementoenamel junction which separates the crown of the tooth and its root from the bone (Fig. 9.1). From a functional viewpoint, the periodontium is a unique, very dynamic and adaptable tissue. The periodontal ligament has one of the fastest turnover rates of connective tissue in the body and maintains its dimensions even if the teeth are moved or the ligament is regenerated. At the same time, the periodontium provides support for the tooth, resists biting forces, and, importantly, provides a seal around the tooth. It is important to recognize that the tooth is a solid structure that extends from inside the body to outside the body. The biologic "seal" provided by the periodontium is under constant microbial challenge from more than 300 microbial species.

Periodontal disease is a chronic infection of the periodontium that results in the loss of the periodontal ligament and surrounding alveolar bone [45]. In a susceptible host, it is caused by the bacterial biofilm (plaque) that adheres to the tooth surface. Susceptibility has been associated with genetic polymorphisms and factors such as smoking and diabetes. The plaque bacteria initiate an inflammatory/immune reaction that may be limited to the gingival tissues (resulting in gingivitis) or may spread to the supporting periodontal tissues and bone, causing their loss and periodontitis. Bone loss, which compromises tooth support, is indicative of periodontitis. In more severe cases, bone loss causes the teeth to become mobile, to become more susceptible to infection, and ultimately to be lost. In the past, periodontal therapy aimed at stopping the progression of the disease and at reducing infection. This was generally accomplished by mechanically cleaning the teeth and roots (by scaling and root planing) to disrupt the microbial biofilm. In areas of greater tissue loss, surgery is needed to access the root and bone loss areas. When bone loss has been severe, bone grafting procedures have been employed in an attempt to replace the bony component of the periodontium. More recent surgical efforts have focused on regenerating all the components of the periodontium including the bone, periodontal ligament, and cementum. When a tooth is lost, there generally remains a space where the tooth root was located. New bone will therefore fill that space. However, depending on how the tooth was lost, the contour of the bony ridge can be compromised, leading to a bone defect. This makes restoration difficult, whatever the tooth replacement. This problem is magnified when two or more contiguous teeth are lost. When all teeth are lost in either the mandible or the maxilla, other problems exist. One is that with time, whether dentures are partial or complete, the bone below the dentures is lost progressively and little alveolar bone remains to support the denture. A large nerve and blood vessel run

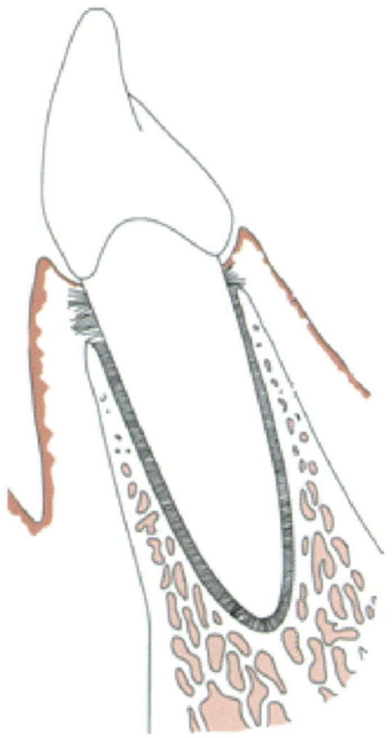

**Figure 9.1.** Schematic cross-sectional representation of the tooth-supporting tissues of the periodontium.

through the mandible, innervating and supplying blood to the lips and face. As a result, only a limited amount of vertical bone is left. When teeth are lost from the maxilla, the sinuses located above the roots of the back teeth enlarge and thus come close to the remaining alveolar ridge crest. This also weakens denture support.

Tooth replacement has been revolutionized by dental implant therapy, which over the last three decades has become the optimal form of tooth replacement [15]. Implants are endosseous titanium screw devices that, after careful preparation of the bone, are screwed into the bone tissue to a depth that ranges from 8 to 12 mm. The procedure is successful in more than 90% of cases. Moreover, a small number of implants can replace many teeth. Implant therapy depends on bone dimensions adequate to allow placing the implant in the bone. Following tooth extraction, alveolar bone undergoes remodeling typical of tissue injury and inflammation. Some systemic interactions may impact upon the remodeling process. Studies of the relationship between bone mineral density (BMD) and alveolar ridge resorption following extraction [30, 32, 48, 69] indicate that the degree of resorption is a function of gender and age. Further, because the tooth extraction site involves both soft and hard tissues, it represents a unique situation for evaluating bone healing and regenerative devices and molecules.

The healing of an extraction site is affected by events in two regions of the healing site: the socket space where the root was located and the residual alveolar bone that supported the tooth. Histologic evaluation of the healing tooth socket has shown that the resultant bone formation fills most of the space. Initial clot formation is followed by cellular infiltration with the formation of highly vascularized granulation tissue. Osteoid becomes evident 7 to 14 days after extraction, and by 30 days the majority of the socket space has become filled with mineralized tissue [1, 12]. Even though a cortical "bridge" covers the coronal aspect of the socket space, the newly formed alveolar bone within the socket continues to remodel, with an increasing percentage of marrow space developing over time. Furthermore, residual tissues from the disrupted periodontal ligament following tooth extraction appear to have little effect on this healing process [12].

The second region involved in the extraction site is the retained alveolar bone that previously supported the tooth. Numerous studies have demonstrated that significant dimensional losses, involving as much as 50% of the buccolingual dimension of the alveolar ridge, occur within the first 3 to 4 months of healing. This dimensional change can amount to 5 to 7 mm of horizontal bone loss and can have a great impact on subsequent dental implant therapy [3, 13, 49, 56]. The patterns of resorption of the residual bone are unique, with greater resorption occurring along the facial than the lingual aspect of the extraction site [2]. This may be because the facial alveolar wall is thinner than the lingual aspect and may therefore be more susceptible to losses of ridge height and width that occur in the course of bone remodeling during the healing process.

Recent attempts to modify the relatively extensive resorption of bone that occurs after tooth extraction have included placement of a dental implant into the extraction site immediately after tooth removal and bone grafting of the sites following the principles of guided bone regeneration (GBR), as will be discussed.

One investigation found no benefit of immediate implant placement in the extraction site. The levels of bone resorption were not altered by the presence of the dental implant [2]. In contrast, the use of xenogenic grafting of the extraction site may to some extent limit the dimensional changes associated with the osseous healing [12]. Both of these approaches are recent, and experience with them is limited. It is hoped that the limitations and benefits of these approaches will become known.

In summary, bone loss around teeth or bone loss after teeth have been removed can present major challenges for dental rehabilitation. Periodontists have focused on restoring the bone around the teeth, and virtually all dentists are involved with dental implants (either in the surgical placement of the endosseous screw or in the tooth replacement on top of the screw). In what follows we will focus on the techniques and strategies that enhance bone formation around teeth and in areas where teeth are missing and on bone formation prior to and at the time of dental implant placement.

## 9.2 Bone Formation Around Teeth

Periodontal disease results in the loss of periodontal ligament and bone adjacent to the tooth. Dental plaque adheres to the side of the tooth, and if the connective tissue attachment to the tooth is lost, a periodontal pocket forms. The bacteria associated with these periodontal pockets change from a largely aerobic to a largely anaerobic flora, one considered to be pathogenic. Over time, the plaque becomes calcified and is termed calculus. Calculus forms along the root surface deep into the pocket and periodontal tissues. Bone loss associated with periodontal pocket formation may involve only the tooth with the pocket and result in vertical bone loss adjacent to the affected tooth. Alternatively, bone loss can also involve the adjacent tooth and lead to horizontal bone loss between teeth. In either case, the optimum choice of treatment would be to remove the plaque and calculus on the surface of the tooth root and regenerate the lost bone, the periodontal ligament, and the cementum on the affected tooth root surface.

The formation of all three tissues is important, because the connective tissue fibers of the periodontal ligament insert into both the cementum and the bone tissue. Regeneration must therefore include tissue formation that is coordinated and involves all components of the periodontium. Two significant complicating factors are that tissue formation must occur in the presence of tooth mobility and the constant challenge from plaque bacteria, since plaque formation is both instant and continuous. Salivary proteins are deposited rapidly on the cleaned surfaces, and bacteria immediately adhere to what is called the acquired pellicle. With continued growth of the microbial flora, plaque is formed.

The tissues of the periodontium (bone, periodontal ligament, cementum, and gingiva) grow at different rates. This can complicate periodontal regeneration, because coordinated growth is required. In fact, the expected outcome for surgical procedures where only scaling and root planing are performed is the formation of a so-called long junctional epithelium, where the epithelium from the gingiva proliferates along the diseased root surface, and no bone, periodontal ligament, or cementum is formed. For this reason, therapies have been devised to inhibit the epithelial proliferation and to favor the formation of the bone, periodontal ligament, and cementum. Because the growth of some tissues is favored over that of other tissues, the term applied to these strategies is "guided tissue regeneration" (GTR), a term reserved for the formation of new periodontal tissue.

In the past, some type of barrier membrane was used to promote preferential growth of selected tissues [44]. The rationale is that if the epithelium is excluded, the slower-growing periodontal ligament cells and the bone and cementum cells will fill the defect and regenerate the periodontium (Fig. 9.2). This was first demonstrated with a membrane filter placed around a mandibular anterior tooth. A nonresorbable expanded polytetrafluoroethylene membrane was manufactured and used as GTR procedures gained popularity. However, in 40% to 60% of the procedures, the gingival tissues over the top of the membrane receded, and the membrane became exposed to the oral cavity and contaminated with bacteria. As a result, tissue regeneration was compromised. This, and the fact that a second surgical procedure

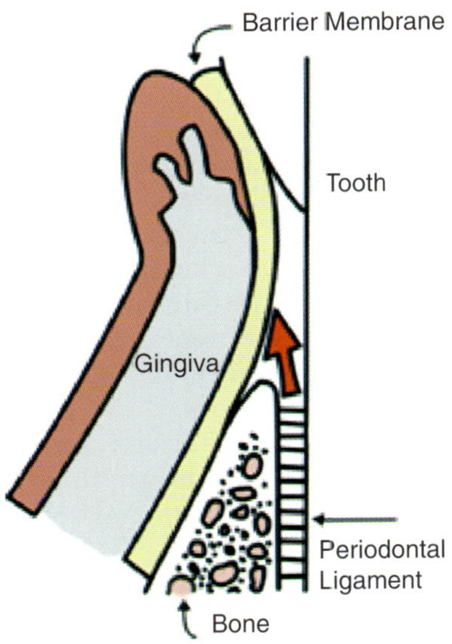


**Figure 9.2.** Schematic cross-sectional representation of the placement of a barrier membrane subgingivally as done for guided tissue regeneration (GTR). This is thought to allow for repopulation of the wound space with cells derived from the bone, periodontal ligament, and cementum.

was needed to remove the membrane, led to the development of several types of resorbable membranes that exclude the gingiva and epithelium. Most of these are made from polylactic and/or polyglycolic acids or collagen. Collagen membranes favor tissue ingrowth and become exposed. However, all membrane barriers are technique-sensitive and require time to shape and place.

In more recent efforts to stimulate GTR, various proteins were added to stimulate growth of the periodontal structures. Early efforts focused on factors that enhance cellular competence and progression through the cell cycle. Because a combination of platelet-derived growth factor (PDGF) and insulin-like growth factor (IGF) stimulated skin wound healing in animals, this combination was used to successfully stimulate periodontal regeneration in dogs and monkeys [29]. However, this combination has not yet been commercially developed. Another approach has used recombinant human bone morphogenetic protein 2 (rhBMP-2) to stimulate periodontal regeneration in humans. In a pilot trial, rhBMP-2 was placed in a collagen sponge around teeth with periodontal disease that were to be extracted. After healing, the tooth and surrounding tissue were removed by block section and examined histologically. The results of this multicenter clinical trial (unpublished) showed that the rhBMP-2 and collagen sponge did not stimulate periodontal regeneration, even though studies in dogs had demonstrated a partial effect on bone growth [71]. In the dog study, however, areas of ankylosis and root resorption had been observed; this was not the case when the PDGF/IGF combination was used.

The newest approved protein stimulant of periodontal regeneration is a heterogeneous mixture of proteins extracted from enamel harvested from developing tooth buds in pigs [18]. The predominant component in this mixture is amelogenin; however, other proteins in this mixture have also been shown to contribute stimulating activities. Extracellular matrix (ECM) proteins stimulate periodontal regeneration to the same extent as GTR procedures that utilize membrane barriers [62]. The advantage of the enamel proteins is that they are less technique-sensitive and affect proliferation and differentiation in epithelium, periodontal ligament cells, and bone cells [60]. rhBMP-2, on the other hand, acts only as a differentiation agent for bone cells. The enamel proteins stimulate proliferation of less differentiated bone cells and differentiation of mature bone cells. These proteins also inhibit the growth of epithelial cells and stimulate periodontal ligament fibroblasts. Also of significance is the fact that the enamel proteins enhance the attachment and growth of bone cells and periodontal ligament cells [31]. The mechanism for attachment does not appear to involve integrin binding (RGD) sequences, but it does require divalent cations. Other studies have demonstrated that the enamel proteins can inhibit anaerobic but not gram-positive growth [66]. Although all factors stimulate periodontal regeneration, the importance of each individual attribute is not presently known.

For many years, clinicians have filled the osseous defects around teeth with some type of bone-replacement graft material in the hope of causing bone tissue to form around the tooth. Even though bone typically is the largest component of the missing periodontal structure, it is not known whether stimulation of cementum or the periodontal ligament is also required to achieve optimal regeneration. For example, when enamel proteins are used, large amounts of cementum are formed, and it seems logical to conclude that this is needed, since the periodontal ligament fibers attach to both cementum and bone.

Bone graft materials have been used to fill the void in the bone around the tooth created by periodontal disease. Many different types of materials have been employed, but demineralized freeze-dried bone allograft (DFDBA) is the best documented for stimulating periodontal regeneration [42]. This material (considered by many to be osteoconductive) contains bone morphogenetic and other proteins, but their specific role in stimulating bone formation is not known. For many years it was assumed that all components of DFDBA had an equal role in stimulating periodontal tissue formation. However, we were able to show that commercial DFDBA varies in its osteoinductive activity, whether derived from the same or from different tissue banks [58]. Osteoinductive activity was also greater when the tissue came from younger donors, with gender making no difference [59]. Furthermore, the addition of exogenous bone morphogenetic protein (BMP) enhanced the osteoinductive activity of the DFDBA preparations. Whether or not variance in osteoinductive activity is of clinical significance, the principal value of this material is to stimulate periodontal regeneration, which it does far better than some other commercially available graft materials [38].

Many materials have been combined to stimulate periodontal regeneration. In larger periodontal defects, some type of bone graft material is required to prevent the gingiva from collapsing into the bone defect, an event that severely limits periodontal regeneration. It is generally thought that a combination of materials will be synergistic and facilitate regeneration. An osteoconductive scaffold combined with factors that stimulate cellular activity is likely to bring about more effective periodontal regeneration [57]. Commercially available enamel matrix proteins have therefore been combined with bone graft material [5]. In one such study with baboons, periodontal defects treated with autogenous bone grafts combined with enamel proteins were compared with untreated periodontal defects [22]. Significant new amounts of cementum and bone were formed, particularly in the narrower lesions (Fig. 9.3). Regeneration occurred by formation of new cementum, periodontal ligament, and of bone that took place beyond a mark that had been placed at the apical (lower) aspect of the original periodontal defect.

The above discussion makes it evident that current therapeutic efforts to treat periodontal disease are aimed at regenerating the lost periodontal tissues, including bone, the periodontal ligament, and cementum.

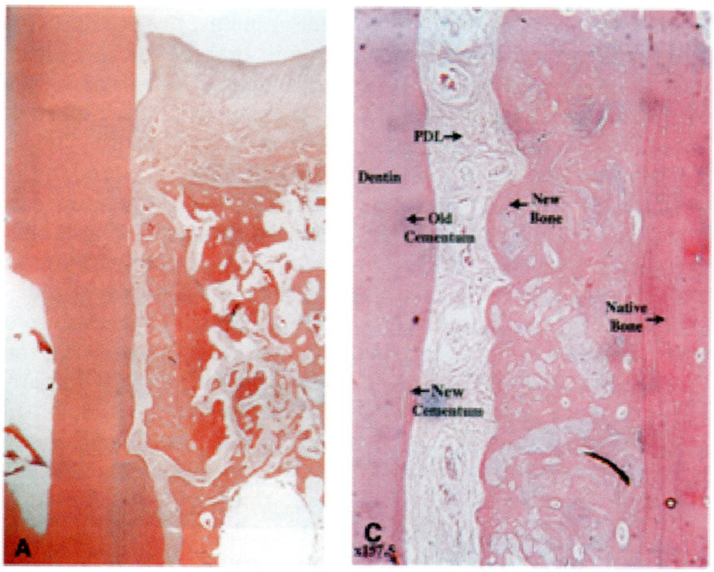


**Figure 9.3.** Histologic view of periodontal regeneration in response to enamel matrix proteins. These slides demonstrate the reformation of supporting bone, periodontal ligament (PDL), and new cementum along the root surface representative of periodontal regeneration in a narrow bony defect.

## 9.3 Bone Formation Around Dental Implants

When teeth are missing or have to be removed, endosseous titanium implants can be inserted in the alveolar bone and the missing tooth or teeth attached to the implants by means of a screw or cement. This procedure has a success rate of over 90%. Early implant placement procedures required waiting periods of 6 to 9 months for bone to surround the implant surface following the osteotomy and insertion of the implant [8, 9]. The amount of time needed was predicated on healing taking place against a relatively smooth titanium surface of the machined implant. Interestingly, success rates with implants with machined surfaces were much lower when the bone was composed of a larger fraction of cancellous and a lesser fraction of cortical bone [33]. Such bone is typically found in the maxilla and in the posterior mandible, with the denser bone found in the anterior mandible. To establish implants in bone that was less dense required significant advances in implant therapy.

Subsequent experimental studies have indicated that bone apposition is more complete and faster if implants have surfaces that are less smooth than when the titanium surface is machined [16]. For this reason, much effort has been spent in trying to optimize the roughness of the titanium surface. Studies in mini-pigs have found that a titanium surface produced by sandblasting and acid etching induces the best bone apposition in comparison with four other roughness procedures [11]. The advantage of this procedure was confirmed in an in vivo canine study [21]. This implant, with the sandblasted and acid-etched surface, was then utilized in a large multicenter international prospective human clinical trial and was shown to induce bone healing in half the time required in the conventional procedure [17]. The new procedure made it possible to insert teeth only 6 weeks after bone drilling, with success rates in the 97% to 99% range. This significant advance in the dental implant field has been confirmed in other studies in which rough surfaces were prepared differently.

Recent efforts to obtain even faster rates of bone healing have centered on changing the surface chemistry of the titanium. For example, in one case, the sandblasted and acid-etched surface was manufactured so as to maintain the chemical activity of the titanium oxide. This approach makes the surface hydrophilic, whereas most other implant surfaces are hydrophobic. The hydrophilic surface reacts quickly with the blood, and the response of bone cells to the surface is enhanced. A study of the minipig maxilla found that bone apposition on the chemically modified, sandblasted, acid-etched implant surface was significantly greater after 2 to 4 weeks than that on the control surfaces [10] (Fig. 9.4). Another study found that the amount of torque required to remove the modified implant was greater than that needed for the control [27]. With these studies as a basis, a randomized, controlled clinical trial was initiated. Sites for two implants were prepared in the posterior areas of either jaw bone (where the bone quality is relatively low), and the two types of implant were inserted according to a random schedule [47]. Implant stability was measured by means of resonance frequency. This technique uses a device that stimulates a transducer on top of the implant, which provides a value for the relative stability of the implant. Weekly readings were obtained on each implant for 6 weeks and again at 3 months. Preliminary results indicated that the implant with the modified surface became stable significantly more quickly than the control. Thus, the animal and human studies gave comparable results, indicating how implant surface characteristics enhance bone apposition to the implant surface. This procedure also improves patient care because the implant becomes stable more quickly. Consequently, the patient has less opportunity to interfere with early healing during function.

Another strategy to enhance bone apposition around implants is to place a bone-replacement graft or bone-stimulating agent around the implant. This is particularly indicated if a space or gap exists between the implant and the bone, a situation that tends to arise if the implant is inserted into the extraction socket immediately after tooth removal (Fig. 9.5). It also arises if an implant is inserted into an area where bone has healed but where new bone has not filled the tooth socket completely. This happens more frequently as dental implant therapy becomes more widespread and the patient is pushing to have missing teeth replaced more quickly.

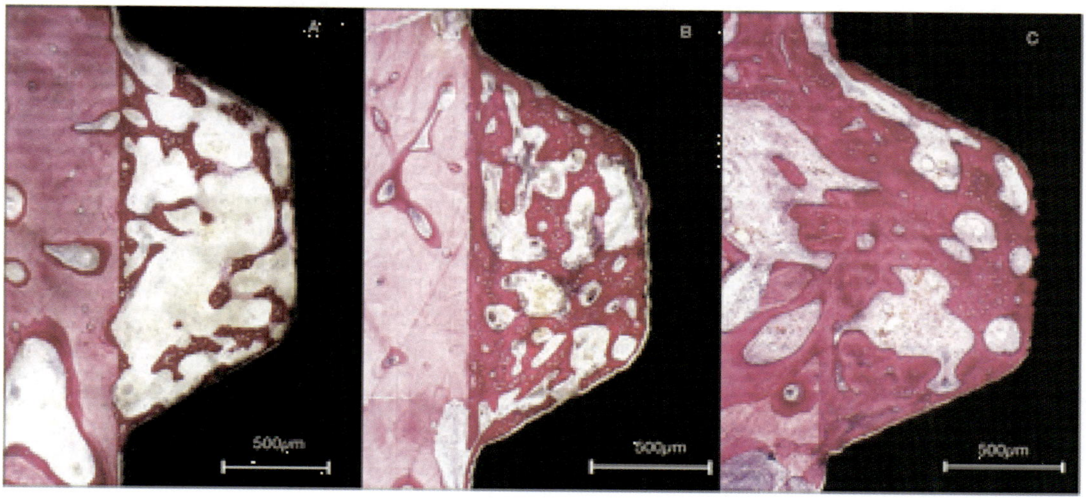


**Figure 9.4.** Histologic view of enhanced bone healing adjacent to a chemically modified implant surface with bone formation (A) after 2 weeks of healing, (B) after 4 weeks of healing, and (C) after 8 weeks of healing. Reproduced from Buser et al. [10], with permission of the International and American Associations for Dental Research.

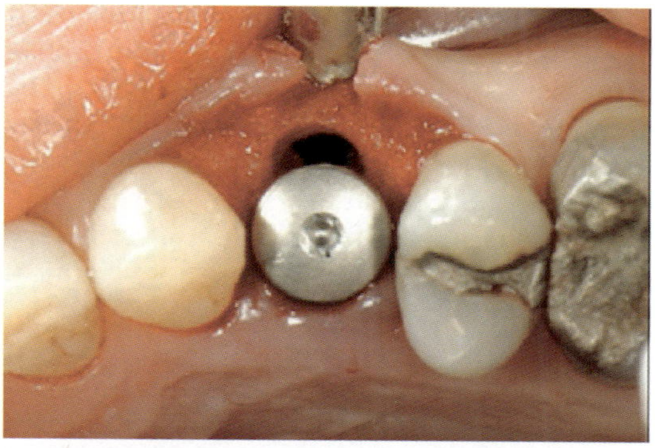

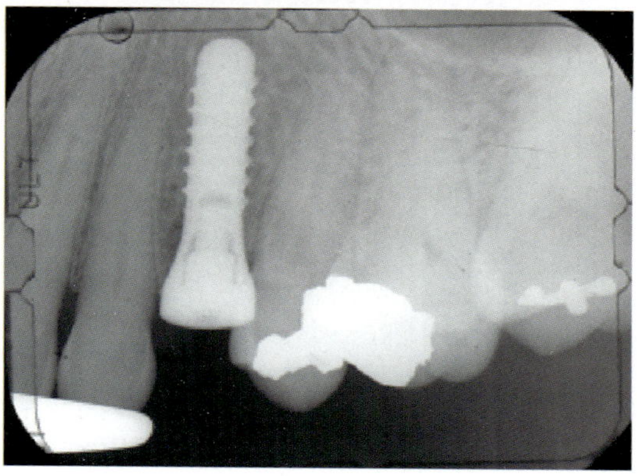

**Figure 9.5.** Clinical view (top) and radiographic view (bottom) of the placement of a dental implant into a tooth extraction socket. The clinical view shows the residual space between the implant and alveolar bone requiring bone grafting.

One of the other materials being tested around implants is BMP. In one study, rhBMP-2, either as part of a collagen sponge or mixed with a polylactide-glycolide polymer, was placed around dental implants inserted in the partially edentulous mandible of dogs [19, 20]. Half of the defects were covered with a nonresorbable membrane. In control sites only carrier was used. Sites that contained BMP had significantly more bone in the defects and against the implant surface than control sites. Early in the study, the membrane-covered sites had less new bone than the uncovered sites. At later times, the membrane-covered sites had more bone. Furthermore, sites with collagen had more bone than sites with the synthetic carrier.

## 9.4 Bone Regeneration in Areas Insufficient for Implant Placement

As implant therapy develops, more implants are inserted into sites that lack sufficient bone to support an implant. It therefore has become necessary to regenerate lost alveolar bone tissue with the aid of guided bone regeneration (GBR).

GBR developed out of earlier attempts to regenerate the supporting tissues of the periodontium, that is, guided tissue regeneration (GTR) [23, 24, 61]. Both approaches typically use membrane barriers to control the cellular repopulation of a maintained wound space [6, 41, 46]. Cells that repopulate the wound space control the resulting tissues. In the case of GBR, the objective is to develop new bone tissue [39, 54, 55]. Early studies demonstrated that the osseous healing that occurs in conjunction with GBR techniques paralleled the healing within a tooth socket following extraction [55]. The initial clot that is formed is followed by granulation tissue with vascular ingrowth, osteoid tissue formation, and mineralization that begins on the edges of the wound surface.

For GBR to be successful, clots must be stabilized and the wound space preserved to allow for cellular and vascular ingrowth and for selective repopulation by osteogenic cells. Each of these objectives is addressed by the use of barrier membranes that serve to define the borders of the osseous defect, restrict fibrotic tissue formation, and provide stability for the ensuing clot. Available biocompatible membranes are either nonresorbable or resorbable, each condition having advantages and disadvantages.

Nonresorbable membranes have proven quite effective at limiting cellular ingrowth but may require a second surgical procedure for removal. An additional limitation of nonresorbable membranes is the increased likelihood that soft-tissue complications will arise during the healing period [63]. Typically, GBR techniques aim to retain the barrier membrane for at least 4 to 6 months to allow for optimal bone growth. Premature loss or degradation of the membrane may compromise bone formation or cause loss of the wound space [65]. Nonresorbable membranes have been modified by reinforcement with titanium. Titanium reinforcement prevents soft-tissue collapse and thus improves maintenance of the desired wound space [35].

The decreased likelihood of soft-tissue complications during the healing process is a major reason for the increased use of resorbable membranes. These membranes are usually made from copolymers of polylactide and polyglycolide or from collagen. A major concern in using these membranes is that the barrier membranes may be resorbed prematurely, thereby diminishing osseous regeneration [72]. A second limitation is their tendency to be easily deformed, which may lead to collapse of the membrane into the wound space. To minimize this problem, the space under these membranes is maintained with the use of bone-grafting materials.

Bone-grafting materials in GBR include autogenous bone, which is considered the gold standard, as well as allogenic, xenogenic, and alloplastic grafting materials. These materials not only support the barrier membrane, but also provide osteogenic cells for autogenous grafting and osteoinductive molecules such as BMPs. They also constitute an osteoconductive scaffold that supports the growth of osseous tissue within the wound space [60, 68].

The use of GBR techniques to augment alveolar bone has proven more successful in lateral ridge than in vertical ridge augmentation procedures. The increased intraoral functional demands on the vertical ridge augmentation procedures are thought to make it more difficult to maintain the wound space necessary for regeneration of the crestal bone [63, 64].

Given these limitations, an alternative technique that has found increased intraoral application is distraction osteogenesis [14, 67]. The development of intraoral fixation devices has made it possible to use distraction osteogenesis as an alternative to gain vertical ridge height in the anterior mandible [51, 52]. However, this procedure is often accompanied by clinical complications that limit its use [26].

## 9.5 Current Trends and Future Applications

The application of bone biology to dental therapy is probably best typified by extending implant therapy to the treatment of recent tooth extractions. Often implants can be placed during the same patient visit at which tooth extraction was carried out. This is termed "immediate implant placement." Although the advantages, disadvantages, and specific methodologies are still under debate, the attempt to use implant therapy to improve patient care will undoubtedly lead to greater utilization of these approaches.

Immediate implant placement developed directly from GBR techniques to augment bone volume. Originally, these approaches were highly dependent upon the methodologies of GBR, that is, membrane placement with a submerged implant and bone grafting. Recently, more reliance has been placed on natural bone healing in the tooth extraction site. The use of barrier membranes has been questioned, and grafting into residual defects to overcome dimensional differences between the implant and tooth socket has been shown unnecessary, if the defects are 2 mm or less (see clinical view in Fig. 9.5). However, it is still uncertain whether placement of a dental implant alters the normal healing pattern. Early studies have indicated that healing in the tooth socket is not significantly affected by implant placement [2].

A second therapeutic approach that has dramatically altered current use of dental implant therapy is maxillary sinus grafting. This procedure relies on GBR techniques to promote bone regeneration in the inferior region of the maxillary sinus. After tooth extraction in the posterior maxilla, sinus pneumatization frequently extends inferiorly as the alveolar ridge resorbs superiorly. As a result, the volume of bone that remains is minimal. This may severely limit the use of implant therapy. To correct these deficiencies in the posterior maxilla, the sinus is augmented with the aid of bone grafting. Sinus augmentation for implant therapy seems to be as successful as implant placement in native alveolar bone [70]. However, the specific approach taken by the clinician makes an important difference. For example, a particulate bone graft provides greater implant success than does block grafting. Also, implants with a roughened surface at the osseous interface assure much greater clinical success, as does the use of a barrier membrane to occlude the osteotomy.

Functional support becomes truly critical for tooth replacement in the posterior maxilla and mandible. Whereas previously there were clear limitations regarding the use of dental implants in the maxillary posterior because of the poor bone quality attributed to maxillary sinus pneumatization, current technology makes implant therapy in these regions of the mouth highly successful and has shortened the time of healing for osseointegration to 6 to 8 weeks, periods previously thought impossible [4, 17].

Current therapeutic approaches can create very nice esthetic and functional results, but this does not mean that progress should not continue toward a more natural result. Such a result might involve an attachment that provides a more natural feel and that is able to adapt better to occlusal forces. One can also envisage an attachment that better accommodates the scalloped architecture of the gingival tissues and recreates the gingival papillary tissue, thus leading to better long-term results.

One of the most studied molecules related to the therapeutics of bone growth is BMP-2, one of over 40 structurally similar proteins that make up the transforming growth factor β (TGF-β) superfamily. BMP-2 is one of several members of the BMP family that are true differentiation factors, capable of triggering the differentiation of mesenchymal stem cells (MSCs) into an osteoblastic lineage that leads to bone formation. The therapeutic effects of BMP-2 are concentration-dependent. As currently used, concentrations of BMP-2 far exceed physiologic levels. To maintain high concentrations locally, a carrier is needed and may also be important to promote regeneration in the extracellular environment. This ECM may also provide an appropriate environment for the response of the progenitor cells to the BMPs. The need for high concentrations may be related, in part, to signals or interactions with other components of the microenvironment.

BMP-2 was used in maxillary sinus grafting in 12 patients whose healing response was evaluated over a period of 4 months [7]. Bone height after grafting was assessed by tomography. The study found a mean gain in bone height of 8.5 mm, sufficient to allow for implant placement in 11 of the 12 patients, a clearly promising result.

As discussed earlier, remodeling of the alveolar bone following tooth extraction can often compromise or complicate dental implant therapy. A recent study of the use of BMP-2 in extraction sites found that BMP-2 significantly increased alveolar bone volume following tooth extraction and that the ability to place dental endosseous implants was greatly enhanced [28]. BMPs have been utilized in GBR procedures to augment alveolar bone in order to allow for direct placement of implants into the dental surface [25, 34, 37, 40, 50].

Carriers that can support the soft tissues more rigidly are still being sought, usually to be combined with some type of growth stimulant. Other proteins and carriers are based on synthetic materials. One such study has examined the combination of a polyethylene glycol (synthetic) carrier with parathyroid hormone (PTH). This mixture significantly stimulated bone formation in standardized defects around implants in a canine model [36].

## 9.6 Summary and Outlook

Advances in our understanding of bone biology and applications of that understanding to dental therapies have led to dramatic changes in our paradigms for patient care. The application of growth factors has thus far been directed at identifying the most appropriate signaling molecules to stimulate the desired biologic response. With progress there will be a need better to understand the dose response, environmental interactions, and time-dependent nature of these biologic mediators. Specifically, intracellular regulation of growth-factor signal transduction on a cell-specific basis presents an intriguing possibility in the future [43, 53]. The extension of tissue-engineered regeneration into the oral environment will allow for improvements in both tooth retention and tooth replacement with the aid of implant therapy. Studies are needed to define combinations of proteins, intracellular and extracellular signal regulators, and carriers that can be used to enhance bone and tissue formation around implants, periodontally compromised teeth, or extraction sites. The oral environment, together with the demands and needs of patients, continues to constitute a unique and complex challenge that necessitates extending the knowledge and applications of bone biology, remodeling, and regeneration.

## References


1. Amler MH (1969) The time sequence of tissue regeneration in human extraction wounds. Oral Surg Oral Med Oral Pathol 27:309–318.
2. Araujo MG, Sukekava F, Wennstrom JL, Lindhe J (2005) Ridge alterations following implant placement in fresh extraction sockets: an experimental study in the dog. J Clin Periodontol 32:645–652.
3. Atwood DA, Coy WA (1971) Clinical, cephalometric, and densitometric study of reduction of residual ridges. J Prosthet Dent 26:280–295.
4. Bornstein MM, Lussi A, Schmid B, Belser UC, Buser D (2003) Early loading of nonsubmerged titanium implants with a sandblasted and acid-etched (SLA) surface: 3-year results of a prospective study in partially edentulous patients. Int J Oral Maxillofac Implants 18:659–666.
5. Boyan BD, Weesner TC, Lohmann CH, Andreacchio D, Carnes DL, Dean DD, Cochran DL, Schwartz Z (2000) Porcine fetal enamel matrix derivative enhances bone formation induced by demineralized freeze-dried bone allograft in vitro. J Periodontol 71:1278–1286
6. Boyne PJ (1969) Restoration of osseous defects in maxillofacial casualties. J Am Dent Assoc 78:767–776.
7. Boyne PJ, Marx RE, Nevins M, Triplett G, Lazaro E, Lilly LC, Alder M, Nummikoski P (1997) A feasibility study evaluating rhBMP-2/absorbable collagen sponge for maxillary sinus floor augmentation. Int J Periodontics Restorative Dent 17:11–25.
8. Brånemark P-I, Breine U, Adell R, Hansson O, Lindstrom J, Ohlsson A (1969) Intraosseous anchorage of dental prostheses. Scand J Plast Reconstr Surg 3:81–100.
9. Brånemark P-I, Hansson B, Adell R, et al. (1977) Osseointegrated implants in the treatment of the edentulous jaw: experience from a 10-year period. Scand J Plast Reconstr Surg 11:1–132.
10. Buser D, Broggini N, Wieland M, Schenk RK, Denzer AJ, Cochran DL, et al. (2004) Enhanced bone apposition to a chemically modified SLA titanium surface. J Dent Res 83:529–533.
11. Buser D, Schenk RK, Steinemann S, Fiorellini JP, Fox CH, Stich H (1991) Influence of surface characteristics on bone integration of titanium implants. A histometric study in miniature pigs. J Biomed Mater Res 25:889–902.
12. Cardaropoli G, Araujo M, Hayacibara R, Sukekava F, Lindhe J (2005) Healing of extraction sockets and surgically produced augmented and non-augmented—defects in the alveolar ridge. An experimental study in the dog. J Clin Periodontol 32:435–440.
13. Carlsson GE, Bergman B, Hedegard B (1967) Changes in contour of the maxillary alveolar process under immediate dentures. A longitudinal clinical and x-ray cephalometric study covering 5 years. Acta Odontol Scand 251:45–75.
14. Chin M, Toth BA (1996) Distraction osteogenesis in maxillofacial surgery using internal devices: review of five cases. J Oral Maxillofac Surg 54:45–53.
15. Cochran DL (1996) Implant therapy I. Ann Periodontol 1:707–790.
16. Cochran DL (1999) A comparison of endosseous dental implant surfaces. J Periodontol 70:1523–1539.
17. Cochran DL, Buser D, ten Bruggenkate CM, Weingart D, Taylor TM, Bernard J-P, Peters F, Simpson JP (2002) The use of reduced healing times on ITI implants with a sandblasted and acid-etched (SLA) surface: early results from clinical trials on ITI SLA implants. Clin Oral Implants Res 13:144–153.
18. Cochran DL, Jones A, Heijl L, Mellonig JT, Schoolfield J, King GN (2003) Periodontal regeneration with a combination of enamel matrix proteins and autogenous bone grafting. J Periodontol 74:1269–1281.
19. Cochran DL, Nummikoski PV, Jones AA, Makins SR, Turek TS, Buser D (1997) Radiographic analysis of regenerated bone around endosseous implants in the

canine using recombinant human bone morphogenetic protein-2. Int J Oral Maxillofac Implants 12: 739–748.
20. Cochran DL, Schenk R, Buser D, Wozney JM, Jones A (1999) Recombinant human bone morphogenetic protein-2 stimulation of bone formation around endosseous dental implants. J Periodontol 70: 139–150.
21. Cochran DL, Schenk RK, Lussi A, Higginbottom FL, Buser D (1998) Bone response to unloaded and loaded titanium implants with a sandblasted and acid-etched surface: a histometric study in the canine mandible. J Biomed Mater Res 40:1–11.
22. Cochran DL, Wennstrom JL, Funakoshi E, Heijl L (2003) Chapter 2. The Biologic Concept: biomimetics. In: Periodontal Regeneration. Rationale and Clinical Use of Enamel Matrix Derivative. Quintessence Publishing Co, Carol Stream, IL, pp 7–10.
23. Dahlin C, Gottlow J, Linda A, Nyman S (1990) Healing of maxillary and mandibular bone defects using a membrane technique. Scand J Plast Reconstr Hand Surg 24:13–19.
24. Dahlin C, Linda A, Gottlow J, Nyman S (1988) Healing of bone defects by guided tissue regeneration. Plast Reconstr Surg 81:672–676.
25. Dunn CA, Jin Q, Taba M Jr, Franceschi RT, Bruce Rutherford R, Giannobile WV (2005) BMP gene delivery for alveolar bone engineering at dental implant defects. Mol Ther 11:294–299.
26. Enislidis G, Fock N, Millesi-Schobel G, Klug C, Wittwer G, Yerit K, Ewers R (2005) Analysis of complications following alveolar distraction osteogenesis and implant placement in the partially edentulous mandible. Oral Surg Oral Med Oral Pathol Oral Radiol Endod 100:25–30.
27. Ferguson SJ, Broggini N, Weiland M, de Wild M, Rupp F, Geis-Gerstorfer J, Cochran DL, Buser D (2005) Biomechanical evaluation of the interfacial strength of a chemically modified SLA titanium surface. J Biomed Materials Res: Part A (submitted for publication).
28. Fiorellini JP, Howell TH, Cochran D, Malmquist J, Lilly LC, Spagnoli D, Toljanic J, Jones A, Nevins M (2005) Randomized study evaluating recombinant human bone morphogenetic protein-2 for extraction socket augmentation. J Periodontol 76:605–613.
29. Giannobile W (1999) Chapter 14. Periodontal tissue regeneration by polypeptide growth factors and gene transferin. In: Lynch SE, Genco RJ, Marx RE, eds. Tissue Engineering. Applications in Maxillofacial Surgery and Periodontics. Quintessence Publishing Co, Carol Stream, IL, pp 231–243.
30. Hirai T, Ishijima T, Hashikawa Y, Yajima T (1993) Osteoporosis and reduction of residual ridge in edentulous patients. J Prosthet Dent 69:49–56.
31. Hoang AM, Klebe RJ, Steffensen B, Ryu OH, Simmer JP, Cochran DL (2002) Amelogenin is a cell adhesion protein. J Dent Res 81:497–500.
32. Humphries S, Devlin H, Worthington H (1989) A radiographic investigation into bone resorption of mandibular alveolar bone in elderly edentulous adults. J Dent Res 17:94–96.
33. Jaffin RA, Berman CL (1991) The excessive loss of Branemark fixtures in type IV bone: a 5-year analysis. J Periodontol 62:2–4.
34. Jennissen HP (2002) Accelerated and improved osteointegration of implants biocoated with bone morphogenetic protein-2 (BMP-2). Ann NY Acad Sci 961: 139–142.
35. Jovanovic SA, Nevins M (1995) Bone formation utilizing titanium reinforced barrier membranes. Int J Periodont Rest Dent 15:56–69.
36. Jung RE, Cochran DL, Domken O, Seibl R, Jones A, Hammerle CHF (2005) The effect of matrix bound parathyroid hormone on bone regeneration. J Dent Res (submitted for publication).
37. Jung RE, Glauser R, Scharer P, Hammerle CH, Sailer HF, Weber FE (2003) Effect of rhBMP-2 on guided bone regeneration in humans. Clin Oral Implants Res 14:556–568.
38. Lallier TE, Yukna R, St Marie S, Moses R (2001) The putative collagen binding peptide hastens periodontal ligament cell attachment to bone replacement graft materials. J Periodontol 72:990–997.
39. Linde A, Thoren C, Dahlin C (1993) Creation of new bone by an osteopromotive membrane technique. Int J Oral Maxillofac Surg 51:892–897.
40. Liu Y, de Groot K, Hunziker EB (2005) BMP-2 liberated from biomimetic implant coatings induces and sustains direct ossification in an ectopic rat model. Bone 36:745–757.
41. Melcher AH, Dreyer CJ (1962) Protection of the blood clot in healing circumscribed bone defects. J Bone Joint Surg 44B:827–831.
42. Mellonig JT (1999) Chapter 16. Freeze-dried bone allografts in periodontics. In: Lynch SE, Genco RJ, Marx RE, eds. Tissue Engineering. Applications in Maxillofacial Surgery and Periodontics. Quintessence Publishing Co, Carol Stream, IL, pp 259–268.
43. Mumford JH, Carnes DL, Cochran DL, Oates TW (2001) The effects of platelet-derived growth factor-BB on periodontal cells in an in vitro wound model. J Periodontol 72:331–340.
44. Murphy KG, Gunsolley JC (2003) Guided tissue regeneration for the treatment of periodontal intrabony and furcation defects. A systematic review. Ann Periodontol 8:266–302.
45. Novak MJ (2002) Chapter 4. Classification of diseases and conditions affecting the periodontium. In: Newman MG, Takei HH, Carranza FA, eds. Carranza's Clinical Periodontology. 9th ed. Saunders, Philadelphia, pp 64–73.
46. Nyman S, Lang NP, Buser D, Bragger U (1990) Bone regeneration adjacent to titanium dental implants using guided tissue regeneration. Int J Oral Maxillofac Implants 5:9–14.
47. Oates TW, West J, Jones J, Kaiser D, Cochran DL (2002) Long term changes in soft tissue height on the facial surface of dental implants. Implant Dent 11:272–279.
48. Ortman LF, Hausmann E, Dunford RG (1989) Skeletal osteopenia and residual ridge resorption. J Prosthet Dent 61:321–325.
49. Pietrokovski J, Massler M (1967) Alveolar ridge resorption following tooth extraction. J Prosthet Dent 17:21–27.
50. Rachmiel A, Aizenbud D, Peled M (2004) Enhancement of bone formation by bone morphogenetic protein-2 during alveolar distraction: an experimental study in sheep. J Periodontol 75:1524–1531.
51. Raghoebar GM, Heydenrijk K, Vissink A (2000) Vertical distraction of the severely resorbed mandible. The Groningen distraction device. Int J Oral Maxillofac Surg 29:416–420.

52. Raghoebar GM, Meijer HJA, Stegenga B, Van't Hof M, van Oort RP, Vissink A (2000) Effectiveness of three treatment modalities for the edentulous mandible. A five-year randomized clinical trial. Clin Oral Implant Res 11:195–201.
53. Ray AK, Jones AC, Carnes DL, Cochran DL, Mellonig JT, Oates TW Jr (2003) Platelet-derived growth factor-BB stimulated cell migration mediated through p38 signal transduction pathway in periodontal cells. J Periodontol 74:1320–1328.
54. Sandberg E, Dahlin C, Linde A (1993) Bone regeneration by the osteopromotive technique using bioabsorbable membranes. An experimental study in rats. Int J Oral Maxillofac Surg 51:1106–1114.
55. Schenk RK, Buser D, Hardwick WR, Dahlin C (1994) Healing pattern of bone regeneration in membrane-protected defects. A histologic study in the canine mandible. Int J Oral Maxillofac Implants 9:13–29.
56. Schropp L, Wenzel A, Kostopoulos L, Karring T (2003) Bone healing and soft tissue contour changes following single-tooth extraction: a clinical and radiographic 12-month prospective study. Int J Periodontics Restorative Dent 23:313–323.
57. Schwartz Z, Carnes DL, Pulliam R, Lohmann CH, Sylvia VL, Liu Y, et al. (2000) Porcine fetal enamel matrix derivative stimulates proliferation but not differentiation of pre-osteoblastic 2T9 cells, inhibits proliferation and stimulates differentiation of osteoblast-like MG63 cells, and increases proliferation and differentiation of normal human osteoblast NHOst cells. J Periodontol 71:1287–1296.
58. Schwartz Z, Mellonig JT, Carnes DL, De La Fontaine J, Cochran DL, Dean DD, Boyan BD (1996) Ability of commercial demineralized freeze dried bone allograft to induce new bone formation. J Periodontol 67: 913–926.
59. Schwartz Z, Somers A, Mellonig JT, Carnes, Dean DD, Cochran DL, Boyan BD (1998) Ability of commercial demineralized freeze-dried bone allograft to induce new bone formation is dependent on donor age but not gender. J Periodontol 69:470–478.
60. Schwartz Z, Weesner T, van Dijk S, Cochran DL, Mellonig JT, Lohmann CH, Carnes DL, Goldstein M, Dean DD, Boyan BD (2000) Ability of deproteinized cancellous bovine bone to induce new bone formation. J Periodontol 71:1258–1259.
61. Seibert J, Nyman S (1990) Localized ridge augmentation in dogs: a pilot study using membranes and hydroxylapatite. J Periodontol 61:157–165.
62. Silvestri M, Sartori S, Rasperini G, Ricci G, Rota C, Cattanco V (2003) Comparison of intrabony defects treated with enamel matrix derivative versus guided tissue regeneration with a nonresorbable membrane. A multicenter controlled clinical trial. J Clin Periodontol 30:386–393.
63. Simion M, Baldoni M, Rossi P, et al. (1994) A comparative study of the effectiveness of e-PTFE membranes with and without early exposure during the healing period. Int J Periodontics Restorative Dent 14:166–180.
64. Simion M, Jovanovic SA, Tinti C, et al (2001) Long-term evaluation of osseointegrated implants inserted at the time or after vertical ridge augmentation. Clin Oral Impl Res 12:35–45.
65. Simion M, Trisi P, Piattelli A (1994) Vertical ridge augmentation using a membrane technique associated with osseointegrated implants. Int J Periodontics Restorative Dent 14:496–511.
66. Spahr A, Lyngstadaas SP, Boeckh C, Andersson C, Podbielski A, Haller B (2002) Effect of the enamel matrix derivative Emdogain on the growth of periodontal pathogens in vitro. J Clin Periodontol 29: 62–72.
67. Urbani G, Lombardo G, Santi E, Consolo U (1999). Distraction osteogenesis to achieve mandibular vertical bone regeneration: a case report. Int J Periodontics Restorative Dent 19:321–331.
68. Urist MR (1965) Bone formation by autoinduction. Science 150:893–899.
69. Wactawski-Wende J, Grossi SG, Trevisan M, Genco RJ, Tezal M, Dunford RG, Ho AW, Hausmann E, Hreshchyshyn MM (1996) The role of osteopenia in oral bone loss and periodontal disease. J Periodontol 67:1076–1084.
70. Wallace SS, Froum SJ (2003) Effect of maxillary sinus augmentation on the survival of endosseous dental implants. A systematic review. J Periodontol 8:328–343.
71. Wikesjo UME, Hanisch O, Sigurdsson TJ, Caplanis N (1999) Chapter 17. Application of rhBMP-2 to alveolar and periodontal defects. In: Lynch SE, Genco RJ, Marx RE, eds. Tissue Engineering. Applications in Maxillofacial Surgery and Periodontics. Quintessence Publishing Co, Carol Stream, IL, pp 269–286.
72. Zellin G, Gritli-Linde A, Linde A (1995) Healing of mandibular defects with different biodegradable and non-biodegradable membranes: an experimental study in rats. Biomaterials 16:601–609.

# 10. Multiscale Computational Engineering of Bones: State-of-the-Art Insights for the Future

*Melissa L. Knothe Tate*[a]

## 10.1 Introduction

Computational models provide a platform that is equivalent to an in vivo, in vitro, and in situ or ex vivo model platform. Indeed, the National Institutes of Health have made the development of predictive computational models a high priority of the "Roadmap for the Future" (http://nihroadmap.nih.gov/overview.asp; see especially "New Pathways to Discovery"). The power of computational models lies in their usefulness to predict which variables are most likely to influence a given result, simulation of the system response to changes in that variable, and optimization of system variables to achieve a desired biological effect. Typically, these models are computer representations of the actual system, based on experimentally determined parameters and system variables; increasingly these computer models are referred to as in silico models (Fig. 10.1).

For example, if one were interested in the role played by the intramedullary blood supply on bone regeneration in a segmental long-bone defect, it might be possible to substitute an in silico model for the numerous experimental studies that would be necessary to predict how loss of the blood supply affects cell viability, recruitment of progenitor cells, etc. In reality, one model rarely suffices to unravel the research question at hand. Although in vivo studies are needed to measure the relevant parameters, they are inherently limited due to financial and time constraints. A computational model based on realistic parameters can forecast responses to changes in experimental variables and determine the experimental approaches that will most likely answer a specific research question. Hence, the first goal of this chapter is to describe the **strengths of computational modeling approaches, when used in tandem with experimental approaches, to unravel the most enigmatic research questions of bone biology.**

In addition to increasing understanding of biological system dynamics, in silico computer models provide an ideal approach to optimizing experimental design prior to implementation and testing. This not only improves efficiency, but may lead to engineering more functional tissue prototypes. Tissue models are designed to optimize a specific function that is to be replaced; by building predictive computational models, it is possible to determine key parameters that influence the specific function(s) to be replaced. The second goal of this chapter is to **outline the process of rational tissue design and optimization, using as example how a**

[a] Based on the work from Dr. Knothe Tate's research team (carried out by former and current students, including Eric J. Anderson, Steven Kreuzer, Hans-Jörg Sidler, Adam Sorkin, Roland Steck, and Andrea Tami) and the clinical and research practice of Ulf R. Knothe, M.D., D. Sc. This chapter is dedicated to my team.

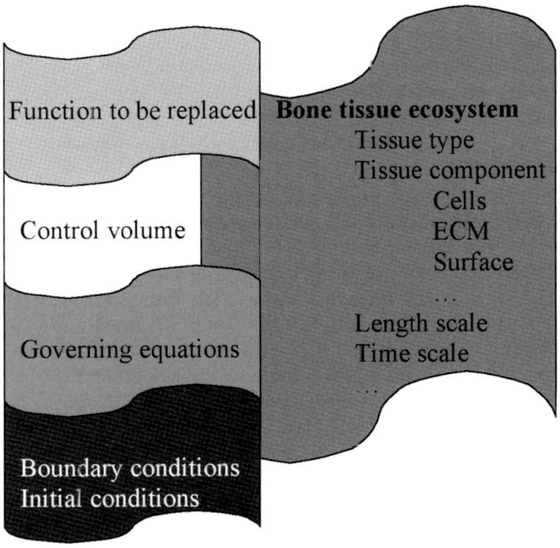


**Figure 10.1.** Pieces of the "bone model puzzle." Depending on the system of interest, bone can be modeled in a variety of ways. Key elements that are common to all models include the function that is to be replaced (which defines the goal of the model), the control volume (an abstract representation of the highly idealized model that aids in reducing the system to one with a finite, determinate set of variables), governing equations that provide mathematical predictions of model behavior in response to changes in system variables, boundary conditions, and initial conditions. The size and boundaries of the model system or control volume are determined by a variety of factors, including the tissue type and component to be modeled, as well as the length and time scale to be addressed in the system of interest. ECM (extracellular matrix).

**problem can be approached at multiple length and time scales.** The chapter is not designed as a cookbook for computer modeling; rather it is intended to encourage tissue engineers to utilize modeling, thereby increasing the power of their research.

## 10.2 Nature's Design Solution: the Biological Ecosystem cum Gold Standard for Tissue Design Specifications

Nature's designs are full of complexities and redundancies, yet designing a multifunctional tissue such as bone is a technical tour de force, and tissue engineers have yet to match Nature's success. In first approaching the problem, it is helpful to consider the critical functions that the engineered tissue needs to replace. Bone, as designed by Nature, is a remarkably resilient and multifunctional, dynamic, and self-healing structure. Bone is one of the few tissues in the human body that heals without scarring, and it represents as such an ultimate smart material. If one sets out to design the ultimate smart material success may be elusive; however, if one considers the mechanisms behind the remarkable capacity of bone not to scar, it is possible to gain insight into one of the most powerful intrinsic material properties of bone, i.e., the capacity to adapt its structure to its prevailing function over time. It is not the material of bone per se that imparts this remarkable property to the tissue, but the cells within bone that are the biological machines continually building and rebuilding structure in response to the prevailing dynamic environment. Hence, any rational tissue-engineering approach must consider the cells and the fact that they migrate within the dynamic environment of the tissue in times of tissue modeling, homeostasis, and disease.

The role of fluids in bone tissue engineering has been receiving increasing attention. In fact, from the perspective of fluids alone, bone is an ecosystem. The transport of life-supporting substances and the removal of waste are basic requirements for the maintenance and survival of any ecosystem. The human body and subsets of the body, e.g., organ systems, are also ecosystems with water, gases, nutrients, waste products, and regulatory substances such as hormones and cytokines in constant flux within the system (Fig. 10.2). Different hierarchical levels define organ, tissue, and cell physiology within the entire system. Materials are transported into these mostly in the form of water solutes. Except for the case of respiratory gas exchange with the external environment, fluid convection represents the most powerful transport mechanism throughout the human body [18] and in bone in particular [8].

By considering the mechanisms that underlie specific functions of bone, tissue engineers can make use of specific aspects of the structure of bone when designing tissue replacements. The functions of bone are many. At a systems level, the skeleton provides mechanical support, making it possible for a

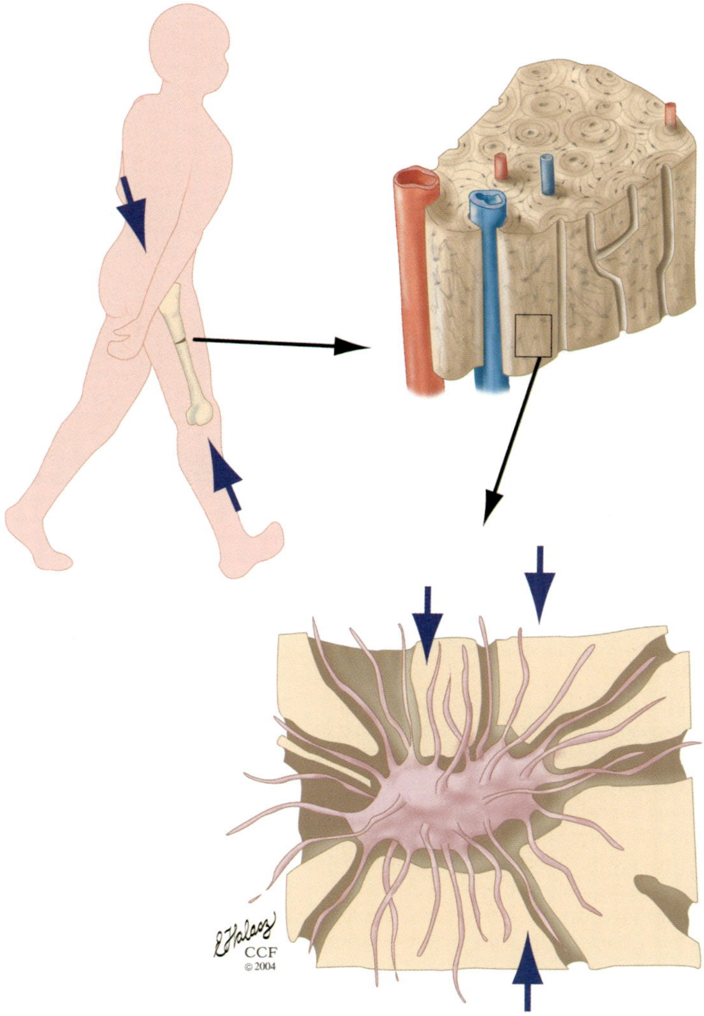


**Figure 10.2.** The ecosystem of bone. This concept is depicted at three length scales, including systemic level (human walking [m]), organ level (femur on thigh overlay [$m^{-1}$]), tissue level (wedge of cortical bone [$m^{-2}$]), and cell level (osteocyte in lacunocanalicular system [$m^{-6}$ to $m^{-7}$]). Of particular note, the ecosystem includes both the fluid environment of the vascular and pericellular space and the organic and inorganic structural elements comprising the matrix, as well as the living component of the tissue (i.e., the cells). Reprinted with the permission of the Cleveland Clinic Foundation.

system of organs to move along the ground. The system is composed of soft tissues, such as the brain (central command system), heart (central supply system), muscles (power for movement), and digestive system (provides materials that allow all organs to survive and maintain their structure). Bones also provide protection from impact. The ribs protect the heart and lungs, and the cranium protects the brain. At the tissue level, bone is a living electrophoretic and ion-exchange column; in this role, bones provide a reservoir and mobilization surface for calcium, a key signaling molecule. In addition, bone provides an enclave for the bone marrow, where hematopoiesis takes place.

## 10.3 Concept of Engineering Bones at Multiple Time and Length Scales

Engineers create implants, artificial joints, protective plates, etc., that function for a defined period of time, under static or dynamic loading conditions. An example is the standard hip replacement, which generally lasts 15 years. The challenge to tissue engineering, however, is that function varies with time and age. At steady physiological state, a histological image of bone at a given time, may, like a snapshot,

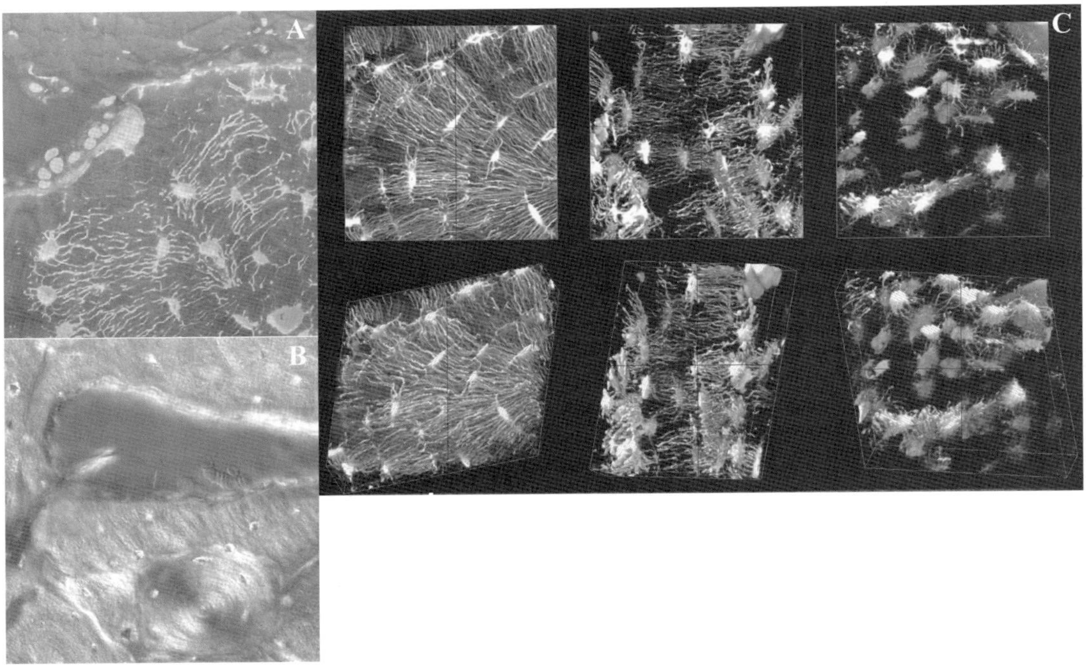


**Figure 10.3.** Bone remodeling processes histologically visualized at the tissue and cellular levels. Although histology is helpful for observing the physiology of the system at one moment in time, remodeling and adaptation are inherently dynamic. Hence, it is impossible to understand the dynamics of remodeling from a static image depicting one point in time. (A) After 20 weeks of immobilization, osteoclasts resorb the surface of the bone, as evidenced by the crater on the top right edge of the cross section of the ulna. Upon remobilization, osteoblasts subsequently lay down new osteoid, which fluoresces highly in its unmineralized state (infilled crater: osteoblasts are visible along the upper edge of the bone). The osteocyte network is observable across the tissue, linking every cell in the tissue with the blood supply, with bone surfaces to which forces are imparted, and with the marrow cavity. (B) Further into the cortex, remodeling is observed as a classic cutting cone. Because this image is taken after remobilization, osteoblasts are observable along the edges of the cutting cone. Red blood cells are visible in the resorption cavity. As the osteoblasts fill in new bone, an osteon is formed, as is visible in the same micrograph, orthogonal to the cutting cone in the plane of the image (see osteonal cross section below the cutting cone). (C) As bone is resorbed and as bone degrades because of aging or disease, the cellular network changes, as observed through changes in network connectivity, cell shape. and cell size. Reprinted with permission from *Advances in Osteoporotic Fracture Management*, Volume 2, M. L. Knothe Tate, A. E. G. Tami, T. W. Bauer, and U. Knothe, "Micropathoanatomy of osteoporosis: indications for a cellular basis of bone disease," pp. 10, 11, 2002. Copyright 2002, Remedica Medical Education and Publishing.

represent the system appropriately. However, as soon as the equilibrium is disturbed, the steady-state assumption no longer applies. Further, depending on the function to be addressed, the time scale of the system may vary from fractions of a second, as in cell signaling, to periods as long as 1 month, the time it takes osteoclasts to resorb a cavity and osteoblasts to fill it in with fresh osteoid (Fig 10.3, Fig 10.4), to months and years, the time it may take a bone to regain its prior mechanical strength after fracture.

One can conceive of a bone replacement that is virtually indestructible yet self-healing, is fully integrated with the biological tissue, and is immune to the biological consequences of aging (loss of bone mineral density, cross-linking of proteoglycans in the extracellular matrix, and cell senescence), but such a bone replacement does not exist. Rather, normal, healthy bone will be considered the "gold standard" for design specifications. For the purposes of this chapter, we will examine bone and its constituents within the engineering concept of a control volume as well as within the context of biological machines and materials. Furthermore, we will consider how surgeons harness Nature's endogenous strategies to replace and promote healing in missing or failed bones. Then we will follow up the design goal with development of computer models, first modeling actual tissue properties and leading into the rational design and optimization of tissue-engineered scaffolds. Finally, we will discuss the experimental validation of in silico models, a step necessary to determine whether model predictions are borne out.

## 10.4 Computational Cell and Tissue Models at Multiple Length Scales

When bone is remodeled as a system, four themes need to be addressed. First, **structure and function of bone are interdependent** and one cannot be addressed without affecting the other; this is true across length and time scales throughout the life of the animal. Second, **cells are the living component of bone**, and it is the movement and activity of cells that enables bone to adapt to its dynamic environment; hence cells must be considered in any model of engineered bone tissue. Even if "just" a scaffold is modeled, it is necessary to remember that the scaffold provides a surface for cell adhesion, migration, and proliferation. The properties of the scaffold will determine the maximum number of cells residing in and on the scaffold as well as the cells' mechanobiological milieu. Third, although idealizations are made in model development at one particular length scale, the **implications of these idealizations at other length scales** need to be addressed. Fourth, any model must **incorporate more than one function**, such as the interplay between the mechanical role of bone and its structural organization. This mandates a transdisciplinary approach to modeling bones in silico.

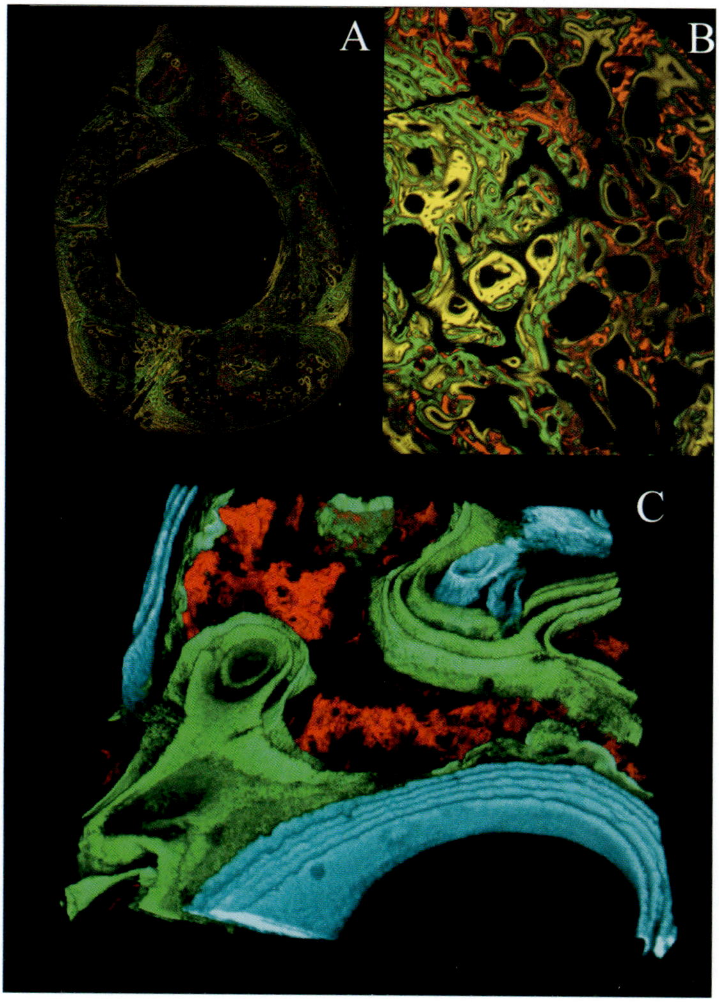


**Figure 10.4.** Regeneration of bone assessed by fluorochrome integration. Fluorochromes are fluorescent agents that are injected intravitally, i.e., into the living animal. The agents are integrated (biochemically through chelation) into the mineralized matrix at the time that new bone is being laid down. They allow for elucidation of the timing of bone apposition when fluorochromes with different excitation and emission spectra (imparting different colors in the micrographs) are administered at different time points. For this particular case, the time points are captured during the regeneration of bone within a segmental bone defect in the femur of a sheep. (A) Cross section showing robust regeneration in the previously empty space of the defect zone. (B) Alizarin red was administered first, followed by calcein green (2 weeks later) and tetracycline (yellow, 2 weeks thereafter). It is likely that the disorganized woven bone that was first laid down during the rapid proliferation stage of healing was remodeled and replaced by more organized lamellar bone during the remodeling phase (green and yellow in B, green and blue in C). (C) Confocal imaging allows for addition of the third dimension, which reveals in more detail the volume and time course of bone generation in a particular volume of interest.

Before modeling a particular bone tissue engineering system, several examples of in silico models at different length scales will be described. This will give the reader a sense of what can be (and has been) done, as well as indicate the inherent limitations of the approaches described. This will be followed by a guide to building models. In reviewing in silico models of bone, emphasis will be placed on models developed in the author's group over the past decade.[b] These examples illustrate the process of developing and analyzing a computational tissue model. Each of the models presented below was developed so that it can be validated experimentally and to achieve insights across length scales.

## 10.5 Organ to Tissue Scale In Silico Models

Until approximately 40 years ago, computational modeling of bones implemented a solid mechanics approach, to explore structure–function relationships on the basis of the structural components of bone, e.g., the trabecular architecture and mineralized matrix [14, 29]. This changed when Maurice Biot adapted the theory of poroelasticity, originally developed for fluid-saturated soil mechanics studies, to model bone as a stiff, fluid-saturated "sponge" [4]. Yet the inclusion of the fluid component of bone (25% of bone's total volume) in computational models has only recently become widespread. Slowly, in the past two decades, bone physiologists and mechanical engineers adopted the concepts of poroelasticity to investigate the interplay between mechanics and fluid transport in bones subjected to mechanical loads. Bassett [3] in 1966 Piekarski and Munro [19] more than 10 years later postulated that pressure gradients developing in mechanically loaded, fluid-saturated bone drive the fluid from areas of high pressure to areas of low pressure, carrying solutes such as nutrients and waste products to and from the cells. Convective transport (compared to diffusive transport) efficiently provides bone cells, including osteocytes, their basic metabolic needs and removes waste products. In the past 15 years, advances in endothelial cell research [22] and new computational methods [6, 11] indicated the need for and possibility of incorporating convective transport in computational models of bone with recent work incorporating the concept for the engineering of functional bone replacement tissue [16, 23].

The first models of bone as a fluid-filled structure showed that Piekarski and Munro's postulate was feasible [6, 11]. The models in turn led to a series of in vivo, ex vivo, and in vitro experiments that, although novel in approach, often raised more questions than they answered [9, 10]. The reason was that the state of all variables in the biological system was difficult to determine. This underscored the need for predictive computational models that would identify the parameters having the greatest effect on transport into bone. This led to the development of highly idealized models of the rat tibia and ulna that would show the effect of mechanical loading on global fluid flow, based on specific tracer distributions observed histologically. We now describe two models designed to increase understanding of the rationale of the modeling approach.

The end-loading model of the rat ulna (Fig. 10.5C), first described by Lanyon et al. [13], imparts a cyclic compressive load to the distal and proximal ends of the ulna via a mechanical testing machine that controls the magnitude and rate of load. Because of the inherent curvature of bone, compressive loading induces a combination of compressive and bending loads within the bone. Interestingly, load transfer is shared by the ulna and radius through the interosseous membrane [27], the mechanics of which are only beginning to be understood. We can predict these loads locally using finite element modeling, in which the bone is meshed into a finite number of elements and the stress and strain generated through loading are calculated for each element in order to simplify the complex problem (Fig. 10.5B). Such predictions are validated with the aid of strain gauges that are glued to the bone prior to loading and that deform under loads. The deformation alters the resistive properties of the wire mesh

[b] It should be noted that computational modeling of bone is a thriving research area and numerous research groups apply different approaches to the problem; a review of all previous approaches is beyond the scope of the specific goals addressed in this chapter, but a PubMed search with the keywords "computational," "model," and "bone" yielded more than 200 examples at the time of publication.

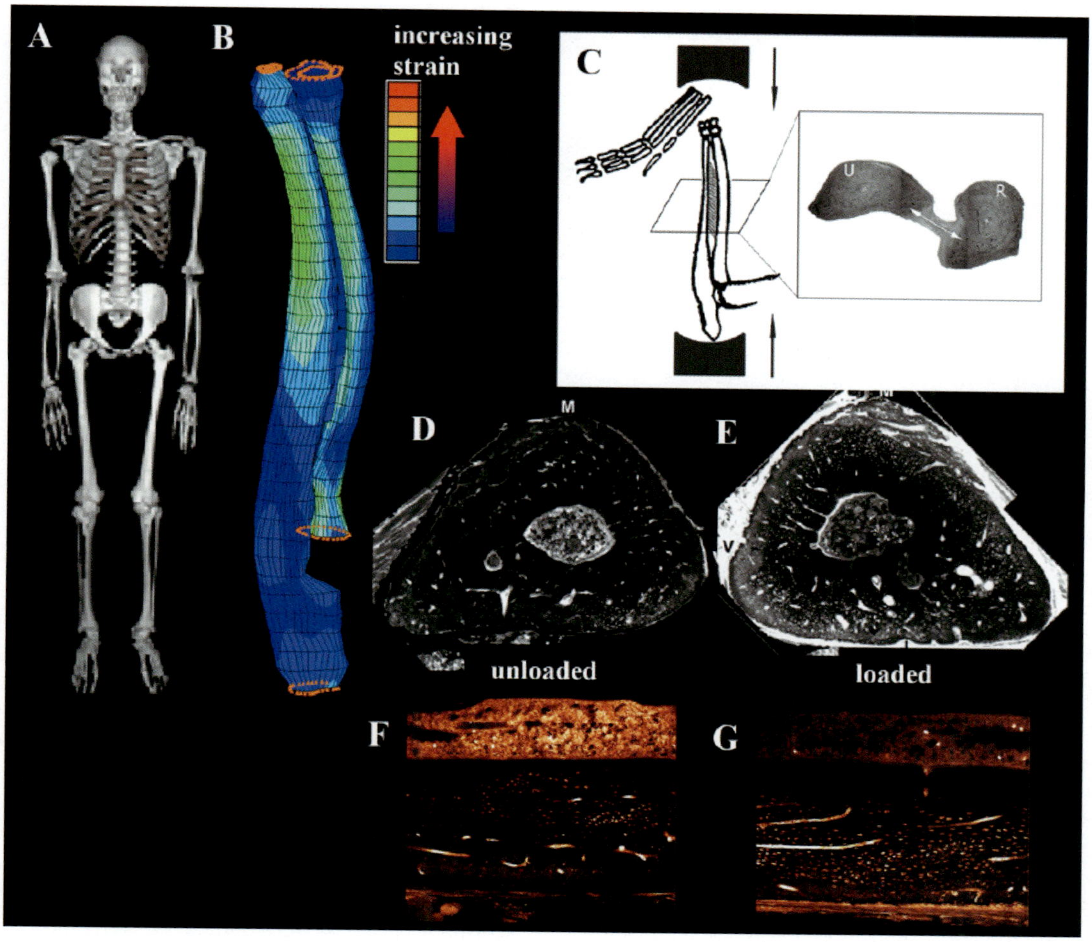


**Figure 10.5.** Organ-tissue-length scale model of the rat ulna. The model relates solid and fluid mechanical behavior of bone to transport within bone, based on a loading scenario after the ulna compression model of Lanyon and colleagues [13]. (A) The human radius and ulna is a length scale larger than that of the rat (B–D). (B) Finite element model depicting strain in the radius, R and ulna, U, etc. Reprinted from *Biorheology*, Volume 40, A. E. Tami, M. B. Schaffler, and M.L. Knothe Tate, "Probing the tissue to subcellular level structure underlying bone's molecular sieving function," pp. 583, 584, 586, 2003, with permission from IOS Press.

embedded in the gauge, which can be measured, and with the aid of circuit theory, the strain is calculated. In this way, one can check whether the strains predicted by the finite element model are appropriate for the model at hand. It would of course be cumbersome to glue thousands of gauges to the surface of the bone to obtain values for the actual strains at every point for every loading condition. Instead, this information is provided by the predictive model, which also predicts the strain within the bone interior.

It is possible to observe the displacement or flow of fluid in bone by using visualization methods, as in a fluid mechanics experiment in which dye is added to the flow stream to visualize streamlines and/or turbulence. In a living being an intravital fluorescent tracer (of a specific molecular size and shape) is injected into the blood stream prior to the application of loads. After the ulna has been loaded, the bone can be examined histologically and the tracer distribution in the loaded ulna can be compared to that in the contralateral, unloaded control (Fig. 10.5D–G).

Observations of histological cross sections taken from the mid-diaphysis show that mechanical loading pushes the tracer fluid out of the medullary cavity, through the cortex, and toward the periosteum (Fig. 10.5D and E).

Longitudinal bone sections demonstrate this effect well; in the unloaded ulna, the contents of the medullary canal are highly fluorescent, and isolated blood vessels in the cortex show the fluorescent tracer as well; the periosteum shows little or no tracer (Fig. 10.5F). In contrast, in the loaded ulna, the medullary cavity is much less fluorescent, the blood vessels of the cortex show fluorescence, and many periosteocytic spaces also exhibit fluorescence, as does the periosteum. Qualitatively, the effect of mechanical loading is clear. However, to understand the interplay between loading and transport in bone, loading magnitudes and durations must be correlated with tracer concentrations locally and throughout the skeleton.

## 10.6 On Choosing Models and Relationships Appropriate for Length Scale

We chose the rat ulna model to understand the problem in terms of organ and tissue distribution of fluids. However, the length of the rat ulna (~3 cm) is almost an order of magnitude less than that of the human ulna (~25 cm). Differences in scale between a model and the biological system of interest may present challenges to carrying out experiments and to interpreting the model results for the human situation. Experimental challenges typically involve the inherent difficulties in achieving spatial resolution (in strain gauge measurements or bone structure imaging) in such tiny bones. Even more confounding may be the fact that rat and mouse bones (like those of other small animals with high metabolic rates) do not have the osteonal structure of human bone [16, 17]. Recent studies point to the role the mean transport path distance plays in the organization of bone. Mammals with thick cortices (up to several centimeters in humans and more than 10 cm in elephants) require a two-tiered transport structure for long-distance fluid transport, just as a circulatory system is needed once organisms reach a size that can no longer be served by diffusional transport alone.

In osteonal bone, two systems assure distribution of fluid and solutes locally and in the organ as a whole. The osteonal layers are organized around a central Haversian canal that contains a blood vessel. Haversian canals, which run along the length of the long bone, are connected to one another by Volkmann's canals, which run orthogonal to the long bone axis. Each cell is connected to a blood vessel (within a Haversian or Volkmann canal) via the pericellular space or lacunocanalicular system, which constitutes the cell's "circulatory system." In this way, the nutritional needs of every cell can be met in large bones.

In smaller bones the transport network is simpler, in that blood vessels are dispersed in the cortex. Because the cortex is relatively thin and the blood vessels are quite close, no cell in the cortex, the periosteum, or the medullary cavity is more than 200 μm away from a blood vessel. In this way, nutrients are distributed to the cells through the lacunocanalicular system, and no branching transport system is needed.

Having shown by modeling and experimentation [7, 10] that load-induced fluid flow involves convective transport, we then studied how magnitude, mode (e.g., compression versus tension), and duration of loading affected convective transport in bone. We used the four-point-bending model of the rat tibia (Fig. 10.6A) for this purpose [28]. Our virtual model was based on three-dimensional data obtained from microcomputed tomographic (μCT) images of a rat tibia (Fig. 10.6B). The model volume was that of the tibial cortex, without the distal and proximal joint surfaces or the fibula (Fig. 10.6B and C). The tibia model was loaded with a four-point approach similar to what had been applied experimentally. The model was meshed into 7200 elements (pieces), comprising 20 node pore pressure elements. This allowed for sufficient computational sensitivity to calculate pressure fields and the resulting fluid velocities within the poroelastic material that had been chosen to simulate the solid-fluid material properties of bone. In the model we represented bone as a **continuum**. This means that bone is a stiff, fluid-filled sponge or poroelastic material. In a **discrete** model (see Table 10.1 for a comparison of the two approaches), the bone structure would be represented with a specific microarchitecture and defined porosities. Applying the continuum assumption effectively "smears" local properties to effective tissue values and does not account for microscopic detail. Furthermore, we defined the material of our virtual model to have limited

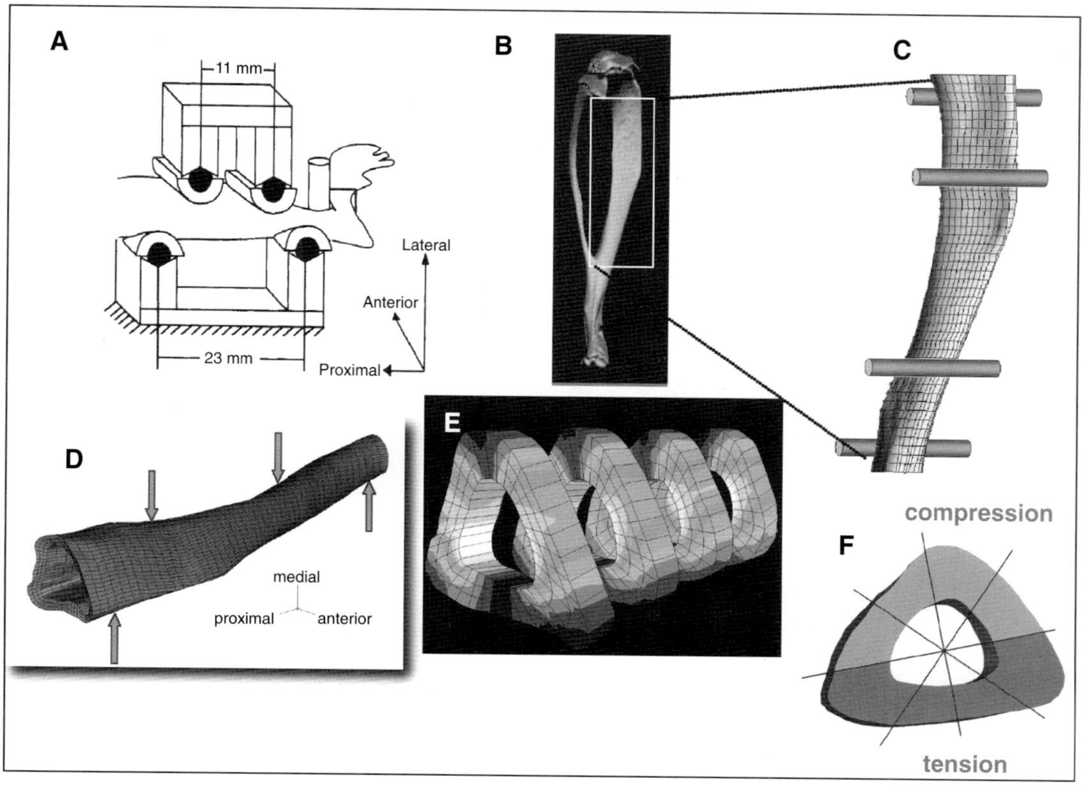


**Figure 10.6.** Organ-tissue-length scale model of the rat tibia. The finite element (C, D, E) model reveals the interplay between solid and fluid mechanics. The loading mode used simulated the in vivo four-point bending model [28] where external loading of the tibia (A). In the area between the distal and proximal junctions with the fibula (B) results in bending loads (F). Fig. 10.6B, C, F reprinted from *Journal of Theoretical Biology*, Volume 2003, R. Steck, P. Niederer, and M.,L. Knothe Tate, "A finite element analysis for the prediction of load-induced fluid flow and mechanochemical transduction," p. 251 (Fig. 10.6B and C), p. 253 (Fig. 10.6F), 2003, with permission from Elsevier.

**Table 10.1. Comparison of the continuum and discrete approaches to computational modeling**

| Continuum approach | Discrete |
|---|---|
| Properties and variables are averaged over a given volume. This results in **effective parameters** | Idealistic or real representation of system to be modeled |
| Justified when the length over which significant inhomogeneities occur is small | |
| Advantages<br>Utilizes classical engineering approach<br>Relatively "lean" computing | Advantages: a priori examination of system properties including<br>Effects of structure<br>Effects of site-specific pore distribution |
| Disadvantages<br>Does not reflect changes in structural characteristics of the system to be modeled<br>Gives only average, "effective" results | Disadvantages<br>High computational effort |

**anisotropic** properties, including elasticity (which describes the material's deformational behavior under mechanical load) and permeability. Anisotropy describes properties that vary as a function of orientation.

In addition to specifying the control volume and the assumptions underlying the model, the **boundary conditions** of the model need to be defined. For instance, if one is interested in determining the degree of hypoxia in a volume of tissue as a function of oxygen transport to and from the cells within that volume, the boundary condition, i.e., the flux of oxygen into the tissue from surrounding tissues, needs to be specified. If, on the other hand, the presence of injury or disease causes the control volume to become disconnected functionally from its surrounding tissue, then a sealed boundary condition with a constant zero or low flux may be more appropriate. The mass balance of oxygen within the model will be significantly affected by which boundary condition is chosen. In the studies at hand, we defined a stiffness and permeability of the material at all edges (boundaries). To model a long bone (Fig 10.7), we defined the endosteum, i.e., the boundary between the cortex and the medullary cavity, as material that exhibits low stiffness but high permeability. This definition reflects the high degree of vascularization and permeability of the endosteum, but does not significantly reflect the structural strength of bone. In contrast, the bone cortex contributes significantly to the structural strength and stiffness of bone, but is less permeable than the soft endosteal tissue, because it is made up of relatively impermeable mineralized matrix. Finally, the periosteum or outer surface of the bone was assumed to exhibit both low stiffness and permeability. A parametric study was carried out to determine the degree of influence that each of these variables has on the flow field within the bone cortex. Interestingly, it was shown that variation in periosteal permeability exerts the greatest influence on pore pressure distribution, which drives flow within the cortex (Fig. 10.7). This was particularly surprising, given that the periosteum is often assumed to be a "sealed surface" in modeling bone as a poroelastic material. [12, 26].

Four-point bending loads were applied to the model (Fig. 10.6D), with the loading conditions the same as those applied in vivo. Using the equations of poroelasticity [4] embedded in the finite element program, we calculated pressure gradients that are shown in Fig. 10.6E. Each cross section of the tibia (in Fig. 10.6E) was then depicted with one aspect under compression and one under tension, with the neutral axis in between (Fig. 10.6F).

We then calculated mass transport with the aid of the heat transfer package of the finite element software. Mass and heat transport are governed by the same equations, provided inertial terms can be neglected, as here. This calculation led to the magnitudes (Fig. 10.9A) and directions for every velocity vector at every element in the model. We had expected that fluid would be squeezed out of segments under compression and taken up by segments under tension. This was not the case in our original model (Fig. 10.9B) and led us to examine critically the definition of the material parameters in the new model (see below). To calculate mass

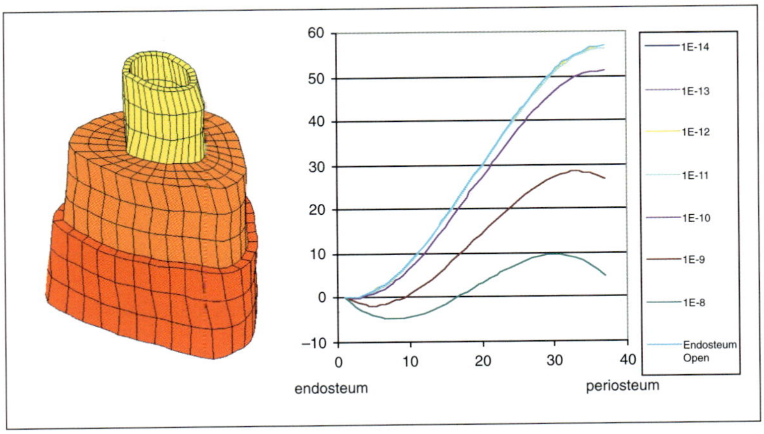


**Figure 10.7.** Effect of boundary conditions on the development of pore pressure in the cortex. The finite element mesh is divided into three concentric sections to define independently material properties of the endosteum (yellow, sheath closest to the medullary cavity, low stiffness and high permeability), the cortex (orange, high stiffness and medium permeability), and the periosteum (red, outer sheath, low stiffness, permeability varied). The pore pressure (kPa, y-axis) in the cortex between the surface closest to the endosteum and periosteum, respectively, is plotted as a function of periosteal sheath permeability (colored lines represent permeability, as defined in the sidebar). Reprinted from Journal of Theoretical Biology, Volume 220, R. Steck, P. Niederer, and M.L. Knothe Tate, "A finite element analysis for the prediction of load-induced fluid flow and mechanochemical transduction in bone," p. 252, with permission from Elsevier.

**Setup 1**

| Site | Porosity | Void Ratio | Permeability $k_{11}$ | Permeability $k_{22}$ | Permeability $k_{23}$ |
|---|---|---|---|---|---|
| 1 | 7% | 0.075 | 2.47E-20 | 2.47E-21 | 2.47E-22 |
| 2 | 6% | 0.064 | 1.58E-20 | 1.58E-21 | 1.58E-22 |
| 3 | 5% | 0.053 | 1.00E-20 | 1.00E-21 | 1.00E-22 |

**Setup 2**

| Site | Porosity | Void Ratio | Permeability $k_{11}$ | Permeability $k_{22}$ | Permeability $k_{23}$ |
|---|---|---|---|---|---|
| 1 | 10% | 0.111 | 1.00E-19 | 1.00E-20 | 1.00E-21 |
| 2 | 8% | 0.087 | 3.71E-20 | 3.71E-21 | 3.71E-22 |
| 3 | 5% | 0.053 | 1.00E-20 | 1.00E-21 | 1.00E-22 |

**Setup 3**

| Site | Porosity | Void Ratio | Permeability $k_{11}$ | Permeability $k_{22}$ | Permeability $k_{23}$ |
|---|---|---|---|---|---|
| 1 | 15% | 0.176 | 1.00E-18 | 1.00E-19 | 1.00E-20 |
| 2 | 10% | 0.111 | 1.00E-19 | 1.00E-20 | 1.00E-21 |
| 3 | 5% | 0.053 | 1.00E-20 | 1.00E-21 | 1.00E-22 |

**Setup 4**

| Site | Porosity | Void Ratio | Permeability $k_{11}$ | Permeability $k_{22}$ | Permeability $k_{23}$ |
|---|---|---|---|---|---|
| 1 | 5% | 0.053 | 1.00E-20 | 1.00E-21 | 1.00E-22 |
| 2 | 5% | 0.053 | 1.00E-20 | 1.00E-21 | 1.00E-22 |
| 3 | 5% | 0.053 | 1.00E-20 | 1.00E-21 | 1.00E-22 |

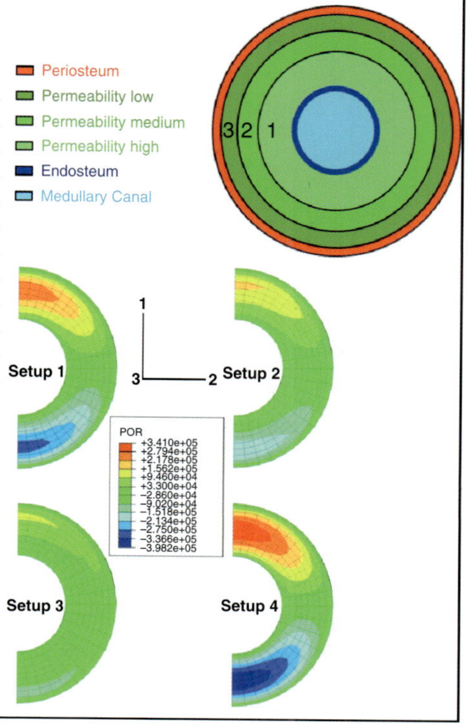


**Figure 10.8.** Parametric model exploring relationships between fluid velocity magnitudes, directions, resulting tracer concentrations, and adaptation.

transport, a diffusion constant for our molecule of interest; fluid velocities were calculated in the first step of the model and tracer concentrations were calculated as a function of location and time with the aid of the heat transfer equations, which reduce to the **general diffusion convection equation**

$$\frac{\partial C}{\partial t} = \frac{\partial}{\partial x_i}\left( k_i \frac{\partial C}{\partial x_i} - u_i C \right) + Q - KC$$

where $C(x_i, t)$ is the concentration (dependent variable), $x_i$ is the index form for cartesian coordinates, $t$ is the time, $k_i$ are the diffusion coefficients, $u_i$ are the components of the velocity vectors calculated in the first step, $Q$ is the source or sink coefficient (positive for source and negative for sink), and $K$ is the reaction rate for the molecule or chemical species of interest. Owing to the extremely slow flow rates that prevail in bone, acceleration (or inertial) effects can be neglected. Thus, for the two-dimensional case, this equation can be written in the following form, where $u$ and $v$ are the location-dependent components of the average velocity vector:

$$\frac{\partial C}{\partial t} + u\frac{\partial C}{\partial x} + v\frac{\partial C}{\partial y} - \frac{\partial}{\partial x}\left[ k_x \frac{\partial C}{\partial x} \right] - \frac{\partial}{\partial y}\left[ k_y \frac{\partial C}{\partial y} \right] - Q + KC = 0$$

When we calculate molecular tracer concentration across the tibia cortex, the areas of highest concentration correspond to the areas of lowest fluid velocity. This makes sense when one considers that the molecules will dwell longest in areas of low flow and will be transported rapidly through areas of high flow. Interestingly, the areas of highest adaptation in response to the four-point bending loads applied to in vivo models (Fig. 10.9D) co-localized better with those areas with the highest molecular concentrations and the lowest flow velocities [12]. This at first was puzzling, inasmuch as, according to the prevailing mechanotransduction hypotheses of the time, increasing shear stress through increasing flow velocity should have exerted a dominant effect, analogous to the influence of flow on endothelial cells in blood vessels. However, as stated above,

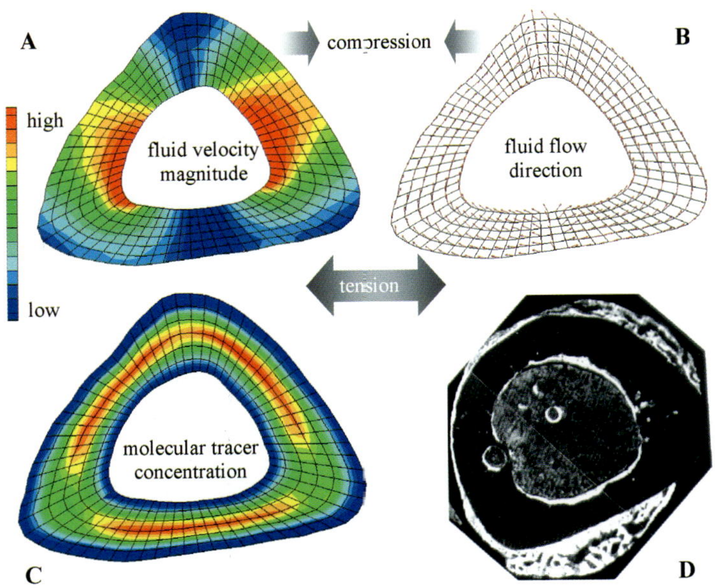


**Figure 10.9.** Areas if endosteal bone apposition in the four-point bending model of the rat tibia (D) correspond to areas of increased molecular tracer concentration (C) rather than areas of highest fluid velocity magnitudes (A) or specific fluid flow directions (B). Fig. 10.9A–C reprinted from *Journal of Theoretical Biology*, Volume 2003, R. Steck, P. Niederer, and M. L. Knothe Tate, A finite element analysis for the prediction of load-induced fluid flow and mechanochemical transduction, p. 254 (Fig. 10.9B), p. 255 (Fig. 10.9A and C), 2003, with permission from Elsevier. Fig. 10.9D reprinted with the permission of Charles Turner.

areas of endosteal bone apposition colocalized with areas exposed to higher concentrations of molecular tracer and not to areas of high fluid velocity magnitudes or specific fluid flow directions.

The power of computational modeling to elucidate biological systems is illustrated by a parametric study evaluating how definitions of site-specific material properties may influence model predictions (Fig. 10.8). Because a reliable experimental method has only recently been reported [5], the value of predictive modeling becomes compelling: using predictive models, we can determine which system parameters influence relevant biological effects. This, in turn, helps set priorities in planning experimental studies. One parametric model (Fig. 10.8) accounted for concentric layers of bone that show differences in porosity, void ratio, and permeability in three dimensions. Histological examination provided the rationale for this model as follows: in cross section, rat cortical bone exhibits "zones" or concentric layers that form as the bone grows and that show marked differences in the number of vascular canals and cells and in matrix density. We hypothesized that these differences would influence the distribution of pore pressures in the cross section and thereby influence transport through the bone. When parameters were varied by orders of magnitude and the corresponding pore pressures across the model sections were calculated, obvious differences in pore pressure distributions that depend on site-specific material definitions became apparent. This simulation guided future studies along two paths, first, there was a need for better definition of material parameters in our models, i.e., the need to make them also site-specific; and second, the poroelastic approach in which the whole cross-section was treated as a continuum needed to be readdressed. Our solution was to build a discrete model of bone at the tissue level to define locally relevant effective permeabilities that could later be implemented in the continuum model.

## 10.7 Tissue to Cell to Molecular Scale

Changes in local and tissue-level permeability influence the transport of nutrients and waste products to and from the osteocytes, as well as the transport of signaling molecules throughout the bone cell network. These changes in permeability are caused in part by changes in the pericellular transport network that result from aging and disease. Due to the inherent limitations of the continuum approach, and because our goal was to determine site-specific permeabilities as input parameters for an organ-tissue-level continuum model, we looked for an alternative approach in building discrete models that were virtual representations of bone at three organizational levels: tissue, cell, and molecule. Stochastic models lend themselves to the study of effects of structural and compositional changes on the flow of interstitial fluid through the pericellular network. We applied this approach, which is used extensively in chemical engineering [15], to develop a stochastic network model to simulate flow through the pericellular network and through the matrix microporosity, and to determine the influence of decreasing osteocyte density on cortical bone permeability [24].

Network modeling involves two steps. First, the random network of nodes and connecting bonds is constructed for optimal representation of the structure to be simulated, in this case the cellular network of bone (Fig. 10.10A and B). Second, the flow through this network is calculated. Both steps are repeated several times until statistical significance is achieved. In the first step, a three-dimensional, cubic-lattice network model, with the dimensions $L \times L \times L$ ($L = 15$), is developed according to methods described by Meyers and Liapis [15]; this simulates the properties of the matrix microporosity. Two different bond diameters, representing the pores between the apatite crystals and the pores between the collagen fibers, respectively, are distributed randomly with defined probabilities across the network. This maintains the overall porosity of the matrix. Next, osteocytes are distributed randomly across the nodes of the network. For every osteocyte, the distance to the neighboring osteocytes is determined. If the distance is smaller than a predefined threshold value, the osteocytes are connected by a canaliculus. Finally, since the network represents the tissue and is not an isolated entity, **periodic boundary conditions** are implemented for the microporosity bonds and the canaliculi (Fig. 10.10B and C). In the second step, the actual flow through the network is calculated. The driving force for this flow is a pressure gradient $p = p_{in} - p_{out}$ between the upper and the lower surfaces of the network. Therefore, all nodes on these surfaces are assigned either $p_{in}$ or $p_{out}$. The flow rate, $Q_{ij}$ through the bond between two nodes can be calculated as a function of the pressure gradient between the two nodes:

$$Q_{ij} = \frac{(p_i - p_j)d^3}{\left(\left(\frac{128l}{\pi d} + 24\right)\mu\right)}$$

where $d$ is the bond diameter, $l$ is the distance between the two nodes, and $\mu$ is the fluid viscosity [15]. The pressure at each node is calculated by solving a system of linear equations for the flow balance at each node. When the pressure at each node is known, the flow through the entire network can be calculated, $\kappa$, and, by using Darcy's law, the permeability of the network can be determined:

$$\kappa = \frac{Q_{tot}}{\Delta p}$$

In order to demonstrate the effect of osteocyte density on tissue permeability, we utilized data that quantify the change in osteocyte density in trabecular bone of patients 30 to 60 years old [20, 21]. The permeability is calculated as a mean value from the outcome of 20 calculations of the model for every osteocyte density (Fig. 10.10D). Whereas the osteocyte density is assumed to vary almost linearly [20, 21], the loss in permeability must be approximated with a power law ($R^2 = 0.98$).

These calculations illustrate the profound effect of declining osteocyte density on tissue permeability. The data predict that a 5% decrease in osteocyte density between the ages of 30 and 40 years will decrease bone permeability by almost 50%. Such a reduction is likely to have a marked effect on transport to and from bone cells.

On the basis of microscopic observations, a logical next step in model development is to determine the influence of osteocyte connectivity on tissue permeability. Osteocytes in close proximity to each other are typically connected by canaliculi that decrease in number with increasing distance from the blood supply; they also decrease in the presence of bone disease. Furthermore, by taking into account the preferred spatial orientation of the lacuno-canalicular network, it is possible to detect anisotropic differences in the permeability of bone tissue, which will be important for the development of more accurate, continuum-level finite element models. Finally, by excluding pores that are too small to allow the passage of a given molecule, we have been able to simulate the molecular sieving properties of bone tissue in preliminary studies. Our discrete models were designed to bridge the level between tissue and cell, but also to bridge to the molecular level. This approach is therefore useful to examine the transport of specific

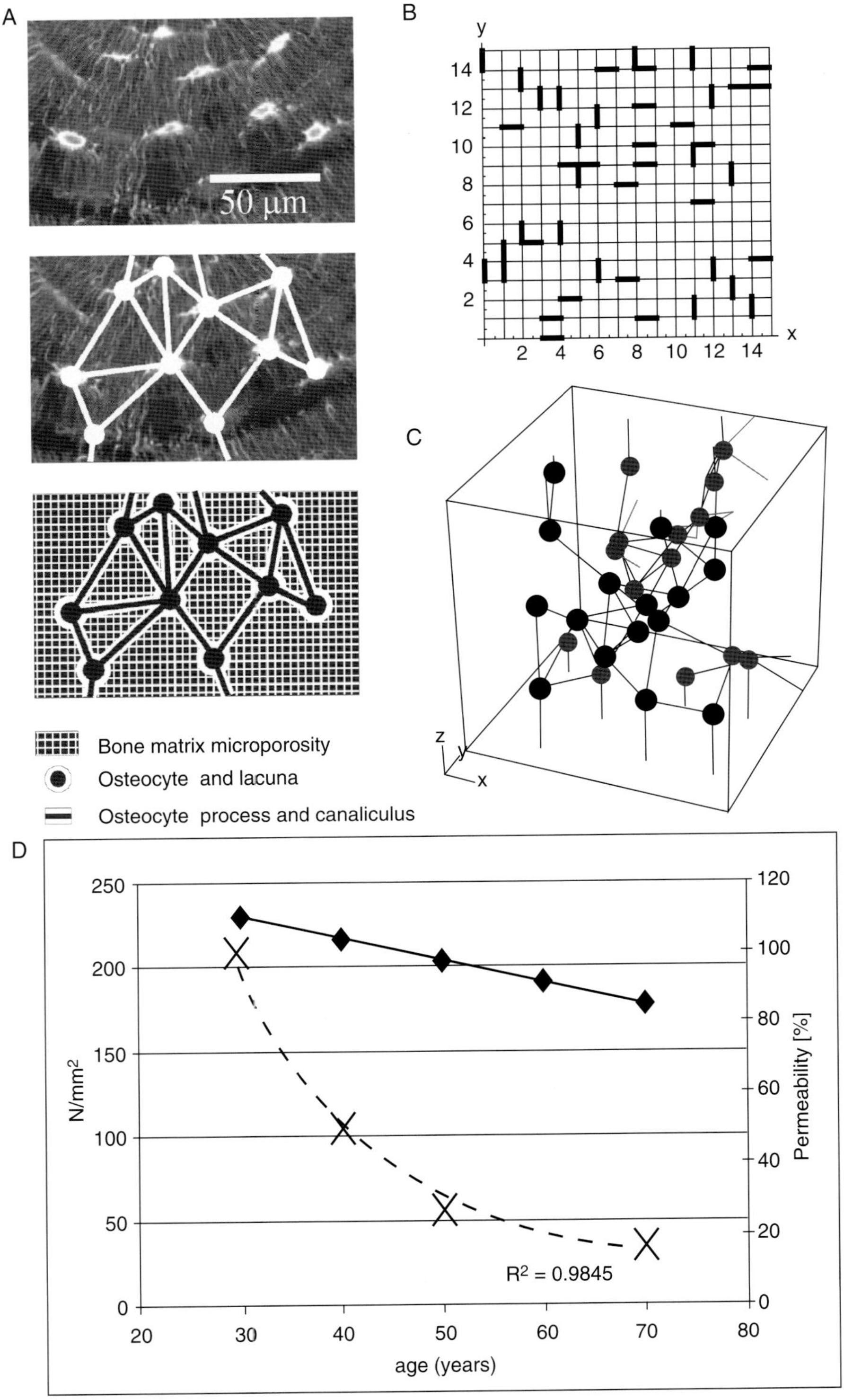


**Figure 10.10.** A stochastic network model (A, B, C) is built to represent the exact conformation and organization of the pericellular network, including the cells and their processes, as well as matrix microporosity (A). (D) Based on published data [19, 20] of osteocyte number (N) decline with age (diamonds), a profound concomitant decrease in tissue permeability (x's) is predicted using stochastic network models. Adapted from Steck R, Knothe Tate ML (2005) In Silico Stochastic Netowork Models that Emulate the Molecular Sieving Characteristics of Bone. *Annals of Biomedical Engineering*, 33(1): 87–94, used with permission from Springer.

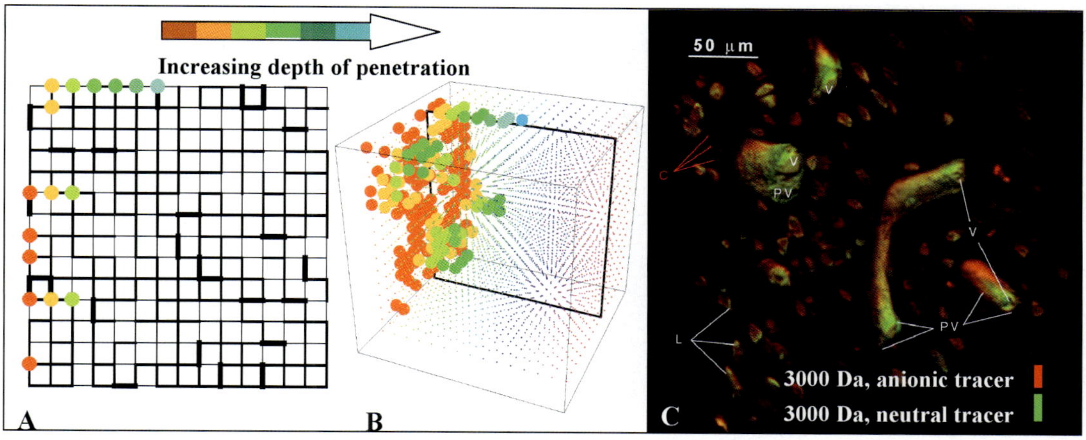


**Figure 10.11.** Stochastic network modeling (A, B) as a means to study delivery of drugs and molecular agents in actual microvolumes of bone (C). Fig. 10.11C reprinted from *Biorheology*, Volume 40, A.E. Tami, M. B. Schaffler and M. L. Knothe Tate, Probing the tissue to subcellular level structure underlying bone's molecular sieving function, p. 586, 2003, with permission from IOS Press.

molecules through the pericellular network within a defined tissue volume (Fig. 10.11). The model predicts the depth of penetration of specific molecules, and the predictions can be validated experimentally with the aid of fluorescently tagged molecules. The predictions apply to the perivascular space (PV, Fig. 10.11), the lacunar pore (L), and canaliculi (C) and can be validated experimentally in scaled-up models that are produced by stereolithographic methods.

## 10.8 Cell to Subcellular Scale

Yet another modeling approach lends itself to study of the mechanobiological effects of solid and fluid interactions in bone. Specific computational fluid dynamics (CFD) programs have been developed to study mechanics and transport in nano- and microelectromechanical systems. We utilized such a program to develop a computational model of an osteocyte in situ to understand the mechanical milieu of the cell and the role of fluid flow in mechanotransduction from the system as a whole to the cellular level. Fluid flow was explored at the length scale of the cell by developing a model of the fluid space around an osteocyte (Fig. 10.12). Flow through the microporosity was not included in initial models. The CFD program was run to calculate the pressure gradient, fluid velocity, and maximum shear and radial stresses imparted to the cell by the fluid (Fig. 10.12). The model predicted that osteocytes are subjected primarily to sustained hydrodynamic pressure and low stresses, whereas cellular processes are subjected primarily to shear gradients [1]. Increasing the number of canaliculi in the virtual model had a minimal effect on the magnitudes of pressure and stress. Because these effects cannot currently be measured at the cellular level, a computational model becomes essential for engineering design, as in the development of scaffolds, where cell recruitment, migration, and adhesion are essential.

Obviously it is important to check the validity of the assumptions that have gone into model construction. Since the CFD program uses the Navier-Stokes equations as the governing equations for flow field calculations, the validity of the **continuum assumption** underlying the Navier-Stokes equations was tested to ensure that the approach was appropriate at the length scale of our system. Validation studies have shown that the simulation is appropriate to lengths of approximately 10 nm, a length that is just below the minimum postulated dimension of the annular flow channel that surrounds the osteocytes [1].

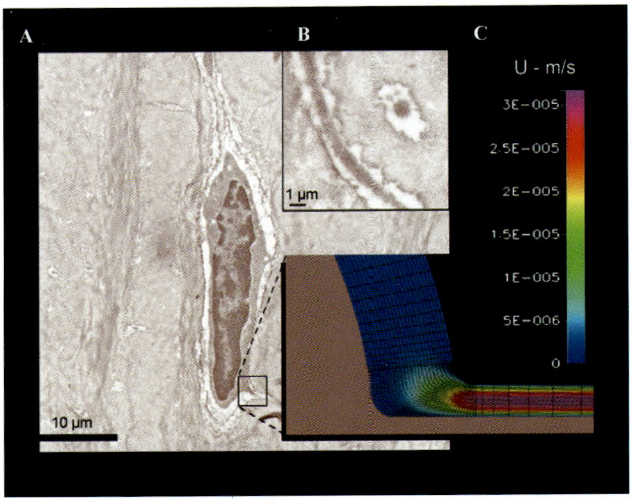


**Figure 10.12.** Transmitted electron micrographs of an osteocyte. Osteocyte (A) with inset showing osteocyte process in the plane and perpendicular to the plane of the micrograph, respectively; a computational fluid dynamics model of a portion of the osteocyte and its process predicts the fluid velocity magnitudes within the pericellular space. Reprinted from *Annals of Biomedical Engineering*, Volume 33, E. J.Anderson, S. Kaliyamoorthy, J. I. D. Alexander, and M. L. Knothe Tate, "Nano-microscale models of periosteocytic flow show differences in stresses imparted to cell body and processes," p. 54 and cover image, 2005, with permission from Springer.

## 10.9 Design Approaches: The Tissue Engineer's Computational Toolbox

The previous sections addressed variations in length scale as they affect model building at the level of the whole organ, the tissue, the cell, and the molecule. Also discussed was whether to approximate the system as a **continuum** or to consider its structure **discrete** (Table 1).

Although progress has been made in the field of tissue engineering, significant challenges remain in the regeneration and repair of bone, a tissue that serves mostly structural and mechanical functions. A major stumbling block has been the lack of understanding of the mechanisms of transport to and between bone cells. The importance of fluid flow for the promotion of cell viability and tissue health has been described [14, 16] in the earlier portions of this chapter. The following case study illustrates one example in which computational modeling can be used to optimize engineered tissue design.

## 10.10 Case Study: Design Optimization of a Tissue-Engineering Scaffold

The case study has used a scaffold designed by Dr. Lorna Gibson (MIT) and kindly provided by Dr. David Dean, in collaboration with Dr. Antonio Mikos. The scaffold is a prototype developed with stereolithography and made of poly(propylene fumarate) (PPF), a photopolymerizable, biodegradable resin. The scaffold is a three-dimensional, layered cylinder with nine circular and four semicircular channels in the longitudinal direction (Fig. 10.13A and B); all channels are connected through seven transverse rectangular channels. Scaffold geometries are created using a solid modeling program (Pro/Engineer, PTC) and then fabricated from poly(propylene fumarate), PPF, with the aid of a stereolithographic, rapid prototyping machine (Viper si2™, 3D Systems, Valencia, CA). By using µCT imaging methods (Scanco, Bassersdorf, Switzerland), actual geometries for the prototypes (Fig. 10.13B) can be compared with the target geometries. To predict flow through the target design scaffold and through actual rapid prototyped scaffolds, computational fluid dynamics methods were applied analogous to those outlined in the previous section.

First, a fluid mesh is created and fluid flow is calculated by using a CFD software package (CFD-ACE, CFDRC, Huntsville, AL). In these studies, we first estimate the effects of nonconformance with specifications, i.e., the variance between the target and the actual geometries, by reducing iteratively (from 0% to 100% in 25% increments) the through-channel diameters in the scaffold. Flow is induced by a pressure gradient for a fluid medium idealized as water (density $= 1000\,kg/m^3$, viscosity $= 0.001\,kg/ms$). On the basis of the mass flow rate calculated through the fluid mesh of each computational scaffold, permeability is determined by Darcy's law,

$$k = \frac{\dot{m}\mu L}{A_{cs}\rho\Delta P}$$

where $k$ is permeability ($m^2$), $\dot{m}$ is mass flow rate, μ is fluid viscosity, $L$ is scaffold length, $A_{cs}$ is cross-sectional area, $\rho$ is fluid density, and $\Delta P$ is the applied pressure gradient. Permeability in the longitudinal direction is calculated in this way and validated experimentally by using the same mass flow rate (Fig. 10.13C) [2].

This preliminary study demonstrates the potential of using nano-micro fluid dynamics to predict and optimize scaffold performance parameters, including fluid flow and permeability, prior to scaffold manufacture [2]. Interestingly, several rapid prototyped SLA scaffolds were shown to be impermeable on the basis of μCT and experimental measurements; this was due to lack of continuity in the scaffold through-channels. Had we had our predictive model and equations prior to rapid prototyping of the pilot scaffolds, we could have optimized our design for function prior to the prototyping phase.

In a further step, we have used our model to predict the mechanobiological milieu of cells seeded on the scaffold. For this purpose, we again employ CFD to simulate fluid flow through a tissue-engineered scaffold based on Navier-Stokes equations for steady flow induced by a pressure gradient. Flow is simulated in the longitudinal direction of the cylindrical scaffold, where the top and bottom of the cylinder are set as the inlet and outlet, respectively (Fig. 10.14A–E). **Boundary conditions** are defined in such a way that the rounded sides of the scaffold are sealed, restricting flow from entering or leaving in the transverse direction. The perfusate medium is treated as if it were water. A pressure gradient of 100 Pa is applied along the length of the scaffold, even though the relationship between fluid velocities applies to a wide range of pressure gradients. When flow is simulated in the longitudinal direction (top to bottom, Fig. 10.14C), differences in longitudinal velocity between the through-channels (v, Fig. 10.14A) and the alternating transverse fluid layers (w, Fig. 10.14B) are observed. In the through-channel geometry, velocity profiles were similar to parabolic pipe flow regimes (Poiseuille flow). When the through-channels enter the transverse layers, the profiles are similar to jet-flow expansion. Owing to the increase in the volume of the transverse layer that is due to jet flow expansion, an order-of-magnitude difference was found between through-channel and transverse layer velocities.

From the perspective of a cell that is attached to the walls of the channels, shear stresses

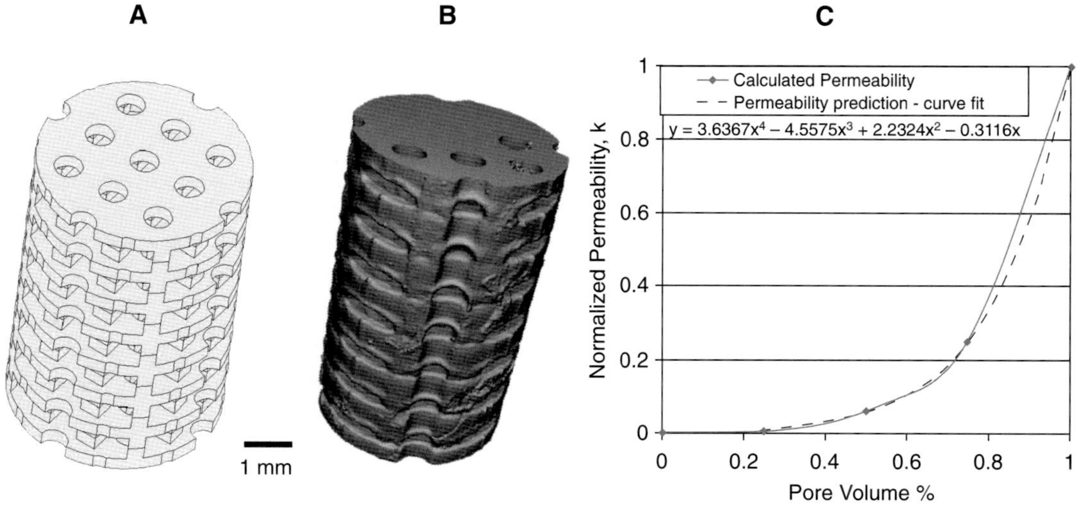


**Figure 10.13.** Comparison of scaffold parameters. (A) Computer-aided design (CAD) drawing of target scaffold geometry; (B) microcomputer tomographic (μCT) image of actual manufactured prototype geometry; (C) predicted permeability (k) of scaffold. Adapted from [2].

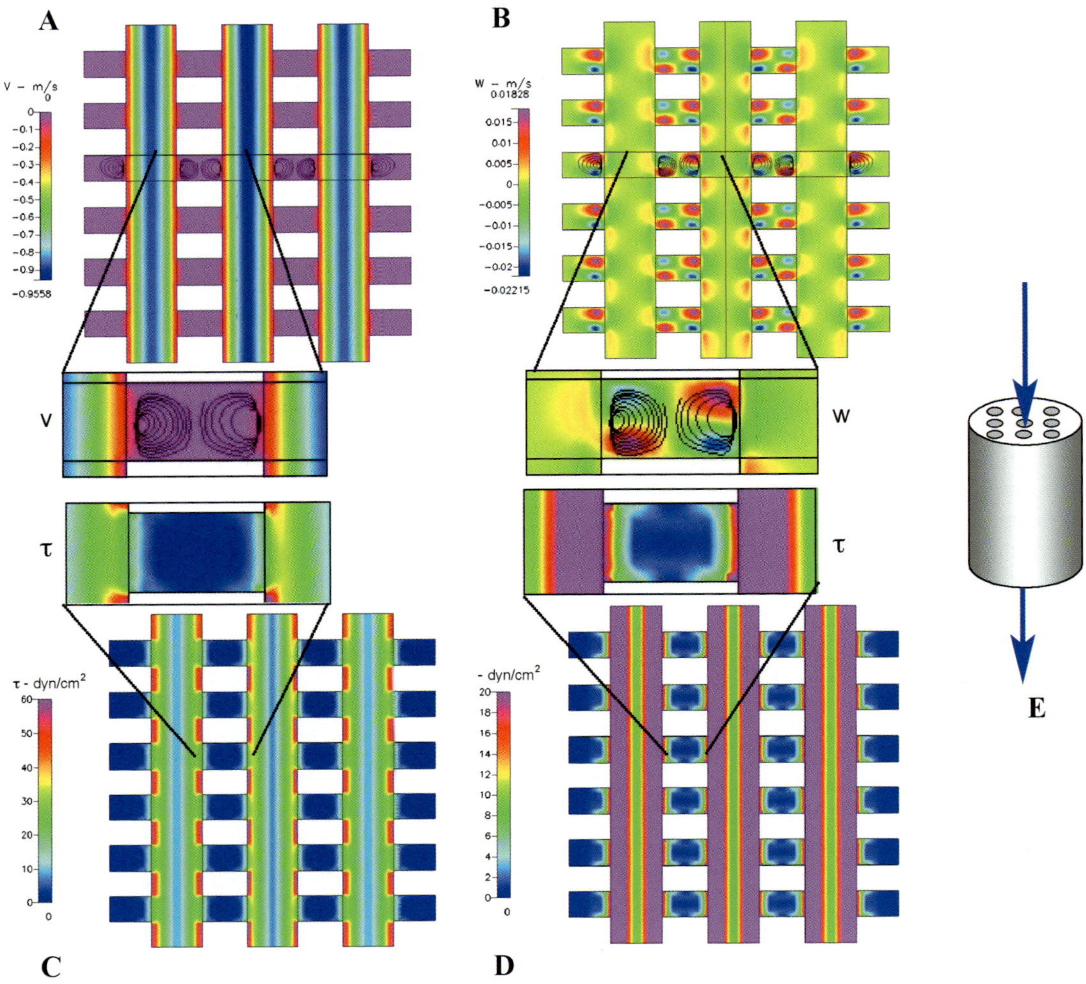


**Figure 10.14.** Predictive computational model of flow through a tissue-engineering scaffold. (A) Flow (v) in longitudinal direction with close-up of streamlines in inset. (B) Flow (w) in transverse direction, with close-up in inset. (C and D) Shear stresses (t) in the same area of the scaffold. (E) Schematic diagram showing flow direction through the cylindrical scaffold, whereby the sides of the cylinder are sealed. Adapted from [2].

differ markedly between the through-channels and the transverse layers as well (Fig. 10.14C and D). Wall shear stress is calculated from the laminar viscosity and the wall strain rate, determined through the solutions of Navier-Stokes equations within the scaffold

$$\tau_w = \mu\gamma$$

where $\tau$ is the shear stress at the wall, $\mu$ is the fluid viscosity, and $\gamma$ is the strain rate determined from the second invariant of the stress tensor. The high-flow environment of the through-channels produces high shearing stresses along the longitudinal walls. In contrast, the low-velocity transverse layers provide low-level shear stresses that promote cell adhesion, as well as mechanical stimuli conducive to osteoblastic differentiation.

# 10.11 Epilogue

We have addressed the two goals of the chapter:

- Describing the strengths of computational modeling approaches, when used in tandem with experimental approaches, to unravel the

most enigmatic research questions of bone biology; and
- Outlining the process of rational tissue design and optimization by using virtual in silico models to approach the problem at multiple length and time scales.

At the risk of exhausting the reader, the only thing to add is this imperative: "Go forth, find a collaborator, and model!"

# References


1. Anderson EJ, Kaliyamoorthy S, Iwan J, Alexander D, Knothe Tate ML (2005) Nano-microscale models of periosteocytic flow show differences in stresses imparted to cell body and processes. Ann Biomed Eng 33:52–62.
2. Anderson EJ, Savrin J, Cooke M, Dean D, Knothe Tate ML (2005) Evaluation and optimization of tissue engineering scaffolds using computational fluid dynamics. In: Annual Meeting of the Biomedical Engineering Society, Baltimore.
3. Bassett CAL (1966) Electromechanical factors regulating bone architecture. In: Fleisch H, Blackwood HJJ, Owen M, eds. Third European Symposium on Calcified Tissues. Springer Verlag, New York.
4. Biot M (1955) Theory of elasticity and consolidation for a porous anisotropic solid. J Appl Phys 26:182–185.
5. Fernandez-Seara MA, Wehrli SL, Takahashi M, Wehrli FW (2004) Water content measured by proton-deuteron exchange NMR predicts bone mineral density and mechanical properties. J Bone Miner Res 19:289–296.
6. Knothe Tate ML (1994) Diffusive and convective transport in the osteon. M.S. thesis, Divisions of Applied Mechanics and Engineering Design, Department of Mechanical and Process Engineering, Institute of Biomedical Engineering and Medical Informatics, Swiss Federal Institute of Technology, Zurich.
7. Knothe Tate ML (1997) Theoretical and experimental study of load-induced fluid flow phenomena in compact bone. Ph.D. thesis, Mechanical and Biomedical Engineering, Swiss Federal Institute of Technology, Zurich.
8. Knothe Tate ML (2003) Whither flows the fluid in bone? An osteocyte's perspective. J Biomech 36: 1409–1424.
9. Knothe Tate ML, Knothe U (2000) An ex vivo model to study transport processes and fluid flow in loaded bone. J Biomech 33:247–254.
10. Knothe Tate ML, Knothe U, Niederer P (1998) Experimental elucidation of mechanical load-induced fluid flow and its potential role in bone metabolism and functional adaptation. Am J Med Sci 316:189–195.
11. Knothe Tate ML, Niederer P (1998) A theoretical FE-based model developed to predict the relative contribution of convective and diffusive transport mechanisms for the maintenance of local equilibria within cortical bone. Adv Heat Mass Transfer Biotechnol 40:133–142.
12. Knothe Tate ML, Steck R, Forwood MR, Niederer P (2000) In vivo demonstration of load-induced fluid flow in the rat tibia and its potential implications for processes associated with functional adaptation. J Exp Biol 203:2737–2745.
13. Lanyon L, Mosley J, Torrance A (1994) Effects of the viscoelastic behavior of the rat ulna loading model. Bone 25:383–384.
14. Maurer B, Lehmann C (2006), Die Statik von Knochen. In: Karl Culmann und die graphische Statik. Zeichnen, die Sprache des Ingenieurs. Ernst und Sohn, Berlin.
15. Meyers JJ, Liapis AI (1998) Network modeling of the intraparticle convection and diffusion of molecules in porous particles pack in a chromatographic column. J Chromatogr A 827:197–213.
16. Mishra S, Knothe Tate ML (2003) Effect of lacunocanalicular architecture on hydraulic conductance in bone tissue: implications for bone health and evolution. Anat Rec A Discov Mol Cell Evol Biol 273: 752–762.
17. Mishra S, Knothe Tate M (2004) Allometric scaling relationships in microarchitecture of mammalian cortical bone. 50th Annual Meeting of the Orthopaedic Research Society, San Francisco, 29:0401.
18. Niederer PF, Knothe Tate ML, Steck R, Boesiger P (2000) Some remarks on intravascular and extravascular transport and flow dynamics. Int J Cardiovasc Med Sci 3:21–31.
19. Piekarski K, Munro M (1977) Transport mechanism operating between blood supply and osteocytes in long bones. Nature 269:80–82.
20. Qiu S, Rao DS, Palnitkar S, Parfitt AM (2002) Age and distance from the surface, but not menopause, reduce osteocyte density in human cancellous bone. Bone 31:313–318.
21. Qiu S, Rao DS, Paltnitkar S, Parfitt AM (2002) Relationships between osteocyte density and bone formation rate in human cancellous bone. Bone 31: 709–711.
22. Reich KM, Frangos JA (1991) Effect of flow on prostaglandin E2 and inositol trisphosphate levels in osteoblasts. Am J Physiol 261(3 Pt 1):C428–432.
23. Sikavitsas VI, Bancroft GN, Lemoine JJ, Liebschner MA, Dauner M, Mikos AG (2005) Flow perfusion enhances the calcified matrix deposition of marrow stromal cells in biodegradable nonwoven fiber mesh scaffolds. Ann Biomed Eng 33:63–70.
24. Sidler H, Steck R, Knothe Tate ML (2006) Site-Specific Porosity and its Impact on Load-Induced Fluid Movement in Cortical Bone, 52nd Annual Meeting of the Orthopaedic Research Society, Chicago, 31: 1591.
25. Steck R, Niederer P, Knothe Tate ML (2003) A finite element analysis for the prediction of load-induced fluid flow and mechanochemical transduction in bone. J Theor Biol 220:249–259.
26. Steck R, Knothe Tate ML (2005) In silico stochastic network models that emulate the molecular sieving characteristics of bone. Ann Biomed Eng 33:87–94.
27. Tami AE, Niederer P, Steck R, Knothe Tate ML (2003) New insights into mechanical loading behavior of the ulna-radius-interosseous membrane construct based on finite element analysis of the ulnar compression

model. 49th Annual Meeting of the Orthopaedic Research Society, New Orleans. 28:1196.
28. Turner CH, Forwood MR, Rho JY, Yoshikawa T (1994) Mechanical loading thresholds for lamellar and woven bone formation. J Bone Miner Res 9: 87–97.
29. Wolff J (1892) Das Gesetz der Transformation der Knochen. Berlin: Herschwald Verlag.

# Index

Printed in Singapore